The Chemistry of Allelopathy

ACS SYMPOSIUM SERIES **268**

The Chemistry of Allelopathy
Biochemical Interactions Among Plants

Alonzo C. Thompson, EDITOR
U.S. Department of Agriculture

Based on a symposium sponsored by
the Division of Pesticide Chemistry
at the 187th Meeting
of the American Chemical Society,
St. Louis, Missouri, April 1984

American Chemical Society, Washington, D.C. 1985

Library of Congress Cataloging in Publication Data

The chemistry of allelopathy.
  (ACS symposium series, ISSN 0097-6156; 268)

  "Based on a symposium sponsored by the Division of Pesticide Chemistry at the 187th Meeting of the American Chemical Society, St. Louis, Missouri, April 1984."

  Includes bibliographies and indexes.

  1. Allelopathic agents—Congresses. 2. Allelopathy—Congresses.

  I. Thompson, Alonzo C. II. American Chemical Society. Division of Pesticide Chemistry. III. American Chemical Society. Meeting (187th: 1984: St. Louis, Mo.) IV. Series.

OK898.A43C44   1984   581.2'3   84–24626
ISBN 0–8412–0886–7

ACS Symposium Series

M. Joan Comstock, *Series Editor*

Advisory Board

FOREWORD

The ACS SYMPOSIUM SERIES was founded in 1974 to provide a medium for publishing symposia quickly in book form. The format of the Series parallels that of the continuing ADVANCES IN CHEMISTRY SERIES except that in order to save time the papers are not typeset but are reproduced as they are submitted by the authors in camera-ready form. Papers are reviewed under the supervision of the Editors with the assistance of the Series Advisory Board and are selected to maintain the integrity of the symposia; however, verbatim reproductions of previously published papers are not accepted. Both reviews and reports of research are acceptable since symposia may embrace both types of presentation.

CONTENTS

PREFACE

THE CLASSIFICATION OF CHEMICAL SUBSTANCES as allelopathic agents requires bringing together the talents of several scientific disciplines. Biologists have obtained a large mass of evidence that documents the presence of substances produced by one plant that interfere, in some way, with the growth of another plant. However, the relationship is not a straightforward, one-step process. As in most natural cause-and-effect observations, we are dealing with a diverse process, including environmental conditions, acting upon plants to produce a given result. We can often measure the results, but we have difficulty isolating which parts are exerted by the individual components in producing those results. Generally, our studies and understanding tend to be based on research of an individual organism.

Through the years, natural product chemists have isolated, identified, and synthesized from plants many compounds that have been subsequently used by the biochemist to establish biosynthetic pathways. Through this cooperative effort, much has been learned about how plants grow and about the mechanisms that mediate growth. These disciplines, however, are closely related; and the respective scientists have little difficulty in understanding one another. A working relationship must be developed between the biologist and chemist to elucidate the biological–chemical relationships of possible allelopathic substances. This relationship necessitates that each become, to some degree, knowledgeable about the other's work.

Once biological activity has been established (in the laboratory or the field) and once the chemical work has been accomplished, we need to confirm allelopathic activity in the natural environment. To accomplish this end, the effects of soil and microbial flora must be considered. Thus, the disciplines of soil chemistry and microbiology are required. The chapters in this volume deal primarily with the biology and chemistry of phytotoxins isolated from plants; however, we hope that these topics will stimulate soil chemists and microbiologists to contribute to solving the problems associated with the study of allelopathy. Thus, the purpose of this volume is not only to bring before the scientific community a representation of research efforts in the area of allelopathy, but also to promote the relationships among the scientific disciplines required to solve and utilize this natural phenomenon for our benefit.

As editor, I am grateful to the authors for their contributions and to the U.S. Department of Agriculture for its support of this effort.

ALONZO C. THOMPSON
U.S. Department of Agriculture
Mississippi State, MS

June 1984

Allelopathic Research in Agriculture
Past Highlights and Potential

ALAN R. PUTNAM

Michigan State University, 105 Pesticide Research Center, East Lansing, MI 48824

Allelopathy produces marked impacts in diverse terrestial and aquatic ecosystems including influences on plant succession and patterning, inhibition of nitrogen fixation and nitrification, and inhibition of seed germination and decay. There are two major challenges to agricultural researchers: To minimize the negative impacts of allelopathy on crop growth and yield, and to exploit allelopathic mechanisms as additional pest control or crop growth regulation strategies. Plant products, microbial products, or their synthetic analogs could comprise the next generation of pesticides and growth regulaators. Joint efforts of chemists, plant physiologists, microbiologists, ecologists, and perhaps others, will be required to achieve maximum progress in this endeavor. Allelopathy research offers unlimited opportunities to solve practical agricultural problems and to contribute fundamental knowledge regarding the chemistry and biology of interspecific relationships.

The term allelopathy was coined by Molisch in 1937 (1). Presently, the term generally refers to the detrimental effects of higher plants of one species (the donor) on the germination, growth, or development of plants of another species (the recipient). Allelopathy can be separated from other mechanisms of plant interference because the detrimental effect is exerted through release of chemical inhibitors (allelochemicals) by the donor species. Microbes associated with the higher plants may also play a role in production or release of the inhibitors (2). Allelopathy is included among a higher-level order of chemical ecology involving interactions among many different organisms. Whittaker and Feeny (3) have defined interspecies

allelochemic effects and classified allelochemicals on the basis
of whether the adaptive advantage is gained by the donor or
recipient. Allomones, which give adaptive advantage to the
producer include repellants, escape substances, suppressants,
venoms, inductants, counteractants, and attractants.
Allelopathic chemicals may be classified as suppressants. Some
inhibitors from plants may also induce intraspecific effects
(autotoxicity).

Impacts in Agroecosystems

The fact that allelopathy can exert detrimental impacts on
agriculture was apparently recognized by Democritus and
Theophrastus in the fifth and third century BC respectively, by
deCandolle in 1832, and more recently by many ecologists and
agronomists (4, 5). Allelopathy has been related to problems
with weed:crop interference (6), with phytotoxicity in stubble
mulch farming (7), with certain types of crop rotations (8), and
with orchard replanting (9) or forest regeneration (10). In some
alleged allelopathic interactions, it is not clear whether
reduced crop growth is a direct result of released toxins, or
whether the toxins precondition the crop plant to invasions by
plant pathogens. Rice (4) indicated that allelopathy may
contribute to the weed seed longevity problem through at least
two mechanisms: chemical inhibitors in the seed prevent seed
decay induced by microbes or inhibitors function to keep seed
dormant, although viable for many years.
     There is extensive evidence that allelopathy may contribute
to patterning of vegetation in natural ecosystems (11). Distinct
zones of inhibition are present under and adjacent to a variety
of woody species, and often toxins from their litter are
implicated (12). One might speculate that aggressive perennial
weed species quickly gain dominance by exploiting allelopathic
mechanisms.

Sources of Allelopathic Chemicals

Chemicals with allelopathic potential are present in virtually
all plant tissues, including leaves, stems, roots, rhizomes,
flowers, fruits, and seeds. Whether these compounds are released
from the plant to the environment in quantities sufficient to
elicit a response, remains the critical question in field studies
of allelopathy. Allelochemics may be released from plant tissues
in a variety of ways, including volatilization, root exudation,
leaching, and decomposition of the plant residues.
     Reports on volatile toxins originate primarily from studies
on plants found in more arid regions of the world. Among the
genera shown to release volatiles are Artemisia, Eucalyptus, and
Salvia (4). When identified, the compounds were found to be
mainly mono- and sesquiterpenes. Work of Muller and associates
(13) has indicated that vapors of these compounds may be absorbed
by surrounding plants, and that the chemicals can be absorbed
from condensate in dew, or by plant roots after the compounds
reach the soil.

A myriad of compounds are also released by plant roots (14). The compounds are probably actively exuded, leaked, or they may arise from dead cells sloughing off the roots. Much of the evidence for root-mediated allelopathy has come from studies where nutrient solutions cycled by the root systems of one plant are added to media containing the indicator species. A recent study by Tang and Young (15) successfully utilized an adsorptive column (XAD-4) to selectively trap organic, hydrophobic root exudates while allowing nutrient ions and other hydrophillic compounds to pass through. They identified 16 compounds exuded from the roots of Bigalta limpograss (Hemarthia altissima) including a variety of benzoic, cinnamic, and phenolic acids.

A variety of chemicals may be leached from the aerial portions of plants by rainwater or by fog-drip (16) including organic acids, sugars, amino acids, pectic substances, gibberellic acids, terpenoids, alkaloids, and phenolic compounds. Colton and Einhellig (17) suggested that leaf leachates of velvetleaf (Abutilon theophrasti) may be inhibitory to soybean (Glycine max). We have recently discovered specialized hairs on the stems of velvetleaf plants which exude toxic chemicals.

After death of the plant, chemicals may be released directly by leaching of the plant residues. A variety of compounds may impose their toxicities additively or synergistically. Along with direct release of compounds from the tissue, microbes in the rhizosphere can produce toxic compounds by enzymatic degradation of conjugates or polymers present in the plant tissue. Examples of this phenomenon are the action by microbes on the cyanogenic glycosides of Johnsongrass (Sorghum halepense), and Prunus species to produce toxic HCN, and the corresponding benzaldehydes (18).

The toxicity arising from plant residues undoubtedly provides some of the more challenging problems and opportunities for agronomists, horticulturists, and weed scientists. Where stubble-mulch farming has been practiced in the plains states for soil and water conservation, toxins from the stubble have proven toxic to certain rotational crops (7). Now in agriculture there is a movement to employ conservation tillage (including no-tillage) practices which preserve surface plant residues. Not only can these residues have an influence on crop emergence, growth, and productivity, but they can also influence similar aspects of weed growth. Our recent work indicates that management of selected crop residues can greatly reduce weed germination and growth (19).

Natural Products Identified as Allelopathic Agents

Inhibitors from plants and their associated microbes represent a myriad of chemical compounds from the extremely simple gases and aliphatic compounds to complex polycyclic aromatic compounds.

The compounds implicated in allelopathy have been divided into chemical classes by recent reviewers (4, 20). They can be arbitrarily classed as (A) hydrocarbons, (B) organic acids and aldehydes, (C) aromatic acids, (D) simple unsaturated lactones,

(E) coumarins, (F) quinones, (G) flavonoids, (H) tannins, (I) alkaloids, (J) terpenoids and steroids and (K) miscellaneous and unknowns. Although many of these compounds are secondary products of plant metabolism, several are also degradation products which occur in the presence of microbial enzymes.

New chemicals are constantly being isolated from plants and microorganisms daily. Swain (21) recently reported that over 10,000 low-molecular weight products have already been isolated from higher plants and fungi. In addition, he proposed that the total number might approximate 400,000 chemicals. Some of these chemicals or their analogs could provide important new sources of agricultural chemicals for the future. There is considerable interest within the agricultural chemical industry on at least two approaches involving allelochemics for weed control. One involves the development of crop cultivars (perhaps through genetic engineering) which can either themselves suppress associated weeds or provide sources of natural product herbicides or their precursors. Another approach is to produce natural herbicides through batch culture with microorganisms.

Challenges in Allelopathy Research

Although allelopathic interactions have been observed for centuries, the science of allelopathy is in its infancy. Much needs to be accomplished, and it will require joint efforts of scientists from several disciplines. Although by no means a complete list, the following areas need intensive study.

Improved Methods for Collection, Bioassay, Isolation, and Characterization of Compounds. Techniques used to characterize natural products are evolving rapidly as more sophisticated instrumentation is developed. Plant physiologists and chemists should work closely together on this aspect, since rapid and reproducable bioassays are essential at each step. There is no standard technique that will work effectively for every compound. Briefly, isolation of a compound involves extraction or collection in a appropriate solvent or adsorbant. Commonly used extraction solvents for plants are water or aqueous methanol in which either dried or live plant parts are soaked. After extracting the material for varying lengths of time, the exuded material is filtered or centrifuged before bioassay. Soil extraction is more difficult, since certain solvents (e.g. bases) may produce artifacts.

Chemical separations may first be accomplished by partitioning on the basis of polarity into a series of solvents from non-polar hexane to very polar compounds like methanol. Compounds may also be separated by molecular size, charge, or adsorptive characteristics, etc. Various chromatography methods are utilized, including columns, thin layer (TLC) gas-liquid (GLC), and more recently, high pressure liquid (HPLC) systems. HPLC has proven particularly useful for separations of water soluble compounds from relatively crude plant extracts. Previously, the major effort toward compound identification involved chemical tests to detect specific functional groups, whereas characterization is now usually accomplished by using a series of spectroscopic analyses. Initially, ultraviolet spectroscopy (UV) is useful in this regard to detect specific functional groups. More recently, infrared (IR) spectroscopy and nuclear magnetic resonance (NMR) have helped immensely in determining natural products structure by indicating the functional groups and relative positions of atoms. Mass Spectrometry is a relatively recent addition to the analytical arsenal that provides additional clues as to molecular size and composition. It can quickly provide confirmation of complex organic molecular structures. Tandem Mass Spectrometry (MS/MS) or GC-MS are more recent developments which also allow analyses of mixtures of compounds. Effective studies of allelopathy must now include natural products chemists who can provide structure elucidation rapidly.

Factors Affecting Allelochemical Production or Release and their Modes of Action. This area of research should prove fruitful for the plant physiologists and biochemists who are interested in regulation of plant metabolism. Studies to date have been limited to only a few compounds.

Plants appear to vary in their production of allelopathic chemicals depending upon the environment in which they are grown and in particular, in response to stresses that they encounter. One practical difficulty faced by researchers is that greenhouse-grown plants may produce limited quantities of inhibitors. Ultraviolet (UV) light is absent in closed greenhouses, and several investigations have shown that UV light greatly enhances the production of allelopathic chemicals (4). For example, when greenhouse light was supplemented by UV, sunflower (Helianthus annuas) produced much more scopolin and chlorogenic acid (22).

Nutrient deficiencies may also influence the production of allelochemics. The compounds studied in great detail have been the phenolic compounds and scopolin-related chemicals. Deficiencies of boron, calcium, magnesium, nitrogen, phosphorus, potassium, and sulfur have all been reported to enhance the concentration of chlorogenic acids and scopolin in a variety of plants (4). In other species, chlorogenic acids have decreased in plants that are deficient in magnesium or potassium.

The type and age of plant tissues are extremely important since compounds are not uniformly distributed in the plant. Among species, there are great differences in ability to produce allelochemics. Within species, differences may exist in the amount of toxin produced by different genotypes. For example, various oat (Avena sativa) lines show differences in their ability to exude scopoletin and related compounds (23). Some cucumber (Cucumis sativus) accessions greatly inhibited weed germination, while others had no effect, or even stimulated growth (24). The implications of all these findings are that plant types may be either selected or bred that are more allelopathic, or that inhibitor production can be enhanced by exerting the proper stresses on the plants.

Mode of action research has caused similar challenges for investigators working with either natural products or synthetic pesticides. The major difficulty is to separate secondary effects from primary causes. Although effects can be measured in

isolated systems, there always remains the critical questions of
whether the inhibitor reaches that site in the plant in
sufficient concentration to specifically influence that reaction,
and whether other processes may be affected more quickly.   At
present, allelochemicals have been reported to inhibit nutrient
uptake by roots, cell division, extension growth, photosynthesis,
respiration, protein synthesis, enzyme activity, and to alter
membrane premeability (4), but little is known about their action
at the molecular level.
       Ecological Studies. Plant succession, particularly in old
fields and cut-over forests has intrigued ecologists for decades.
The appearance and disappearance of species and changes in
species dominance over time has been attributed to numerous
factors including physical changes in the habitat, seed
production and dispersal, competition for resources, or
combinations of all these.   Rice and co-workers (4) have
presented extensive evidence that allelopathy may play an
important role in the disappearance of the pioneer weeds (those
most rapidly invading old fields).   Additional findings in this
area could help us manage vegetation more effectively.
       Certain reforestation problems have also been linked to
allelopathy.   There are logged-over sites on the Allegheny
Plateau in Northwestern Pennsylvania that have remained
essentially treeless for up to eighty years (10).   Several
herbaceous weed species have been shown to produce toxins that
inhibit establishment of the black cherry (Prunus serotina)
seedlings that normally reinfest these sites.   Among the more
active are goldenrods (Solidago) and Aster species.   One wonders
why this idea could not be exploited for vegetation management on
right-of-way lands.
       In many ecosystems, plants tend to pattern themselves as
pure stands or as individuals spaced in rather specific densities
or configurations.   Many desert species show obvious zones of
inhibition around which few, if any, alien species   are able to
invade.   These patterns often cannot be adequately explained by
competition alone, and are probably caused by a combination of
factors including allelopathy.   The phenomenon happens with
herbaceous plants as well as woody shrubs and trees.
       Muller reported that black mustard (Brassica nigra) can form
almost pure stands after invading annual grasslands of coastal
southern California (25).   This was attributed to inhibitors
released from the dead stalks and leaves which do not permit
germination and growth of other plants.   These observations
provide agronomists hope that similar results could be exploited
with crops, specifically to achieve almost pure stands of crops
(over weeds) by use of an allelopathic mechanism.
       Positive and Negative Impacts of Allelopathy for Weed Science.
There is considerable evidence which now suggests that some of
the more aggressive perennial weed species, including quackgrass
(Agropyron repens) (26), Canada thistle (Cirsium arvense) (27),
Johnsongrass (28), and yellow nutsedge (Cyperus esculentus) (29)
may impose allelopathic influences, particularly through toxins
released from their residues. There are also several annual weed
species in which allelopathy is implicated.   Perhaps best

documented is giant foxtail (<u>Setaria faberi</u>) whose residues severely inhibited the growth of corn (<u>Zea mays</u>) (<u>6</u>).

Extracts of several important weed species were found to inhibit the nodulation of legumes by <u>Rhizobium</u> (<u>4</u>). Among those were Western ragweed, large crabgrass, prostrate spurge and annual sunflower. Our recent studies indicate that quackgrass releases compounds that inhibit nodulation and nitrogen fixation on a number of legumes. Adverse effects of weeds on nitrogen fixation appears to be an agricultural problem that deserves much more research attention.

The classic seed burial studies of W. J. Beal and his successors have shown seeds of at least one weed species, Moth Mullein (<u>Verbascum blattaria</u> L.) can remain viable in soil for a peiod of 100 years, whereas three species continued to germinate after 80 years of burial (<u>30</u>). Weed seeds not only resist decay by soil microbes, but they vary in dormancy characteristics. There is considerable evidence that chemical inhibitors are responsible for both phenomena. Unsaturated lactones and phenolic compounds in particular, are potent antimicrobial compounds present in many seeds (<u>4</u>). Fruits and seeds are also known to contain diverse germination inhibitors including phenolic compounds, flavonoids and/or their glycosides and tannins. Unique methods to destroy inhibitors could provide an excellent weed management strategy.

Recently, some weed scientists have attempted to directly exploit allelopathy as a weed management strategy. One approach has been to screen for allelopathic types in germplasm collections of crops, and to transfer this character into cultivars by either conventional breeding or other genetic transfer techniques. Superior weed suppressing types have been reported in cucumber (<u>24</u>), oat (<u>23</u>), sunflower (<u>31</u>), and soybean collections (<u>32</u>). When thoroughly researched, this idea may have potential for crops that are maintained in high density monocultures i.e. turfgrasses, forage grasses, or legumes.

Another approach is the utilization of allelopathic rotational crops or companion plants in annual or perennial cropping systems (<u>19</u>). Living rye (<u>Secale cereale</u> L.) and its residues have been shown to provide nearly complete suppression of a variety of agroecosystem weeds (<u>33</u>). Similarly, residues of sorghums, barley, wheat and oats can provide exceptional suppression of certain weed species (<u>19</u>). Although some progress has been made on identifying the allelochmicals from these plants, much remains to be accomplished.

Allelopathic plants may also provide a strategy for vegetation management in aquatic systems. The diminutive spikerush (<u>Eleocharis coloradoensis</u>) has been reported to displace more vigorous and unwanted aquatic plants i.e. pondweeds (<u>Potamogenton</u> species) and <u>Elodea</u> in canals and drainage ditches. Frank (<u>34</u>) attributed this to allelopathic effects, and more recently the phototoxic compound dihydroactinidiolide (DAD) was isolated and characterized from the spikerush plant (<u>35</u>). This chemical has since been shown to be inhibitory to pondweeds.

An important contribution from allelopathy research may be the discovery of novel chemicals either useful as pesticides or

precursers to pesticides. Both higher plants and microorganisms
are rich sources of diverse chemistry. Some excellent leads have
already been made in this area. For example, a cineole
derivative is now being developed as a herbicide by a major
chemical company. In addition, several potential herbicide
candidates have been isolated from broths of <u>Streptomyces</u>
cultures. Biotechology will undoubtedly complement chemical
synthesis for production of our future agrichemicals.

<u>Literature Cited</u>

1. Molisch, H., "Der Einfluss einer pflanze auf die andere -
    Allelopathie."   G. Fischer, Jena, 1937.
2. Kaminsky, R. Ecol. Monograph. 1981, 51, 365-82.
3. Whittaker, R.H., Feeny, P.P. Science, 1971, 171. 757.
4. Rice, E.E. In Allelopathy, Academic: New York 1983.
5. Putnam, A.R.; Duke, W.B. Ann. Rev. Phytopathology, 1978.
6. Bell, D.T., Koeppe, D.E. Agron. J. 1972, 64, 321.
7. McCalla, T.M., Haskins, F.A. Bacterial Rev. 1964, 28, 181.
8. Conrad, J.P. Amer. Soc. of Agron. 1927, 19, 1091.
9. Borner, H. Contrib. Boyce Thompson Inst. 1959, 20, 39.
10. Horsley, S. B. Can. J. Forest Res. 1977, 7, 205.
11. Bell, D.T., Muller, C.H. Amer. Midl. Natur. 1973, 90, 277.
12. delMoral, R., Muller, C.H. Amer. Midl. Natur. 1970, 83, 254.
13. Muller, C.H. Torrey Bot. Club. 1965, 92, 38.
14. Rovira, A.D. Bot. Rev. 1969, 35, 35.
15. Tang, C.H., Young, C-C. Plant Physiol. 1982, 68, 155.
16. Tukey, H.B., Jr. Bull. Torrey Bot. Club. 1966, 93, 385.
17. Colton, C.E., Einhellig, F.A. Amer. J. Bot. 1980, 67, 407.
18. Conn, E.E. Ann. Rev. Plant Physiol. 1980, 31, 433.
19. Putnam, A.R., DeFrank, J., Crop Protection. 1983, 2, 173.
20. Putnam, A.R. Chem. Eng. News. 1983.
21. Swain, T. Ann. Rev. Plant Physiol. 1977, 28, 479.
22. Koeppe, D.E., Rohrbaugh, L.M., Wender, S.H. Phytochem. 1969,
    8, 889.
23. Fay, P.K., Duke, W.B. Weed Sci. 1977, 25, 224.
24. Putnam, A.R., Duke, W.B. Science. 1974, 185, 370.
25. Muller, C.H. Vegetatio. Haag 18, 1969, 348.
26. Gabor, W.E., Veatch, C. Weed Sci. 1981, 29, 155.
27. Wilson, R.G., Jr. Weed Sci. 1981, 29, 159.
28. Nicollier, G.P., Pope, D.F., Thompson, A.C., J. Agr. Fd.
    Chem. 1983, 31, 744.
29. Drost, D.C., Doll, J.D. Weed Sci. 1980, 28, 229.
30. Kivilaan, A., Bandurski, R.S. Am. J. Bot. 1973, 60, 146.
31. Leather, G.R. Weed Sci. 1983, 31, 37.
32. Massantini, F., Caporali, F., Zellini, G. Symp. on the
    Different Methods of Weed Control and Their integration.
    1977, 1, 23.
33. Barnes, J.P., Putnam, A.R. J. Chem. Ecol. 1983, 9, 1045.
34. Frank, P.A., Dechoretz, N. Weed Sci. 1980, 28, 499.
35. Stevens, K.L., Merrill, G.B. J. Agr. Fd. Chem. 1980, 28, 644.

RECEIVED April 1, 1984

Economics of Weed Control in Crops

J. M. CHANDLER

Texas Agricultural Experiment Station, Texas A&M University, College Station, TX 77843

In United States crops, losses caused by weeds and the cost of their control totals about $15.2 billion annually. Total losses from weed competition with major crops produced in the United States is approximately $8.9 billion. Total expenditures for herbicides used in United States crops are approximately $2.1 billion with application of these herbicides costing $938 million annually. Equipment and labor costs of about $3.1 billion are spent annually for interrow tillage of crops and tillage to control weeds between cropping seasons. Without current control technology the losses from weeds would be double or triple. Expenditures for research studies of herbicides and plant growth regulators have increased during the past 5 years, but only minimal resources have been allocated to studies of allelochemicals.

The recent gains in world population and the pressure they place on natural resources is having an impact on the economics of weed control currently and will have far reaching impact on it in the future. Since the early 1940's the world's population doubled - from less than 2.3 billion to over 4.6 billion in 1983. The annual rate of growth is about 1.7 percent and recent projections indicate the global population will increase to slightly over 6 billion by 2000.

Natural resources affect people and people affect natural resources. Many qualities or attributes of nature are customarily described as "natural resources" - soils, climate, plants, minerals, and others. But these are not really resources for human welfare until appropriate amounts of human energy, resourcefulness, capital, and managerial capacity have been applied to these qualities of nature, to transform them into materials and services usable by people.

Natural resources are scarce in relation to human wants or demands. Many natural resources and their products are traded in markets. The basic cost of natural resources is determined by the willing seller negotiating with the willing buyer, until they

0097–6156/85/0268–0009$06.00/0
© 1985 American Chemical Society

have agreed upon a price. When a price has been arrived at in a
market, and this price is generally known, it is an important
piece of information which can guide both producers and
consumers. Producers can make their production decisions in
light of the price. Actually, they should use the price they
expect when their product is ready for market. Consumers can use
the price as a guide for purchases and consumption decisions.

A weed, which is a form of a natural resource, is simply a
living organism whose presence conflicts with the interests of
people. Undesirable as they may be, weed species are natural
occurrences in the agricultural ecosystems created by man's food
and fiber production. Over 300,000 species of plants inhabit the
earth, but only 30,000 of these are weeds. About 1,800 weed
species cause serious economic losses in crop production, and
about 300 weed species are serious in cultivated crops throughout
the world. Most cultivated crops are plagued by 10 to 30 weed
species that must be controlled to avoid yield reductions (1).

Holm et al. state that across the world about 200 weed
species are involved in 95% of man's weed problems as related to
food production (2). They list 80 species as primary weeds and
120 species as secondary weeds. The United States has seventy
percent of the world's worst weeds; 24% of these are grasses, 67%
broadleaf weeds, and 7% sedges.

Public Welfare as Affected By Weeds

Plants are detrimental in many ways to the health and well-being
of man. It is difficult to quantify the health aspects of weeds.
Weeds known to produce allergenic reactions in humans can be
divided into two major groups. One group produces skin eruption
as a result of bodily contact with the plant, while the other
produces symptoms usually affecting the respiratory tract through
inhalation of pollen grains.

Allergenic plants causing skin eruption by contact include
poison ivy (Rhus radicans L.), poison oak (Rhus
toxicodendron L.), poison sumac (Rhus vernix L.), and
stinging nettle (Urtica dioica L.). In the United States,
poison ivy and poison oak cause nearly 2 million cases of skin
poisoning and skin irritation annually, for a loss of 333,000
working days. In addition, these weeds cause 3.7 million days of
restricted activity among those people who are susceptible to the
toxins (1).

Plants causing allergenic reactions by inhalation of pollen
grains are numerous and widely distributed. Plant families whose
pollen causes allergic reactions include Gramineae,
Amaranthaceae, Chenopodiaceae, Leguminosae, Polygonaceae,
Plantaginaceae, and Compositae. Ragweed (Ambrosia spp.) is the
most important single allergenic plant on the North American
continent and affects more of our citizens than does smog. Hay
fever and similar allergies from weed pollen cause tremendous

absenteeism among school children and working adults, because
over 20 million people in the United States are affected.
    In the United States, more than 700 species of plants are
known to be poisonous. On occasion death occurs from eating
poisonous plants, berries, seeds, or tubers. It has been
estimated that 75,000 cases of plant poisoning occur annually
with children being the prime victims. Adult poisoning often
results from the consumption of therapeutic teas or use of toxic
weed, such as hemp, as hallucinogens and mind-altering drugs.
    Weeds indirectly affect the health of man and animals by
harboring animals or insects. Control of aquatic weeds is
effective for mosquito (Anopheles quadrimaculatus Say)
control by eliminating breeding habitats. In control of the
tsetse fly (Glossian spp.) (vector of sleeping sickness in
Africa), herbicides are involved to reduce growth of the brush so
essential to the survival of the fly.

Classes of Losses Caused By Weeds in Crops

Crop losses due to the presence of weeds may be severe although
the damage caused is not always as obvious as losses caused by
other pests. These losses occur at various stages in the crop
production cycle. Weeds (a) reduce the yield of crops; (b)
impair the quality of the crop; (c) increase costs of hand
tillage, mechanical tillage, fertilizer, and herbicides; (d)
prevent efficient irrigation and water management; (e) reduce the
value of the land and increase the acreage needed for crop
production; (f) serve as hosts and habitats for insects,
nematodes, disease-causing organisms, and rodents.
    Weeds directly compete with the crop for water, nutrients,
light and other growth factors. Competition for water begins
when root systems overlap as they absorb water and nutrients (3).
Competition for water depends on the rate and completeness with
which a plant utilizes the soil water supply (4). Competition
for water usually occurs with other forms of competition. For
example, competition between weeds and peas (Pisum spp.)
centered on light and water depending on weed height (5).
Research has shown that weeds usually absorb fertilizer faster
and in relatively larger amounts than crops and therefore derive
greater benefit (6). Studies in West Texas showed that cocklebur
(Xanthium spp.), Russian thistle (Salasola kali L. var.
tenuifolia Tausch), and puncturevine (Tribulus terrestris
L.) utilized twice the amount of nitrogen as sorghum [Sorghum
bicolor (L.) Moench.] during the growing season, but crabgrass
(Digitaria spp.), buffalobur (Solanum rostratum Dunal), and
barnyardgrass [Echinochloa crusgalli (L.) Beauo.] used only
half as much on a per acre basis (7). Competition between
selected crops and field bindweed (Convolvulus arvensis L.),
under ample soil water and nutrient supply, centered around
available light (8). Purple nutsedge (Cyperus rotundus L.),

classified as the world's worst weed, is intolerant to shade and
forms 10 to 57% fewer tubers and bulbs when shaded continuously
(2, 9).

Estimated Crop Losses from Weed Competition

The increased use of herbicides along with mechanical tillage
over the past three decades in crop production systems has lead
to a drastic reduction in the hand labor required to control
weeds. Even with this shift, we still have weeds as major pests
in our crops that result in a $8,971,130,000 loss annually. This
figure represents losses resulting from weed competition but does
not include losses in crop quality due to weeds, increased costs
for fertilizer and irrigation, losses during harvest through
decreased efficiency, and losses during storage and processing of
crops. It has been estimated that weeds cause losses of about 10
percent in agricultural production, including crops, livestock,
forests, and aquatic resources currently valued at more than $12
billion (10).

The monetary loss due to weeds competing with the first
seven row crops listed in Table I was approximately five billion
dollars annually. Losses in soybeans [Glycine max (L.)
Merr.] and corn (Zea mays L.) were 21 and 19%, respectively,
of the total. The predominant weeds in these crops were pigweeds
(Amaranthus spp.), cocklebur, foxtails (Setaris spp.) and
crabgrass (11). Monetary losses in cotton (Gossypium
hirsutum L.), sorghum, and peanuts (Arachis hypogaea L.)
were 5, 7, and 2% respectively, with the predominant weeds being
johnsongrass [Sorghum halepense (L.) Pers.], nutsedge,
cocklebur, pigweeds and crabgrass. A monetary loss of 1% or less
occurred in sugarbeets (Beta vulgaris L.), sugarcane
(Saccharum officinarum L.), and forage seed crops.

The total monetary loss resulting from weed competition in
the cereal crops was $1.3 billion annually. The most frequently
reported weeds were mustards (Brassica spp.) followed by wild
oats (Avena fatua L.), bromes (Bromus spp.), and wild
garlic (Allium vineale L.) (11). Losses in vegetables was 5%
of the total while in fruit and nuts the loss was 7% of the
total. Crabgrass, bermudagrass [Cynodon dactylon (L.)
Pers.], johnsongrass, pigweeds, and nutsedges were the most
frequently reported weeds.

Weed competition in hay crops, pasture, and rangeland caused
a $1.5 billion annually. The predominant weeds in hay crops are
quackgrass [Agropyron repens (L.) Beauv.], thistles
(Circium spp.), and bromes. In the eastern pastures the
predominant weeds were thistles, ragweeds, and dock (Rumex
spp.), while on the western rangelands the predominant weeds were
sagebrush (Artemisia spp.), weedy bromes, and larkspurs
(Delphinium spp.).

Table I.   Estimated average annual losses due to weed competition
in crops, United States, 1975-1979

Commodity	Commodity Market Value[a] ($1,000)	Monetary Losses from Potential Production[a] ($1,000)
Row Crops:		
Corn	14,962,306	1,702,251
Cotton	3,229,353	475,134
Sorghum	3,825,443	607,340
Soybeans	10,258,870	1,912,100
Peanut	719,714	161,922
Sugarbeets	638,491	89,024
Total	34,163,792	5,029,203
Small Grain Crops:		
Rice	1,072,280	268,895
Wheat	5,958,990	854,689
Other grain	1,735,065	255,396
Total	8,766,335	1,378,980
Vegetables:	4,698,315	441,449
Fruits and nuts:	4,514,334	619,072
Forage Crops:		
Forage seed crops	190,200	37,400
Hay	6,690,022	676,221
Pasture and rangeland	4,960,600	788,805
Total	11,840,822	1,502,426
TOTAL	63,983,600	8,971,130

[a]Commodity market value and monetary loss were constructed from
information given in "Crop Losses Due to Weeds in Canada and the
United States", 1984 J.M. Chandler, A.S. Hamill, and A.G. Thomas,
Special Report. Weed Science Society of America, Champaign, IL.

It has been estimated that without the use of herbicides in the United States' the losses due to weeds in cotton, corn, peanuts, and soybeans would be 40, 25, 90, and 24% of the total crop value, respectively (12). An impact assessment study has shown that the loss of all herbicides would increase the consumers food bill by $8 billion annually (13).

Extent of Herbicide Usage and Cost in Crops

Recognizing that weeds are active participants in the dynamic agricultural environment has led to the current development and use of ecosystem-oriented methods of weed control or integrated weed management systems. These systems involve the judicious integration of cultural, mechanical, and chemical control procedures. The practices of crop rotation, mulching, fallowing, and flooding can be used for weed control in selected crops in conjunction with other control techniques. Mechanical tillage prior to crop establishment and during the growing season between the crop rows is an effective nonselective weed control procedure. Limited amounts of hand hoeing are still used to control weeds that escape other forms of control, but in the United States it is very expensive. Currently the use of selective herbicides applied in a band directly over the crop row or as a broadcast spray is used widely to control weeds.

In 1949 herbicides were used on 9.3 million ha of agricultural land, on 21.5 million ha in 1959, and on 28.7 million ha in 1962 (14). Herbicides were used on 39.6 million ha in 1966 and on 79.7 million ha in 1976 (15). These figures show that the use of herbicides over the past three decades has increased nine fold.

Between 1971 and 1982 the proportion of row crop acreage in the United States treated with herbicides increased from 71 to 91% (Table II). A modest increase of 38 to 44% occurred in small grain crops. During 1981 ninety percent or greater of the acreage in corn, cotton, soybeans, peanuts, and rice (Oryza sativa L.) was treated with a herbicide (16).

A total of 94 and 204.7 million kilograms of herbicides were applied to crops in 1971 and 1982, respectively, in the United States (Table II). Total quantity of herbicides applied in row crops increased by 43% and in small grains by 66% between 1971 and 1982. In 1971 corn, cotton, and soybeans received 49, 9, and 18%, respectively, of the total quantity of herbicides applied while in 1982 they received 54, 4, and 28%, respectively (16).

In 1980 United States herbicide sales of $2.166 billion represented 44% of the total worldwide pesticide sales for crops (17). In the United States herbicides represented 66% of all pesticide sales for crops (Table III). Herbicides for row crops account for 84% of the total United States herbicide sales with 35% being sold in corn and 34% in soybean. Herbicide sales in small grain crops stand at 8% of the total with 7.5% divided

Table II. Percentage of crop acreage treated with herbicides and total herbicide usage in the United States in 1971 and 1982

Commodity	Proportion of Hectares Treated with Herbicide[a]		Herbicide Applied[a] (Active Ingredient)	
	1971	1982	1971	1982
	(Percent)		(Million kg)	
Row Crops:				
Corn	79	95	45.8	110.4
Cotton	82	97	8.9	7.8
Sorghum	46	59	5.2	6.9
Soybeans	68	93	16.6	56.8
Peanuts	92	93	2.0	2.2
Tobacco	7	71	0.1	0.7
Total	71	91	78.6	184.8
Small Grain Crops:				
Rice	95	98	3.6	6.3
Wheat	41	42	5.3	8.2
Other grain	31	45	2.5	2.7
Total	38	44	11.4	17.2
Forage Crops:				
Alfalfa	1[b]	1	0.2[b]	0.1
Other hay	−	3	−	0.3
Pasture and rangeland	1	1	4.8	2.3
Total	1	1	4.0	2.7
TOTAL	17	33	94.0	204.7

[a] Proportion of hectares treated with herbicides and herbicide applied were constructed from information given in "Inputs Outlook and Situation", 1983, Herman Delvo and Michael Hawthorn, U.S. Department of Agriculture, Economic Research Service, Washington, DC.

[b] Included in alfalfa figure.

Table III. Herbicide and total pesticide purchases by crop
in the United States and the World, 1980

Crop	Total Herbicide Purchases[a]		Total Pesticide Purchases[a]	
	United States	World	United States	World
	($ million)		($ million)	
Row Crops:				
Corn	754	1,077	1,002	1,516
Cotton	142	321	366	1,266
Sorghum	59	83	82	165
Soybeans	749	913	799	1,073
Peanuts	31	46	83	162
Tobacco	17	32	44	144
Sugarbeets	26	274	39	400
Sugarcane	14	100	22	174
Other	29	92	52	209
Total	1,812	2,938	2,489	5,109
Small Grain Crops:				
Rice	53	429	69	1,300
Wheat	92	622	112	970
Other grain	29	279	41	421
Total	174	1,330	222	2,691
Vegetables:	60	203	205	970
Fruits and Nuts:	56	182	292	1,642
Forage Crops:				
Alfalfa	16	36	42	88
Other hay & forage crops	9	15	14	41
Pasture and rangeland	30	78	36	99
Total	55	129	92	228
Other crops:	---	109	---	366
TOTAL	2,166	4,891	3,305	11,006

[a]Total herbicide purchases and total pesticide purchases were
constructed from information given in "A Look at World
Pesticide Markets", Farm Chemical, September 1981. Meister
Publishing Co., Willoughby, Ohio.

equally between vegetables, fruits and nuts, and forage crops.
Forty-four percent of all herbicides sold in 1980 were in the
United States. In addition to the sales of herbicides for
agricultural uses, $750 million was expended on herbicides for
use by commercial companies, government, and homeowners (18).
    The value of all herbicides for crops marketed worldwide in
1980 was estimated at about $4.9 billion (17). The sale of
herbicides for weed control in corn in the United States
represented 70 and 49% of the world herbicide and pesticide
purchases in corn, respectively. The sale of herbicides in
soybeans represents 82 and 69% of the world herbicide and
pesticide purchases in soybeans, respectively. The sale of
herbicides for weed control in row crops in the United States
represents 62 and 36% of the world herbicide and pesticide
purchases for row crops, respectively.
    Worldwide herbicide sales were projected to increase by
11.9% from 1980 to 1985 (17). The projected increase in
herbicide sales in the United States from 1980 to 1985 was
estimated at 12.4%. It was estimated that by 1985 herbicide
sales would represent 61% of all pesticide sales in the United
States, whereas herbicides would account for 42% of the pesticide
sales worldwide (17). It is projected that pesticide sales by
1990 will reach $6.14 billion, a 7% per year growth from 1978's
$2.72 billion in sales (19). It is also projected that the
average cost of pesticides will increase from $3.46/kg in 1978 to
$6.88/kg in 1990 because of higher raw-materials costs and a
shift to higher-performance materials.
    An annual estimated cost of approximately $938,835,000 is
required for labor and equipment to apply 204.5 million kilograms
of herbicides to 147.6 million hectares of cropland in the United
States (Table IV). Mechanical tillage to control weeds between
cropping seasons and interrow tillage of crops has been used for
centuries as a very effective method of weed control.
Approximately 50% of all tillage between crops is done to control
weeds. The number of cultivations required for effective weed
control within a row crop varies from two to five during a
cropping season.
    Tillage for weed control in intertilled row crops cost on
the average about $29.65/ha/yr. With approximately 76.9 million
ha of intertilled row crops the cost of tillage is in excess of
$2.28 billion annually. In drill row crops where tillage is less
intent, a cost of about $12.35/ha/yr is required on approximately
70.8 million hectares. The cost in these moderate tilled areas
is in excess of $874 million annually.
    The estimated cost of equipment, labor, and herbicides for
weed control within a specific crop can be very expensive. For
example, in 1980, cotton was planted on 5.2 million ha in the
United States. The total cost for equipment, labor, and
herbicides used to control weeds was estimated to be $547 million
(20). Herbicides account for 32% of the total cost while
equipment and labor costs were 34% each.

Table IV.   Planted acreage in United States, 1981

Commodity	Planted Hectares[a]
	(1,000)
Row Crops:	
Corn	34,056
Cotton	5,799
Sorghum	6,485
Soybeans	27,520
Peanuts	614
Sugarbeets	509
Sugarcane (harvested)	305
Tobacco (harvested)	394
Total	75,682
Small Grain Crops:	
Rice	1,555
Wheat	35,964
Other grain	10,775
Total	48,294
Vegetables:	1,242
Fruits and Nuts:	1,431
Forage Crops:	
Alfalfa	10,681
Other hay (less wild hay)	10,334
Total	21,015
TOTAL	147,664

[a]Planted hectares per crop were obtained from
"Agricultural Statistics 1982", United States Department
of Agriculture, Washington DC.

The total cost for full-season weed control varies considerably with a specific crop or among different crops (21, 22). The production environment, especially the amount of available moisture influences the weed control strategy and the costs incurred. For example, in cotton the total cost for full-season weed control in West Texas was $76/ha, but in the Mississippi Delta the cost was $156/ha (20).

Full-season weed control in peanuts produced in Georgia was $168/ha. In soybeans produced in Mississippi the cost for full-season weed control was $74/ha while $95/ha was spent for full-season weed control in Illinois (21).

In summary, weed competition with crop plants and their control constitute some of the highest costs in the production of food, feed, and fiber. In the United States, losses caused by weeds and the cost of their control totals about $15.2 billion annually. Total losses from weed competition with major crops produced in the United States is approximately $8.9 billion. Total expenditures for herbicides used on croplands in the United States are approximately $2.1 billion with appliation of these herbicides costing $938 million annually. Equipment and labor costs of about $3.1 billion are spent annually for interrow tillage of crops, and tillage to control weeds between cropping seasons. About 71% of the weed losses due to competition occur in field crops, 17% in forage crops, and 12% in vegetables, fruit and nut crops. Without current control technology the losses would be doubled or tripled. Herbicides comprise 66% of all pesticides sold in 1980. Herbicide sales in corn, cotton, and soybeans account for 77% of all herbicides sales in the United States. It is projected that pesticide sales by 1990 will reach $6.14 billion, a 7% per year growth from 1978's $2.72 billion in sales. Herbicide sales will continue to account for nearly 60% of the total. It has been estimated that by the year 2000, no-tillage farming will be used on over 62 million ha, minimum tillage systems will be used on another 62 million ha, and only 20 million ha will be farmed conventionally. An estimated 68 million kg of herbicides beyond the present level of use will be required annually to make the conversion. Expenditures for research studies of herbicides and plant growth regulators have increased during the past 5 years, but only minimal resources have been allocated to studies of allelochemicals (22).

Literature Cited

1. Burnside, O. C. "Introduction to Crop Production;" W.B. Ennis, Jr., Ed.; American Society of Agronomy, Madison, WI, 1979; p 27-38.
2. Holm, L. G.; Plucknett, D. L.; Pancho, J. V.; Herberger, J. P. "The World's Worst Weeds: Distribution and Biology;" The University Press of Hawaii, Honolula, HI, 1977; p 8.

3. Pavlychenko, T. K.; Harrington, J. B. Science Agriculture. 1935, 16, 151.

4. Donald, C. M. Advances in Agronomy. 1963, 15, 1-118.

5. Nelson, D. C.; Nylund, R. E. Weeds. 1962, 10, 224-228.

6. Vengris, J.; Colby, W. G.; Drake, M. Agronomy Journal. 1955, 47, 213-216.

7. Shipley, J. L.; Wiese, A. F. Texas Agricultural Experiment Station. 1969, MP 909, p 8.

8. Stahler, L. M. American Society Agronomy. 1948, 40, 490-502.

9. Hauser, E. W. Weed Science. 1962, 10, 209.

10. Shaw, W.C. Proc. Western Society of Weed Science. 1984, 37, 1-26.

11. "Extent and Cost of Weed Control with Herbicides and on Evaluation of Important Weeds;" U.S. Department of Agriculture, ARS-H-1, 1972, p 1-227.

12. Abernathy, J. R. "Estimated Crop Losses Due to Weeds with Non-chemical Management;" D. Pimental (Ed.) Handbook of Pest Management in Agriculture, Vol I, CRC Press, Boca Raton, Floridy, 1981; p 159-167.

13. Farm Chemical. "If No Pesticides Could Be Used - How Much More Would Consumers Pay for Food?;" Meister Publishing Co., Willoughby Ohio, December, 1982; p 40-41.

14. Shaw, W. C. Proc. Southern Weed Science Society. 1978, 31, 28-47.

15. Eichers, T. R. "Use of Pesticides by Farmers;" D. Pimental (Ed.) Handbook of Pest Management in Agriculture, Vol II, CRC Press, Boca Raton, Florida, 1981; p 3-25.

16. Delvo, Herman and Hanthorn, Michael. "Inputs Outlook and Situation;" U.S. Department of Agriculture, Economic Research Service, 1983; p 3-23.

17. Farm Chemical. "A Look at World Pesticide Markets;" Meister Publishing Co., Willoughby, Ohio, September 1981; p 55-60.

18. Aspelin, A.L. and Ballard, G.L. "Pesticide Industry Sales and Usage - 1980 Market Estimates;" Environmental Protection Agency, 1980; p 1-13.

19. Chemical Week. "Pesticides: $6 Billion by 1990;" McGraw Hill, Hightstown, NJ, May 7, 1980; p 45.

20. Chandler, J. M. "Cotton Protection Practices in the USA and World - Weeds;" American Society of Agronomy, Madison, WI, 1984; p 605.

21. McWhorter, C. G.; Chandler, J. M. "Conventional Weed Control Technology;" R. Charudattan and H.L. Walker (Ed.), Biological Control of Weeds with Plant Pathogens; John Wiley and Sons, New York, 1982; p 5-27.

22. Farm Chemical. "What's Happening in Pesticide Research;" Meister Publishing Co., Willoughby, Ohio, September, 1982; p 23.

RECEIVED August 16, 1984

Assessment of the Allelopathic Effects of Weeds on Field Crops in the Humid Midsouth

C. D. ELMORE

Southern Weed Science Laboratory, Agricultural Research Service, U.S. Department of Agriculture, Stoneville, MS 38776

Weed interference, which consists of the combined effects of competition and allelopathy, is responsible for millions of dollars in crop loss in the humid Mid-South. The allelopathic component of interference has not been clearly determined under field conditions. A literature review indicated allelopathy (as determined by greenhouse and laboratory assay) to be a potential factor in the loss caused by many weeds of the Mid-South; e.g. johnsongrass [Sorghum halepense (L.) Pers.], purple nutsedge (Cyperus rotundus L.), yellow nutsedge (Cyperus esculentus L.), prostrate spurge (Euphorbia maculata L.) and velvetleaf (Abutilon theophrasti Medic.). In addition, wheat (Triticum aestivum L.) straw, prickly sida (Sida spinosa L.), and pigweeds (Amaranthus sp.) have been preliminarily implicated as having allelopathic effects on field crops in the humid Mid-South. Although allelopathy by weeds is a potential factor in crop losses, no published reports have unequivocally shown an allelopathic effect on the weed-crop associations in the field in the Mid-South. Therefore, a needed first approach to the demonstration of allelopathy in the field is to determine if weed residues have an allelopathic effect on Mid-South crops. Then, either a statistical or biological approach could be devised to partition weed interference into its components, competition and allelopathy.

Weeds are an enormous problem affecting field crops in the Mid-South. They reduce yields, increase the cost of production, reduce the quality of the harvested produce, and decrease the value of the marketed product. These effects of weeds are present wherever crops are grown; however, they seem to be more devastating

and pernicious in warm, humid temperate, and tropic regions than in
cooler and drier climates.

How weeds produce their effect is the subject of continuing
research. As rather recently defined, the negative effect of weeds
on crop plants includes both competition and allelopathy (1) and
has been termed interference. Previous weed science research
considered the competition component foremost and, usually, solely
(e.g. 2). Allelopathy, however, has received some attention
recently, with a number of symposia and reviews devoted to
allelopathic effects. Some of these cover the effects of weeds on
crops (3, 4, 5). In this chapter, I review the possible role of
allelopathy in weed interference in crop production in the
Mid-South. A second objective of this review is to indicate where
research is needed and to suggest potential lines of future
research, especially with respect to the developing role of
conservation tillage practices in this region.

Allelopathic Weeds

The principal summer annual row crops in the Mid-South include
cotton (Gossypium hirsutum L.), corn (Zea mays L.) soybeans
[(Glycine max (L.) Merr.], grain sorghum (Sorghum bicolor (L.)
Moench.), peanuts (Arachis hypogea L.), and tobacco (Nicotiana
tabacum L.). Sugarcane [(Saccharum officinarum L.) (not an annual
crop)], the small grains wheat (Triticum aestivum L.) and rice
(Oryza sativa L.) and small-acreage specialty crops such as
sunflower (Helianthus annuus L.) and syrup crops are not considered
in this review.

The common and troublesome weeds present in these six crops in
the 13 Southern states (VA, NC, SC, GA, FL, AL, MS, TN, KY, AR, LA,
TX and OK) are regularly surveyed by State Extension Weed
Specialists. The most recent survey includes 59 weeds which are
common or troublesome in these six field crops (6). Twenty weeds
on that list have been implicated as being allelopathic (Table 1);
eleven are among the 10 most common in at least one of the six
major field crops in the Southern states. Ten of these 11 weeds
are considered to be among the 10 most troublesome, usually for the
same crop in which it was found to be common. There is one weed
[bermudagrass (Cynodon dactylon (L.) Pers] that is on the most
troublesome list for cotton and tobacco, but not on any most common
list. Similarly, giant foxtail (Setaria faberi Herrm.) is a common
weed in tobacco, but apparently not a troublesome one.

An interesting feature of this list is that some of our most
troublesome and difficult to control weeds are not suspected
allelopathics. Many weeds, such as morningglory (Ipomoea spp.),
cocklebur (Xanthium strumarium L.), and sicklepod (Cassia
obtusifolia L.), which make almost all lists of serious weed
species, are strong competitors with little or no evidence of
allelopathic activity. In fact, cocklebur is probably the epitome
of what a competitor can be, being even more devastating than

johnsongrass to soybeans (17). Perhaps the right tests have not
been reported for these weed species (Einhellig, at this symposium,
reports that cocklebur is in fact highly allelopathic to soybeans).
More could be said, but suffice it to say that weed competition is
a strong force; in some cases, it is the only force needed by weeds
to exert their detrimental effect on crops.

One of the features of allelopathy, as pointed out by Rice (18),
is that it is specific. Certain species, but not others, are
affected by the allelochemics produced by a plant. This suggests
that allelopathy by a weed must be positively demonstrated for each
crop. References to specific reports of demonstrated allelopathy
by a weed on a crop are given in Table 1. In some cases [crabgrass
(Digitaria sanguinalis (L.) Scop.), spurge (Euphorbia spp.), etc.],
no crop is listed in the allelopathy column.

Table I provides general information on the identified
allelopathic weeds and the crops they affect. More detailed
information on two of these weeds, johnsongrass and purple
nutsedge, will be presented to show the tenuous nature of the
evidence for allelopathy. These two species are undoubtedly
allelopathic, at least under certain conditions. Rigorous proof
that allelopathy is the agent responsible for even a specific
portion of the interference exerted by them is not easily attained,
however, even though these are the weeds with the best research
data available of those weeds that occur in the Mid-South.

Johnsongrass. Johnsongrass was one of the first weeds to be
implicated as having allelopathic potential (19). Friedman and
Horowitz (20) demonstrated that dried and fragmented pieces of
johnsongrass rhizomes added to soil (50 g/kg) and later (2 and 4
months) extracted with distilled water inhibited seedling
development of barley (Hordeum vulgare L.), mustard [Brassica nigra
(L.) Koch.] and wheat. In a subsequent test, barley sown in the
soil to which the rhizomes had been placed, but removed after 2 or
4 months, also seemed to be inhibited in seedling development (21).
Ethanolic extracts of the soil from which the rhizome fragments
had been removed also showed allelopathic activity. This same
technique worked with cotton as the test species (10). In a
similar experiment, Lolas and Coble (9) showed that johnsongrass
was allelopathic to soybeans. They collected soil from a
johnsongrass-infested field and removed the rhizomes before assay.
A regression analysis was used to test the johnsongrass rhizome
biomass for allelopathic activity.

These tests demonstrate that rhizomes of johnsongrass have
biological activity and that the activity is residual. The
allelochemic usually associated with Sorghum species in general,
and with johnsongrass in particular, is the cyanogenic glucoside
dhurrin (3) or its decomposition product (p-hydroxybenzaldehyde).
Whether these compounds act as allelochemics in field situations is
unknown and somewhat suspect, since they would surely be
immobilized or altered in most soil situations, as was shown for

Table I. Weeds With Reported Allelopathic Activity Found in Row Crops in the Southern United States

Weed[1]	Crops Affected		
	Common[2]	Troublesome[3]	Allelopathic[4]
Large crabgrass Digitaria cilaris (Retz.) Koel	corn, peanuts cotton, sorghum soybeans, tobacco	corn, sorghum	--
Pigweeds Amaranthus spp.	corn, cotton peanuts, sorghum soybeans, tobacco	corn, sorghum soybeans, tobacco	soybeans, corn (7)
Johnsongrass Sorghum halepense (L.) Pers	corn, cotton peanuts, sorghum soybeans	corn, cotton peanuts, sorghum soybeans	corn (8) soybeans (9)
Purple nutsedge Cyperus rotundus L.	cotton, peanuts soybeans, tobacco	corn, cotton peanuts, sorghum soybeans, tobacco	cotton (10)
Yellow nutsedge Cyperus esculentus L.	cotton, peanuts soybeans, tobacco	corn, cotton peanuts, sorghum	corn, soybeans (13)
Spurge Euphorbia spp.	cotton	cotton	--
Giant foxtail Setaria faberi Herrm.	tobacco	--	corn (12)
Sunflower Helianthus annuus L.	--	--	sorghum, soybeans (13)

Species			
Bermudagrass Cynodon dactylon (L.) Pers.	–	cotton, tobacco	cotton (10)
Velvetleaf Abutilon theophrasti Medic.	–	–	soybeans (14)
Ragweed Ambrosia spp.	sorghum, soybeans tobacco	soybeans	–
Kochia Kochia scoparia (L.) Schrad	–	–	–
Prickly sida Sida spinosa L.	cotton, soybeans	cotton, peanuts	soybeans (15)
Broadleaf signalgrass Brachiaria spp.	corn, sorghum soybeans	corn, sorghum	–
Lambsquarters Chenopodium album L.	corn, tobacco	tobacco	–
Sandbur Cenchrus spp.	–	–	–
Common purslane Portulaca oleracea L.	–	–	–

Table I. Continued on next page

Table I. (Continued)

Weed[1]	Crops Affected		
	Common[2]	Troublesome[3]	Allelopathic[4]
Shattercane Sorghum bicolor (L.) Moench	—	—	sorghum, corn (16)
Field bindweed Convolulus arvensis L.	—	—	—
Jimsonweed Datura stramonium L.	—	—	—

[1] A weed is listed if it occurs as a weed in any of the row crops of the Mid-South. (From 1984 SWSS Research Report) (6).
[2] A crop is listed if the weed is one of the top 10 common weeds in that crop in the Mid-South.
[3] A crop is listed if the weed is one of the top 10 troublesome weeds in that crop in the Mid-South.
[4] A crop is listed if a report exists of specific allelopathic activity of that weed on the crop.

ferulic acid (22). Recent evidence would implicate other chemical
classes of compounds not previously associated with allelopathy
(22). Perhaps it is not these compounds, but other undetected
highly toxic compounds or even species with short half-life
periods, that are responsible.

These experiments which have been used to document allelopathy
of johnsongrass to cotton and soybeans are either in vitro tests
using Johnsongrass litter (dead decaying fragments of johnsongrass
rhizomes), or greenhouse studies with litter and live plant
material. In the field with an infestation of johnsongrass, how is
allelopathy distinguished and partitioned from competition? Table
II is an illustration that may be useful.

Within any paired treatment in that experiment where the soybean
herbicide injury was the same, and the only differences were
between presence or absence of johnsongrass, the yield was greater
when johnsongrass was absent, even though the control of the weed
was nearly complete in the johnsongrass plots. This, of course, is
a confounded experiment, because competition for 3 weeks may have
also been responsible. Other research would suggest, however, that
three weeks of competition by johnsongrass is not sufficient to
reduce yield (2). The implication here is that the herbicide
treated johnsongrass plants released some factor(s) into the
environment of the developing soybean plant and reduced the yield
by 135 kg/ha. This 135 kg/ha then is probably the allelopathic
component of the 430 kg/ha reduction caused by interference from
uncontrolled rhizome johnsongrass. This would mean that
allelopathy accounts for approximately 30% of the johnsongrass
interference. Such an analysis is obviously too simplistic, the
allelopathic effect may be much more in larger, healthier plants,
or if injured, dead or metabolically altered plants produce or
leach more allelochemics, even less. Thus, the allelopathic factor
is very difficult to partition from competition, as was suggested
by Fuerst and Putnam (25), Dekker et al. (26) and Elmore et al.
(27).

Purple nutsedge. Purple nutsedge, possibly the world's worst weed
(28), has been similarly tested for allelopathy. Friedman and
Horowitz (20, 29), and Horowitz (10), in experiments similar to
those discussed above for johnsongrass, have shown that fragments
of purple nutsedge tubers and rhizomes inhibit growth of wheat,
mustard, barley and cotton. In a separate study (21) they further
identified the allelochemics as phenolic acids. Other work with
purple nutsedge has shown it to be a very strong competitor,
especially in the humid tropics (30). Season-long interference
from purple nutsedge reduced yields of garlic (Allium sativum L.)
by 89% and other vegetable crops significantly. However, corn
yield was not reduced by a large population of purple nutsedge in
the humid tropics when enough nitrogen fertilizer was added,
suggesting that the principal mode of interference by purple
nutsedge is competition for nutrients (2).

Table II. Indications of Possible Allelopathic Effects of Johnsongrass on Soybeans[1]

Rate of metriflufen (kg/ha)	Time of application after emergence (weeks)	Presence of johnsongrass	Johnsongrass control[ab] (%)				Soybean injury[a]				Soybean yield[a] (kg/ha)
			Late June	Early July	Late July	Early Sept.	Late June	Early July	Late July	Early Sept.	
1.1	3	No	...	...	...	...	13b	0d	0d	0c	1,450a
1.1	3	Yes	89a	77c	72c	68c	13b	0d	0d	0c	1,310cd
2.2	3	No	...	...	...	...	25a	20b	5c	5b	1,350bc
2.2	3	Yes	91a	93a	93b	94b	25a	20b	5c	5b	1,220de
1.1	3 and 6	No	...	...	...	...	13b	15c	24b	7ab	1,270cd
1.1	3 and 6	Yes	89a	88b	99a	99a	13b	15c	24b	7ab	1,130ef
2.2	3 and 6	No	...	...	...	...	25a	35a	34a	10a	1,110f
2.2	3 and 6	Yes	91a	95a	99a	100a	25a	35a	34a	10a	980g
0	Not treated	No	...	...	...	...	0c	0d	0d	0c	1,440ab
0	Not treated	Yes	0b	0d	0d	0d	0c	0d	0d	0c	1,010g

[a] Within a column, means followed by the same letter are not significantly different at the 5% level according to Duncan's multiple range test.

[b] Ellipsis (...) indicates that johnsongrass control was not evaluated because the weed was not present.

[1] reproduced with permission from Weed Science (24). Copyright 1980, Weed Science Society of America.

In a replacement-series study, Elmore et al. (27) have shown that purple nutsedge is a stronger competitor than four other species, including cotton. The data are insufficient to document allelopathy, but nutsedge tuber residue has been shown to affect cotton seedling development (31).

In a study at Stoneville on radish (Raphanus sativus L.) (Table III), purple nutsedge residue did not reduce yield in the field. However, plots with the weed present, even where weed control was attempted with herbicide and a plant feeding insect (Bactra) (32) so that weed growth was reduced, resulted in severe yield reductions. Here we see that residues were not toxic but live, growing nutsedge was. A complicating factor was a heavy rain soon after planting. The results of this study would suggest that field

Table III.   Yield of Radish (Raphanus sativus L.) Growing
in Purple Nutsedge

Treatment	Yield[3]
	-gfw-
Weed and residue free [1]	504[b]
Weedfree /with residue [2]	1150[a]
With weeds/controlled [2]	78[c]
With weeds/uncontrolled	128[c]

[1]Planted nutsedge stand killed. Residue augmented with greenhouse grown ovendried residue.
[2]Planted nutsedge stand control was attempted with postdirected glyphosate (rope wick) and Bactra release. Growth of nutsedge was reduced but not killed.
[3]Means followed by the same letter are not different according to Duncan's Multiple Range Test.

research may produce significantly different results from greenhouse pot studies. This may be because plants have a much larger soil volume to exploit in the field than in small pots in the greenhouse, and rainfall could leach allelochemics out of the soil profile. Therefore, studies with both purple nutsedge and johnsongrass suffer the same deficiencies: lack of rigorous proof of allelopathic effects in the field, although the case for johnsongrass may be better substantiated than for purple nutsedge.

Future research. Ample research has been conducted to demonstrate the allelopathic potential of numerous weedy species, and of their associated crops. Future research should be directed to discerning how much of this potential is realized in the field under humid, warm conditions such as that found in the Southeast United States. Newman (33), in a reflective article, concludes that "--it cannot be said with confidence that allelopathy has any significance for agriculture." The evidence so far certainly substantiates his conclusion. To get a better feel for allelopathy in the field, field studies should minimally include the weed and weed residues.

The experimental design should include a weed-free control, a weedy control, a weed free but with weed residue treatment, and at least one treatment with weeds and residue present. Various levels of weeds and weed residue could be incorporated into the study.

Residue management. In an agriculture based on a full tillage-one-crop-per-year system, crop and weed residues present no management difficulties. Residues are simply incorporated into the soil sufficiently ahead of planting to allow decomposition. Green manure cover crops, which were at one time in vogue, required a management scheme that may have been forgotten today. The cover crop had to be turned under sufficiently ahead of planting to avoid interference from fermenting residues on the germinating crop seed. The stubble mulch studies of the 1940's (34) showed convincingly that crop residues on the soil surface can interfere with production. The current best practice for double cropping wheat and soybeans in Mississippi is to burn the wheat stubble, and then plant soybeans no-till in the residue-free surface (35). Wheat straw may be allelopathic to soybeans (36), but it also interferes with field operations. Management of plant residues may be as simple as determining what residue-crop combinations are compatible or, perhaps as Putnam and de Frank (37) have suggested, the allelopathic potential of a plant residue may be altered by chemical treatment.

In many respects, crop production has advanced little in the last two decades. More tools (herbicides and insecticides) are available for use by producers, but in many cases we have not fully utilized lessons learned earlier. Residue management seems to be one of those cases. In no-till agriculture, residue management techniques will have to be developed--by engineering, chemical, or biological means--for the system to function.

Conclusion

Allelopathy should be studied and the weed-crop combinations that develop with new management constraints should be evaluated. However, easy solutions to pest problems will not be forthcoming by manipulating allelopathy. For allelopathy to operate, the weed or weed residue must be present, and then it does not matter much to the producer whether the detrimental effect is due to competition or to allelopathy. Research on allelopathy may have some serendipitous results, however, because, as Newman (33) points out new herbicides may be discovered that are analogous to phytotoxins produced by higher plants, or allelochemic production may be blocked in weeds, or toxin production may be altered in decaying plant residues by chemical manipulation. Although no agriculturally useful information has surfaced from allelopathy studies thus far, the future is open for innovative solutions.

Acknowledgments. The cooperation of K. E. Frick in supplying
Bactra for purple nutsedge control is appreciated. The helpful
reviews of Drs. S. O. Duke, L. G. Heatherly, G. W. Cathey, and G.
H. Egley for this manuscript were greatly appreciated.

Literature Cited
 1. Harper, J. L.   In "Mechanisms in Biological Competition".
    Milthrope, F. L. Ed., Univ. Press, Cambridge. 1961. pp. 1-40.
 2. Zimdahl, R. L.   "Weed-crop competition. A review."   Int. Plant
    Prot. Center, Oregon State Univ., Corvallis, OR. 1980.
 3. Patterson, D. T.   In "Research Methods in Weed Science", So.
    Weed Sci. Soc., Champaign, IL. 1984.
 4. Putnam, A. R.; W. B. Duke.   Ann. Rev. Phytopathol. 1978, 16,
    431-451.
 5. Putnam, A. R.; Defrank, J.;Barnes, J. P.   J. Chem. Ecol. 1983,
    9, 1001-1010.
 6. Elmore, C. D.   So. Weed Sci. Soc. Research Report. 1984, 37,
    192-198.
 7. Bhowmik, P. C.; Doll, J. D.   Agron. J. 1982, 74, 601-606.
 8. Beltiano, J.; Montaldi, E. R.   Rev. Facultad de Agronomic, 1982,
    3, 265-269.
 9. Lolas, P. C.; Coble, H. D.   Weed Sci. 1982, 30, 589-593.
10. Horowitz, M.   Expl. Agric. 1973, 9, 263-273.
11. Dorst, C. D.; Doll, J. D.   Weed Sci.   1980, 28, 229-233.
12. Bell, D. T.; Koeppe, D. E.   Agron. J. 1972, 64, 321-325.
13. Irons, S. M.; Burnside, O. C.   Weed Sci. 1982, 30, 372-377.
14. Dekker, J.; Meggitt; W. F.   Weed Res. 1983, 23, 91-101.
15. Pope, Dan, personal communication.
16. Guenzi, W. D.; McCalla, T. M.   Soil Sci. Soc. Am. Proc. 1962,
    26, 456-458.
17. McWhorter, C. G.; Hartwig, E. E.   Weed Sci. 1972, 20, 56-59.
18. Rice, E. L.   "Allelopathy"; Academic Press. N.Y. 1974.
19. Abdul-Wahab, A. S.; Rice, E. L.   Bull. Torrey Bot. Club 1967,
    94, 486-497.
20. Friedman, T.; Horowitz, M.   Weed Res. 1970, 10, 382-385.
21. Horowitz, M.; Friedman, T.   Weed Res. 1971, 11, 88-93.
22. Dalton, B. R.; Blum, U.; Weed, S. B.   J. Chem. Ecol. 1983, 9,
    1185-1201.
23. Lehle, F. C.; Putnam, A. R.   Plant. Physiol. 1982, 69,
    1212-1216.
24. Azlin, W. R.; McWhorter, C. G.   Weed Sci. 1980, 29, 138-143.
25. Fuerst, E. P.; Putnam, A. R.   J. Chem. Ecol. 1983, 9, 937-944.
26. Dekker, J. H.; Meggitt, W. F.; Putnam, A. R.   J. Chem. Ecol.
    1983, 9, 945-981.
27. Elmore, C. D.; Brown, M. A.; Flint, E. P.   Weed Sci. 1983, 31,
    200-207.
28. Holm, L. G.; Plucknett, D. L.; Pancho, J. V.; Herberger, J. P.
    "The world's worst weeds". Univ. Press of Hawaii, Honolulu,
    HI.
29. Friedman, T.; M. Horowitz.   Weed Sci. 1971, 19, 398-401.
30. William, R. D.; Warren, G. F.   Weed Sci. 1975, 23, 317-323.
31. Elmore, C. D.; Clarke, L. E.   Proc. 1981 Meeting, Weed Sci.
    Soc. Am. 1981, p. 87.
32. Frick, K. E.   Environ. Entom. 1982, 11, 938-945.
33. Newman, E. I.   Pestic. Sci. 1982, 13, 575-582.

34. McCalla, T. M.; Duley, F. L. Soil Sci. Soc. Am. Proc. 1949,
    14, 196–199.
35. Sanford, J. O. Agron. J. 1982, 74, 1032–1035.
36. Steinsiek, J. W.; Oliver, L. R.; Collins, F. C. Weed Sci.
    1982, 30, 495–497.
37. Putnam, A. R.; DeFrank, J. Crop Prot. 1983, 2, 173–181.

RECEIVED June 12, 1984

Chemistry and Biology of Allelopathic Agents

N. BHUSHAN MANDAVA

U.S. Environmental Protection Agency, Office of Pesticide Programs, Washington, DC 20460

It has been shown that allelopathic agents are responsible for prevention of seed decay, dormancy of seeds
and buds, promotion of pathogenic infections, and resistance of plants to diseases. These agents, biosynthesized by both crop plants and weeds, are also involved
in eliciting inhibition of growth of crop plants by crop
plants, growth retardation, of weeds by crop plants, and
inhibion of weeds by crop plants. The specific chemicals
involved in allelopathy remain obscure, although several
secondary plant products, e.g., simple aliphatics, phenolic compounds, terpenoids, and steroids have been reported.
It appears that these plant products play an important
role in crop production and crop protection. Several
toxins have been reported which not only affect the growth
and development of plants, pests and microorganisms, but
also elicit deleterious health effects in animals and
humans. In order to assess the chemical specificity for
allelopathy, identification of the causative agents is
required. A concerted effort in chemical and biological
research is needed to fully understand the physiological
and biochemical implications in allelopathy. An overview
of allelochemicals produced by crop plants is presented.

I. Terminology

Allelopathy (root words: ALLELON and PATHOS) is derived from
Greek, "allelon" of each other and "pathos" to suffer – the injurious effect of one upon another. The subject matter of this
symposium covers that body of knowledge which concerns the production by one plant of chemicals that induce suffering in another
and may also be called chemical pathogenesis. Although the evidence

on allelopathic action supports its conception as a one-way street, from plant A to plant B, yet there is a possibility that plant B fights back with its own chemicals.

Allelopathy, as Muller pointed out (1), must not be confused with physical competition, such as crowding and shading. The total influence of one plant on another should be termed "interference" which in turn includes both physical and chemical effects.

The term "Allelopathy" was coined by Molisch (2) to refer to both detrimental and beneficial biochemical interactions among all classes of plants, including microorganisms. Because the root word "pathy," however, implies detrimental interactions, Rice (3) defines "allelopathy" as follows: "Any direct or indirect harmful effects of one plant (including microorganisms) on another through the production of chemical compounds that escape into the environment." Perhaps, the term "allelopathy" should be extended to include the manifold mutual effects of metabolic products of both plants and animals. Now Rice includes beneficial interactions (18).

Grummer (4) recommended that special terms be used for the allelopathic chemicals based on where they are produced and which plants are affected by them. These special terms include:

1. Phytonicide – an inhibitor produced by a higher plant and effective against microorganisms.

2. Marasmin – a compound produced by a microorganism

3. Koline – a chemical inhibitor produced by a higher plant and effective against higher plants.

4. antibiotic – a chemical produced by a microorganism and effective against microorganisms.

Following this terminology, all allelopathic agents described in this paper are kolines since we confine the subject to higher plants affected by other plants grown in the immediate environment, e.g., crop plants whose growth is inhibited by weeds.

II. Historical Background

The study of allelopathy has a long history. According to Rice (5), Lee and Monsi (6) found a report by Banzan Kumazawa in a Japanese document some 300 years old that rain or dew washing the leaves of red pine (Pinus densiflora) was harmful to crops growing under the pine. This was substantiated by these workers (6) in a series of experiments. Historically, this is considered to be the first report on allelopathy.

In the early 19th Century, deCandolle (7) reported that root excretions of several plants were injurious to common crop plants. His theory was challenged until Livingston (8) produced a convincing evidence in 1907 that the failure of non-bog plants to grow in peat bogs is due to deleterious chemical substances. Later, Schreiner and his co-workers (9, 10), working on the problem of fatigued soils, found that they could recover substances from crop plants and from previously cropped soil that were deleterious to many crop plants. By the 1950's, new interest in crop plants and allelopathy arose. Bonner found that the residue of guayule (Parthenium argentatum) produced trans-cinnamic acid which is toxic to young guayule plants (11). He also found that the cinnamic acid is slowly decomposed in the soil, so that the effect dissipates with time (12). These reports led McCalla (13), Muller (14) and Rice (5), along with many others (5, 15, 16, 17, 18), to provide renewed interest in allelopathy as an important phenomenon in agriculture. These accounts point out that, although the concept of allelopathy has been around for a long time, the study of allelopathy is nonetheless a young field.

III. Present State of Knowledge on Allelopathic Agents

The causes of plant succession have been the subject of considerable research in the past three or four decades. Ecologists have found that allelopathic interactions play a considerable role in natural ecosystems and their effects are seen in phytoplankton succession, terrestrial succession, inhibition of nitrogen fixation and nitrification, dormancy of seeds, and prevention of seed decay. Allelopathic effects are clearly shown in managed ecosystems (agroecosystems), namely agriculture, horticulture and forestry. Since our interest is mainly in the exploitation of allelopathy for agriculture, we shall restrict our coverage to agroecosystems. The following few examples illustrate the importance of allelopathy that affect the growth and development of agricultural crops.

Davis (19) in 1940 extracted and purified the toxic substance from the hulls and roots of walnut (Juglans) and found it to be identical to juglone (5-hydroxy-1,4-naphthoquinone). This compound proved to be a powerful toxin when injected into the stems of tomato, potato and alfalfa plants. The allelopathic action in the case of juglone (walnut tree and its vicinity) is well established.

Bode (20), working on the exudation of absinthin from Artemisia absinthium, showed that the growth of neighboring plants such as Foeniculum vulgare and others was influenced by these toxins and so, for the first time, produced exact evidence that such metabolic products can in nature influence the development of nearby plants.

Several aromatic compounds, such as caffeic acid, chlorogenic acid, trans-cinnamic acid, p-coumaric acid, ferulic acid, gallic acid, vanillic acid, vanillin and p-hydroxybenzaldehyde have been found in crop residues and many have been isolated from field soil (5, 20). Guenzi and McCalla (21) isolated them from residues of corn (Zea mays L.), wheat (Triticum aestivum L.), oats (Avena sativa L.) and sorghum (Sorghum bicolor L.). The same chemicals were found to inhibit the growth of sorghum, soybeans (Glycine max), sunflower (Helianthus annuus L.) and tobacco (Nicotiana tabacum L.) (22, 23, 24).

In recent years, evidence has accumulated that the chemicals produced by weed plants severely affect the growth of crop plants. It has been reported that giant fox tail (<u>Setaria faberii</u> Herrm.) strongly inhibits the growth of corn (<u>25</u>) and soybeans (<u>26</u>). Peters and his coworkers (<u>27</u>, <u>28</u>) found that the allelopathic agents from tall fescue (<u>Festuca arundinacea</u> Schreb.) affect the growth of rape (<u>Brassica nigra</u>), sweet gum (<u>Liquidambar styraciflva</u> L.), birdsfoot trefoil (<u>Lotus corniculatus</u> L.) and red clover (<u>Trifolium pratense</u> L.). Velvet leaf (<u>Abutilon theophrasti</u> Medic.) is another annual weed of wide distribution in the eastern United States and causes yield reductions in cotton (<u>Gossypium hirsutum</u> L.) and soybeans (<u>Glycine max</u>) (<u>29</u>). Another common weed in India is <u>Parthenium hysterophorus</u> L. which was reported to affect the growth of all other crops and other vegetation (<u>30</u>). The major allelopathic agent in the plant was identified as parthenin, a sesquiterpene lactone (Figure 2).

The foregoing examples summarize the allelopathic agents produced by both crop and weed plants which affect the growth of crop plants. However, we must be extremely cautious when implicating the inhibitory (or stimulatory) action of secondary plant products or their metabolites in terms of allelopathic activity, because of the following contradictory reports.

The hypothesis of Gray and Bonner (<u>31</u>, <u>32</u>) that 3-acetyl-6-methoxybenzaldehyde, which is found in the leaves of <u>Encelia farinosa</u> and in the soil beneath these plants, is responsible for the suppression of other plant species has not been substantiated by Muller (<u>33</u>, <u>34</u>). The same is true for the presumed role of trans-cinnamic acid in "soil sickness" associated with guayule (<u>Parthenium argentatum</u>), for the amygdalin of peach roots, and for the phlorizin of apple roots or its breakdown products (<u>35</u>), which were thought to be associated with the "soil sickness" of orchards.

In addition, we have found that crude extracts of tall fescue (<u>Festuca arundinacea</u>) inhibit seed germination and growth of several plant species including pinto and mung beans under laboratory conditions. These extracts consist of complex mixtures of secondary plant products, including several phenolic compounds. However, purification by chemical fractionation procedures results in the diminution of the inhibitory activity. It is sometimes possible that the allelopathic activity may be ascribed to a complex of different inhibiting factors, and also, no doubt, to the synergistic effect of amounts which separately are below the threshold levels of toxicity. Other factors that contribute to the difficulties include:

(1) Degradative changes of allelopathic substances during storage and handling.

(2) Physico-chemical effects of soil (adsorption).

(3) Weakening or strengthening of activity when in combination with other substances.

(4)   Changes in responsiveness due to endogenous or exogenous factors.

Therefore, it seems that when studying allelopathy, it is best not always to be concerned exclusively with those single compounds contained the plant which are shown to be effective under laboratory conditions. From the foregoing account, it appears that we have not progressed much in understanding the phenomenon of allelopathy except for attributing undesirable effects upon the growth patterns of plants to allelopathy. In other words, we are still right in the beginning of the quest for an explanation for them in chemical terms. A concerted effort should be undertaken by biologists and chemists toward a better understanding of the allelopathic effects which could contribute to increased agricultural productivity.

Recently, the reports of two recent planning conferences (36, 37) that recommended more intensive investigations be pursued on the the effects of allelopathic substances on plant growth dynamics and physiological processes. Such investigations could contribute, among others, to understanding the importance of allelopathy in weed-crop rotations and tillage practices.

IV. Allelopathic Agents

A. Biosynthesis

According to Robinson (38), Whittaker and Feany (39), and Rice (5), a great majority of secondary plant products are biosynthesized from acetate and shikimic acid (38), many of which have been implicated as allelopathic agents, and the main groups of compounds are described below.

1.   Aliphatic compounds: Several water-soluble simple organic acids and alcohols are common plant and soil constituents. They include methanol, ethanol, n-propanol and butanol (40), and crotonic, oxalic, formic, butyric, lactic, acetic and succinic acids (41, 42), all of which inhibit seed germination or plant growth. Under aerobic conditions, however, aliphalic acids are metabolized in the soil and therefore, should not be considered a major source of allelopathic activity (40).

2.   Unsaturated lactones: The most notable lactone identified as a potent allelopathic agent is patulin, produced by several Penicillium sp. growing on wheat and other plants. At 10 μg/ml, patulin completely inhibits the germination and growth of seedings of certain cultivated plant species, including corn (3, 49).   Other lactones such as psilotin, psilotinin and protoanemonin (Table 1) are also powerful growth inhibitors, although their role in allelopathy is not fully ascertained (3, 48, 49).

3.   Fatty Acids and Lipids: Although several fatty acids, esters and alcohols are known to be toxic to plant growth, their role in allelopathy is not fully investigated (3). Dihydroxystearic acid (3, 49) is the classic example known to exhibit allelopathic activity.

4. Cyanogenic glycosides   Amygdalin (or its reduced form,
prunasin) and dhurrin are known to be allelopathic (3,48). Not only
are these glycosides hydrolyzed to produce hydrogen cyanide, but the
benzaldehyde or hydroxybenzaldehyde (produced during hydrolysis) is
oxidized to benzoic acid which itself may be toxic to several species
(3, 48).

5. Terpenoids   There are a number of allelopathic agents
among the terpenoids. They include monoterpenes (Figure 1), many
from arid zones (14), such as α-pinene, β-pinene, camphor and
cineole. Sesquiterpenes, such as caryophyllene, bisabolone and cha-
mazulene are shown to have allelopathic activity (Figure 2) (49).
Several sesquiterpene lactones, namely, arbusculin A, achillin, and
viscidulin C are known to be inhibitors of plant growth (3, 48, 49).

6. Aromatic compounds   Phenols, phenolic acids, cinnamic acid
derivatives, coumarins, flavonoids, quinones, and tannins, all of
which are aromatic compounds, comprise the largest group of secondary plant products. They are often referred to as "phenolics" and
have been identified as allelopathic agents in more instances than
all of the other classes of compounds combined (5).

a. Among simple phenols, hydroquinone and its glycoside,
arbutin, have been identified as allelopathic agents(3, 17,
47, 49).

b. Among phenolic acids, benzoic, gallic, ellagic,
vanillic, salicylic and sulfosalicylic acids are all present
in leaf leachates and soil and are known to have allelopathic
activity (Figure 3) (3, 17, 48).

c. Cinnamic, coumaric, caffeic, chlorogenic and sinapic
acids are widely distributed in the plant kingdom and are
reported to be allelopathic chemicals (Figure 3)(3, 48, 49).

d. Coumarins, the lactones of o-hydroxy cinnamic
acids, occur as glycosides in plants and are readily leached
into the environment. Examples include scopoletin, scopolin,
esculetin, esculin and methylesculin, all of which have
allelopathic potential (Figure 4)(3, 48).

e. Quinones are another large class of phenolic compounds
which are found widely in plants. A classic example in this
group is juglone (5-hydroxy-1,4-naphthoquinone) which occurs only
in the genus Juglans and is reported to be a potent allelopathic
agent (Figure 3)(3, 48).

f. Flavonoids are another large group of phenolic compounds
and a number have been shown to be allelopathic. Phlorizin and
its breakdown products, glycosides of kaempferol, quercetin and
myricetin are known allelopathic agents (Figure 5)(3, 47, 48).

g. Both condensed and hydrolyzed tannins, are reported to
possess allelopathic activity (3, 48).

Table 1.   Unsaturated Lactones with Plant Growth Inhibiting
           Activity

Compound	Structure	Source	Biological Activity
Psilotin (R= β-D-glucose) Psilotinin (R=H)	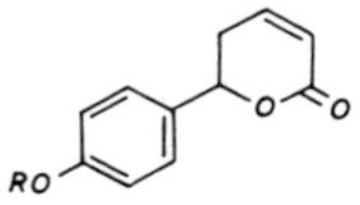	Psilotum nudum and Twesiperis tannensis	Inhibits seed germination and seedling growth
Patulin	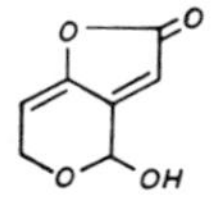	Several Penicillium sp.	Inhibits seed germination and seedling growth
Protoanemonin	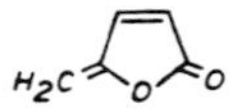	Several Ranunculacea sp.	Inhibits seed germination

1:8-Cineole   Limonene   α-Pinene

β-Pinene   (+) and (-)-Camphor

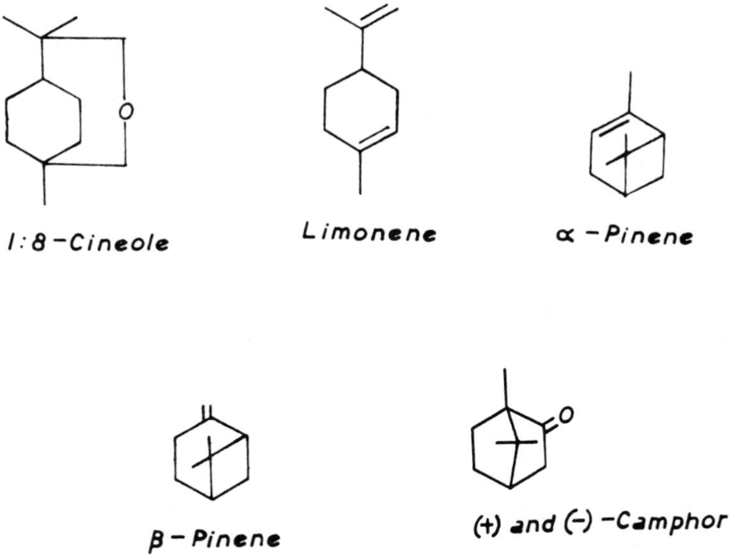

Figure 1.   Volatile Monoterpenes as Allelopathic Agents

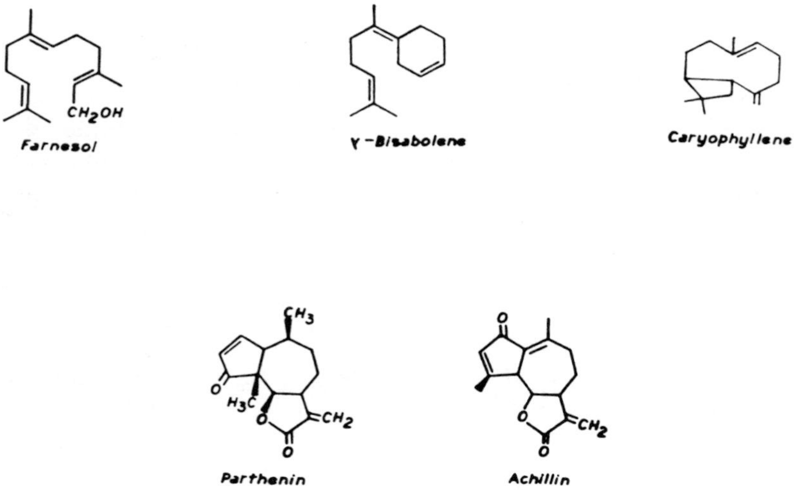

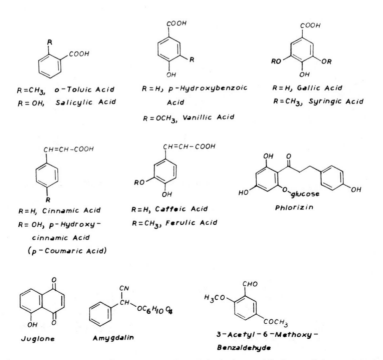

CH2OH
Farnesol
γ-Bisabolene
Caryophyllene
CH3
H3C
O
CH2
Parthenin
Achillin

Figure 2. Sesquiterpenes with Allelopathic Activity

R
COOH
R=CH3, o-Toluic Acid
R=OH, Salicylic Acid
COOH
R
OH
R=H, p-Hydroxybenzoic Acid
R=OCH3, Vanillic Acid
COOH
RO
OR
OH
R=H, Gallic Acid
R=CH3, Syringic Acid
CH=CH-COOH
R
R=H, Cinnamic Acid
R=OH, p-Hydroxy-cinnamic Acid
(p-Coumaric Acid)
CH=CH-COOH
RO
OH
R=H, Caffeic Acid
R=CH3, Ferulic Acid
OH
O
HO
O-glucose
OH
Phlorizin
O
OH
O
Juglone
CN
CH
OC6H10O4
Amygdalin
CHO
H3CO
COCH3
3-Acetyl-6-Methoxy-Benzaldehyde

Figure 3. Aromatic Compounds Eliciting Allelopathic Activity

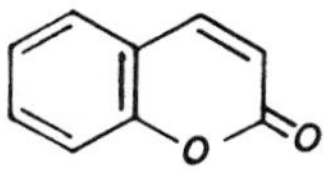

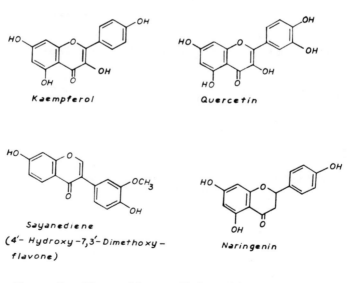

Coumarin

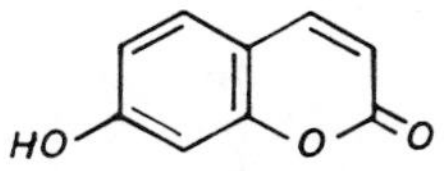

Umbelliferone

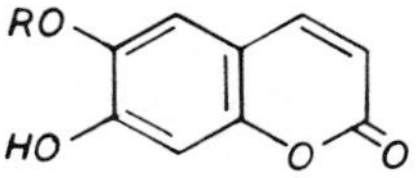

R = H, Aesculetin
R = CH₃, Scopoletin
R = β − D − Glucose,
    Aesculin

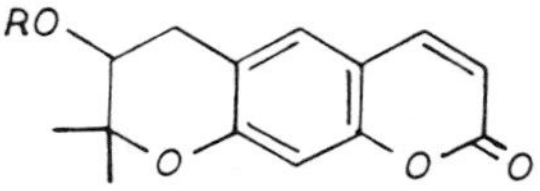

R = H, Decursinol
R = OCCHO(CH₃)₂, Decursin

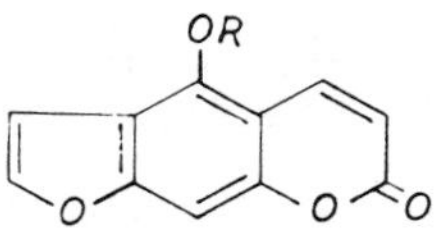

R = H, Psoralin
R = CH₃, Bergaptin

Figure 4.   Coumarins as Allelopathic Agents

Kaempferol

Quercetin

Sayanediene
(4′− Hydroxy −7,3′− Dimethoxy −
flavone)

Naringenin

Figure 5.   Flavonoids as Allelopathic Agents

B. Routes of Entry into the Environment

It has been documented (5) that allelopathic agents are released into the environment by: (a) exudation of volatile chemicals from living plant parts, (b) leaching of watersoluble chemicals from above–ground parts in response to the action of rain, fog or dew, (c) exudation of water-soluble toxins from below–ground parts, and (d) release of toxic chemicals from nonliving plant parts through leaching of toxins from litter, sloughed root cells or tissues decomposed by microorganisms. Audus (46) divides allelochemicals into 4 groups (Table 2) based on their source as shown below.

1. Root exudates   A wide variety of chemicals, such as sugars, amino acids, and aromatics, is excreted by roots of plants. Very little information is available on the allelopathic interaction of root exudates with the higher plants, except for the identification of a few products in isolated cases (46).

Table 2. Allelopathic Agents Isolated From Various Sources (46)

Inhibitors in			
Root Exudates	Leaf Leachates	Volatile Inhibitors	Sick Soil Toxicants
1. Allicin	1. 3-Acetyl-6-methoxybenz-aldehyde	1. Ethylene	1. Decomposition products of amygdalin
2. Chlorogenic acid	2. Genistic acid	2. β-Pinene	2. Phlorizin
3. Melilotic acid	3. Dihydroxy-stearic acid	3. Camphene	3. Phloretin
4. Gallic acid	4. Protocatechuic acid	4. Cineole	
5. o-Coumaric acid	5. Caffeic acid		
6. Piperic acid	6. Syringic acid		
7. 2-Furanacrylic acid	7. p-Hydroxy benzoic acid		
8. Juglone			
9. p-Hydroxy benzaldehyde			
10. Phenylpropio-nic acid			

2. Leaf leachates and decomposition products   Inhibitors
from leaves may be washed into the soil or compounds from
leachates may further decompose into toxic products which
inhibit seed germination and prevent seedling establishment (46).

3. Volatile toxicants   A great many volatile substances,
including gaseous compounds and volatile terpenoids, are
involved in allelopathy (46).

4. "Sick soil" toxicants   Accumulation of toxicants from
previous inhabitants evidently results in growth inhibition of
roots and interfere with seedling establishment (46).

The donor plants which release these chemicals generally store
them in the plant cells in a bound form, such as water-soluble
glycosides, polymers including tannins and lignins, and salts.   It
has been suggested that upon cleavage by plant enzymes or environmental stress, the toxic chemicals are released into the environment from special glands on the stems or leaves (17, 47, 48).   What-
ever may be the mode of formation (e.g., oxidation, reduction, etc.),
it is generally assumed that these chemicals are not toxic to the
donor plants.   Once the chemicals from the donor plants are released
into the environment, they may be either degraded or transformed
into other forms.   The resultant stages of these chemicals may also
be toxic to the host plant.

C. Extraction, Bioassays and Identification

Several procedures have been reported for extraction of the
suspected allelopathic agents from donor plants.   Essentially all
the procedures that were employed attempted to simulate the routes
of entry of toxic substances into the natural environment.   As
shown previously, the allelopathic agents are released through
leaves and roots, or escape into the environment as volatile
materials.   Table 3 summarizes the different extraction and bioassay
procedures employed to isolate and detect the toxic chemicals (17).
For extraction, the investigators used either the plant parts from
the donor plants or the intact donor plants from which the suspected
chemicals were leached through leaves, stems or roots.

Although cold water extraction was generally regarded as the
method of choice, some investigators chose to use boiling water or
organic solvents to ensure that all the toxins are essentially
removed from the plants.   Other investigators question this
approach, because such a drastic treatment would not only extract
the desired chemicals but would also remove a host of other chemicals
which might modify biological activity in the test systems.

In other isolation methods, where the compound(s) was removed
from the donor plants, the plant material was either dried or macerated prior to cold and hot water treatment.   Soxhlet-type extraction
was employed when organic solvents were used.   Leaves and stems from
the intact plants were extracted to collect the suspected volatile
substances and those chemicals likely to be released by rain, mist

Table 3.   Extraction and Bioassay Methods for
Allelopathic Agents (17)

Extraction Methods	Bioassay Methods
I. Plant material removed from plant. 　A. Dried 　　1. Cold water (soaking) 　　2. Hot water (boiling) 　　3. Organic solvents 　　　(Soxhlet extraction) 　B. Live-diced 　　1. Cold water 　　2. Hot water 　　3. Organic solvents II. Plant material intact 　A. Leaf on top 　　1. Rain or fog dip 　　2. Gas trap 　B. Roots 　　1. Water 　　2. Flushing	I. Extracts 　A. Petri dish 　B. Soil 　C. Sand II. Plant material 　A. Cellulose sponge 　B. Soil 　C. Sand III. Intact plant 　interaction A. Sand B. Soil C. Agar

and dew. The chemicals presumed to be released through root systems
were collected as root exudates.   In all cold water extraction procedures, it should be understood that the investigators attempted to
simulate the conditions for natural release of toxic chemicals.

Various bioassay methods have been used to detect the "natural"
release of allelopathic agents. Some authors preferred, after
partial purification, to assay the extracts by petri dish methods
for germination, growth of roots or shoots and other symptoms of
seedlings. The bioassays also included tests in soil or sand and
also in nutrient solution (Table 3).

Chromatographic methods were extensively used to characterize
the allelopathic agents. Because a great majority of the suspected
toxins were already known (and several of them are commercially
available), simple chromatographic methods were thought to be
sufficient for their characterization from the extracts. However,
no detailed investigations were ever undertaken to characterize
those inhibitors that may be present in minute amounts.

Excellent reviews are available (3 ,5, 17) on the procedures for
extraction, bioassays and identification of allelopathic chemicals.

In our efforts to detect and isolate the allelopathic agents from tall fescue and several other grass species, we extracted the detached plant material with water and/or organic solvents. Either solvent extraction method yielded extracts that were inhibitory to the seed germination and seedling growth in our bioassay systems. We noticed that stepwise purification of the crude extracts resulted in gradually diminishing activity. When several of the identified phenolic compounds were individually tested, none of them exhibited biological activity. We concluded that the biological activity was associated with either the threshold levels of chemicals in the crude extracts or synergistic effect of some chemicals. Similar observations were made by others, and it needs to be verified by further experimentation (17, 48).

D. Mechanism of Action

The observation that allelopathic agents affect plant growth and development leads the physiologist to question the modes of action. While we know that diverse secondary plant products exhibit allelopathic activity, it is difficult to understand the mechanism of action of these plant products partly because of:

1. Complications in separating the secondary effects from the primary causes;

2. The uncertainty in translating the observed effects in isolated enzyme and other biochemical systems to intact plant systems; and

3. The lack of understanding of the effect of the allelopathic agents on whole plant photosynthetic processes, namely, changes in stomatal opening, membrane permeability, water content and many other processes that affect the overall photosynthetic processes.

Published data (3) on allelopathic agents, however, indicate that the following parameters affect the growth process due to toxin action:

1. Cell division and cell elongation;

2. Phytohormone-induced action, especially by auxins and gibberellins;

3. Enzyme-mediated action, by sulfhydryl and phenylalanine-lyase (PAL) enzymes as well as by other enzymes such as cellulase, catalase, peroxidase, phosphorylase and pectolytic enzymes;

4. Mineral uptake;

5. Photosynthesis and respiration;

6. Stomatal opening;

7. Protein synthesis and changes in lipid and organic acid
    metabolism;

8. Hemoglobin synthesis; and

9. Membrane permeability.

A detailed discussion on the above parameters that contribute
significantly to mode and mechanism of action, is included in the
monographs by Rice (3, 5) and reviews by Einhellig (48) and Mandava
(49).

E. Exploiting Allelopathy in Agricultural Systems

It was emphasized in the two conference reports (36, 37) that
proper understanding and application of allelochemicals could lead
to potentially increasing the crop productivity by protecting the
crop plants from natural toxins and by increasing the crop yields by
the action of natural stimulants.

Others also have proposed that allelochemicals could prove useful
in crop protection, especially for minimizing the agricultural losses
due to insects and nematodes, and for controlling diseases. The USDA
Research Planning Conference in 1977 estimated that technological
advances could reduce substantially the $30 billion in annual losses
caused by pests and the cost of their control (37). This conference
proposed the following strategies for research in allelopathy.

1. Discover, identify, synthesize and evaluate allelopathic
    compounds for potential use in agriculture to increase the
    control of weeds, diseases, insects and nematodes, and to
    reduce the cost of their control.

2. Determine effects of environment on persistence, activity
    and effectiveness of allelopathy.

3. Determine effects of allelopathic chemicals on ecological
    community, including the use of growing plants as allelo-
    pathic agents.

4. Determine mode of action of allelopathic chemicals against
    target organisms and on crop plants.

5. Investigate the most promising allelopathic agents for harm-
    ful side effects as required by current EPA procedures.

6. Establish pilot test programs where needed to supplement
    small-plot research.

The papers of this present symposium reflect the progress made
in understanding the chemistry of allelopathy following the 1977
USDA Research Planning Conference recommendations. In addition to
these strategies, research efforts on the following aspects should
also be continued to fully exploit allelopathy in agriculture (17).

1. Utilization of beneficial plant associations: Practice of
    mixed cultures, as compared to monoculture adopted in the
    U.S. due to farm mechanization, could lead to utilizing
    the beneficial associations for the long-term benefit of
    agriculture. It has been suggested that such an approach
    is even more profitable for developing countries.

2. Manipulation of weed seed behavior: Proper understanding of
    seed physiology including seed vigor and germination could
    lead to preventing seed decay and controlling germination.
    Increased research efforts should be encouraged toward deve-
    loping methods for increasing the weed seed decay and stimu-
    lating or inhibiting weed seed germination. Appropriately
    included in this approach is the use of microorganisms to
    destroy the weed seed and also use of chemicals such as
    abscisic acid, gibberellins and strigol to control seed
    germination.

3. Minimization of agricultural losses from soil toxins: Toxins
    from soils appear to be responsible for inhibition of nitro-
    gen fixation, metabolism and nodulation in legumes. Removal
    of toxins could be achieved by proper adsorption techniques
    and also by growing companion plants that contribute organic
    matter to microoranisms which help to destroy or degrade
    toxic chemicals.

Proper understanding of allelopathic crop and weed plants including their growth stages at which toxin production occurs and characterization of allelopathic agents from these plants provide new
avenues for developing technologies in weed control, crop efficiency, pest control and plant diseases.

F. Health and Environmental Considerations

Ecologists have long been concerned with the natural chemicals –
whether they come from plant or animal origin – that contaminate the
environment and affect human and domestic animal health. From the
foregoing discussion, we have seen that allelopathic agents from
plant origin alone contribute to the suppression of growth and development of plant communities. In agroecosystems, their action
results in a substantial loss of food and feed production which is
equivalent to a net loss of about 5 billion dollars annually from
weeds alone in the United States(49).

How dangerous to man or animals are the naturally occurring
toxins, including allelopathic chemicals of plant origin? Although
the subject of health and environmental consequences due to allelopathic chemicals has not been studied in detail, available but limited information points out the fact that several natural toxins do
pose hazards to animal and human health, and environment. A single
extensively studied allelopathic agent affecting the health of humans
and animals is parthenin from Parthenium hysterophorus (30). Parthenin from this persistent weed in India causes contact dermititis
among farm workers and also allergic rhinitis for others exposed to

this plant. This weed also causes livestock poisoning, including
deaths (30). Other bitter tasting plants of the Compositae appear to
have severe toxicity in livestock. A sesquiterpene lactone, hymeno-
vin, present in bitter weed (Hymenoxys odorata), has been reported to
be responsible for the deaths of sheep and goats, resulting in a loss
of several million dollars per year in Texas. Another sesquiterpene
lactone, helinalin from Helenium microcephalum (small head sneeze
weed), has been found to be toxic to cattle. Vermeerin, present in
Geigeria species in South Africa is poisonous to sheep. It was
suggested that the sesquiterpene lactones alter the microbial composition of the rumen and thus affect the vital metabolic functions.
Also, plants containing sesquiterpene lactones such as tenulin from
Helinium amarum, when eaten by dairy cattle, impart a bitter taste
to their milk (30, 39).

These adverse health effects are caused not only by sesquiterpene
lactones, but by other groups of allelochemicals as well. Protoanemonin present in buttercups (Ranunculus spp.) and other species has
been reported to be an irritant which can lead to fatal convulsions
in livestock. Neurotoxicity is attributed to such alkaloids as
delphinine present in larkspurs (Delphinium spp.). Steroidal alkaloids such as jervine, cyclopamine and cycloposine present in
Veratrum genus pose a potential teratogen hazard in lambs, resulting
in monkey face disease (50). The milkweed plant family, Asclepiada-
ceae, containing cardenolides, also known as cardiac glycosides, has
been blamed for sporadic livestock poisoning. For example, in a
1975 episode, 250 sheep in Samona County, California died after
eating hay containing Asclepias eriocarpa (51). Other steroid
cardiac glycosides from foxglove (Digitalis purpurea) and certain
other plants can lead to convulsive heart attacks. Toxic chemicals
from the leaves of oleander (Nerium oleander) are potentially lethal
to man. Hypericin, secreted from glands on plants of the genus
Hypericum, can cause fatal blindness and starvation. Tannins which
bind proteins to indigestible complexes can contribute to growth
inhibition in animals. Mustard oils contain allyl isothiocyanates
and their glycosides from Cruciferae are known skin irritants capable
of causing serious injury to animal tissue. Pyrethrins and sesamin
from the Chrysanthamum family affect animal enzyme systems. Pyrethrins also are reported to be toxic to fish (39).

Several volatile terpenes (camphor, cineole, etc.), gaseous
poisons such as hydrogen cyanide (a breakdown product of cyanogenic
glycosides) and mercaptans (e.g., n-propyl mercaptan from onions)
pollute the air. They have been reported to be absorbed from the
air onto soil and to remain in the soil during the dry season. It
has also been stated that other allelopathic agents from leaching
and plant decay remain in the soil for a long period. We know that
several bacteria and fungi release toxins (e.g., the antibiotic
patulin). In other words, irrespective of their route of entry into
the environment, many allelopathic agents remain in the soil for an
extended period, even through several of them are considered to be
biodegradable. A possibility that has not been considered is that
they may enter into our water systems if they survive in the soil
for extended periods. If this is a feasible consideration, then the

allelopathic agents may be found in surface and ground water in a
manner similar to water contaminated by pesticides.

The research strategies recommended by the USDA Research Planning
Conference for EPA-related activities include studies on: (1) allelopathic effects on environment in terms of allelochemical persistence,
activity and effectiveness, (2) ecological consequences due to allelochemicals, and (3) potential risks to human and animal health (37).
In addition to these approaches, we should also pursue studies toward understanding the effect of commercial allelochemicals, alone or
in combination with other chemicals such as pesticides, on human
health, safety and environment.

V. Problems and Perspectives

The concept of allelopathy originated with the ecologists' constant pursuit of understanding of the causes for plant succession in
natural ecosystems. Several decades went by before it was accepted
that the phenomenon of allelopathy exists both in natural and managed
ecosystems. The concept of allelopathy has been built on the basic
assumption that it is a concentration-dependent phenomenon involving
a multitude of diversified chemicals. As a logical pursuit, investigators made it a common practice to determine the threshold levels of
the allelopathic agents that might be required to elicit such an
effect. However, it is difficult to determine the allelopathic quantities that are present and functional in a field environment. As
Einhellig (48) points out, "a persistent question in allelopathy is
whether or not the biochemical agents involved are in sufficient
concentration in the environment to influence germination or growth."
However, very few studies have been designed to prove that the plant
produced a chemical in sufficient quantity over time or that it persisted long enough to affect other plants. Also the present state
of knowledge in allelopathy is such that, "to date, most researchers
have been more concerned with the event and not the casual agent"
(17). In spite of the early work that provided the basis for developing the concept of allelopathy, many plant scientists, especially
plant physiologists, are still skeptical about the realities of
allelopathic potential because of the conflicting and oftentimes
unconvincing reports in allelopathy. A few problems that many of us
still face are:

1. Lack of proper biological test systems to detect the
    causative agents;

2. Lack of proper experimental design and/or setup to deter-
    mine the threshold levels of activity;

3. Lack of proper controls and standards to avoid variation in
    experiments;

4. Inability to translate and transfer the laboratory data into
    field situations where the phenomenon is observed;

5. Complexity of allelopathic chemical mixtures and improper
    assessment of their concentrations;

6. Lack of adequate techniques for detection and quantification
    of suspected chemicals;

7. Improper identification of chemicals likely to be involved
    in allelopathy; and

8. Failure to detect chemicals that might be biologically
    active but might be impossible to detect in the presence of
    other chemicals.

The latter problems are of particular interest to chemists, who
should devise appropriate methods for resolving the complexity of
chemicals, properly identifying them and finally determining their
exact composition and makeup. The participation of chemists is
needed to verify the concept of allelopathy as a concentration-
dependent phenomenon. They should help to reconstitute the chemical
composition as it was found in the original and isolated plant
samples. This systematic approach leads to verification of the
concept as well as to proper assessment of the initial observation
with crude extracts, and to final application to the field
situation. Once the concept is proven, some simulation experiments
need to be performed to maximize the allelopathic effect (toxin
action). The concentration of the toxic chemicals is varied to
where the threshold levels of chemicals prove to be involved in the
exhibition of allelopathy under field conditions.

Another perplexing problem that has often been encountered is
the determination of highly potent compounds which are usually present in minute amounts. They may be masked by other chemicals whose
physical/chemical (chromatographic) properties are so similar that
they may be very difficult to track down with limited quantities of
plant samples. This type of situation is not uncommon in plant
sciences and a typical example is our recent isolation of brassinolide from rape (Brassica napus L.) pollen (52).

Therefore, chemists should concentrate on developing and applying
appropriate analytical instrumentation and other newly-emerging technologies to isolate, separate and identify the complex mixtures
obtained from the crude plant samples. Once all chemicals are
correctly identified, their identity must be confirmed by synthesis
and comparison of biological activity with the natural compounds.
This is an important task for chemists. Until it has been achieved,
no major progress can be made in allelopathy.

Plant physiologists and other biological scientists also have
their important role to play in allelopathy. They must devise suitable bioassays to detect the suspected allelopathic compounds, follow
the biological activity of the individual and associated chemicals,
develop activity profiles for identified chemicals, and determine
the conditions (dose/response) for chemicals to arrive at the threshold levels. They must also determine which chemicals contribute

to activity including any synergism or antagonism and other biological parameters that may be considered vital in allelopathy. After
this assessment under controlled environment, experiments should be
performed under field conditions, not only to verify the allelopathy
but also to find practical solutions to the problem.

As Einhellig (48) correctly stated, "Unless cooperative research
between chemists and physiologists is conducted, we will continue to
be left with the notion that an array of simple phenols common to a
wide array of soils and situations are the major chemical allelopathic agents." As for the importance of the chemical identity, Putnam
and Duke (17) expressed it very well: "after all chemicals are identified, studies can be made on genetics, specificity, environmental
influences and environmental consequences."

Once we know the chemical nature of the allelopathic agents and
their effects on plant growth dynamics, as well as on health and
environment, we can apply genetic manipulation and biotechnology to
develop toxin-resistant plants and to reduce the toxin levels from
the donor plants. These approaches serve a dual purpose because
they contribute to increased agricultural productivity and help to
minimize the potential risks on health and environment.

Acknowledgments

The unpublished work cited in this paper was carried out at the
U.S. Department of Agriculture, ARS, Beltsville, Maryland where the
author was previously employed. Sincere thanks are due to Dr. E.J.
Peters, UDSA, ARS, Columbia, MO and Dr. M.M. Rao, University of
Maryland, College Park, MD for their cooperation in this work.

Literature Cited

1. Muller, C.H. Bull. Torrey Bot. Club, 1966, 93, 332.

2. Molisch, H. "Der Einfluss einer Pflanze auf die audere–Allelo-
   pathie"; Fischer: Jena, 1937.

3. Rice, E.L. "Allelopathy"; Academic Press: New York, 1974, pg.
   353.

4. Grummer, G. "Die gegenseitige Beeinflussung hoherer
   PflanzenAllelopathie"; Fischer: Jena, 1955, pg. 162.

5. Rice, E.L. Biochem. System. Ecol., 1977, 5, 201.

6. Lee, I.K., Monsi, M. Bot. Mag. Tokyo, 1963, 76, 400.

7. Decandolle, M.A.P. Physiologie vegetale. III. Bechet Jeune,
   Lib.Fac. Med. 1832, 1474.

8. Livingston, B.E. U.S. Bur. Soils, Bull. No. 36, 1907.

9. Schreiner, O., Reed, H.S. U.S. Bur. Soils, Bull. No. 40 1907.

10. Schreiner, O., Lathrop, E.C. U.S. Bur. Soils Bull. No. 80, 1911.

11. Bonner, J., Galston, A.W. Bot. Gaz., 1944, 106, 185.

12. Bonner, J. Bot. Gaz., 1946, 107, 343.

13. McCalla, T.M., Duley, F.L. Science, 1948, 108, 163.

14. Muller, C.H. Am. J. Bot., 1953, 40, 1.

15. Evenari, M. Bot. Rev. 1949, 15, 153.

16. Bonner, J. Bot. Rev. 1950, 16, 51.

17. Putnam, A.R. and Duke, W.B. Ann. Rev. Phytopathol., 1978, 16, 431.

18. Rice, E.L. Bot. Rev., 1979, 45, 15.

19. Davis, R.F. Am. J. Bot., 1928, 15, 620.

20. Bode, H.R. Planta, 1940, 30, 567.

21. Guenzi, W.D., McCalla, T.M. Agron. J. 1966, 58, 303.

22. Patterson, D.T. Weed Science, 1981, 29, 53

23. Einhellig, F.A., Kuan, L. Torrey Bot. Club, 1977, 98, 155.

24. Einhellig, F.A., Stille, M.L. Bot. Soc.Amer. Misc. Publ. 1979, 157, 40.

25. Bell, D.T., Koeppe, D.E. Agron. J., 1972, 64, 321.

26. Knake, E.L., Slife, F.W. Weeds, 1961, 26.

27. Peters, E.J. Crop Science, 1968, 8, 650.

28. Peters, E.J., Zam, A.H.B.M. Agron. J. 1981, 73, 56.

29. Colton, C.E., Giuhellig, F.A. Amer. J. Bot. 1980, 67, 1407.

30. Towers, G.H.N., Mitchell, J.C., Rodriquez, E., Bennett, F.D., Rao, P.V.S. J. Sci. Ind. Res. (India), 1977, 36, 672.

31. Gray, R., Bonner, J. Am. J. Bot., 1748, 35, 52.

32. Gray, R., Bonner, J. J. Am. Chem. Soc., 1948, 70, 1240.

33. Muller, C.H. Am. J. Bot. 1953, 40, 53.

34. Muller, C.H. Am. J. Bot. 1956, 43, 354.

35. Bonner, H. <u>Naturaiss</u>, 1958, <u>45</u>, 138.

36. "Biochemical Interactions among plants," National Academy of Sciences; Washington, D.C. 1971, pg. 134.

37. "Report of the Research Planning Conference on the Role of Secondary Compounds in Plant Interactions (Allelopathy)." United States Department of Agriculture, Agricultural Research Service; 1977, pg. 124.

38. Robinson, T. "The Organic Constituents of Higher Plants"; Burgess. Minneapolis, Minnesota, 1963; p. 8.

39. Whittaker, R.H., Feeny, P.P. <u>Science</u>, 1971, <u>171</u>, 757.

40. Hutchinson, S.A. <u>Ann. Rev. Phytopathol.</u>, 1975, <u>11</u>, 223.

41. Takijama, Y. <u>Soil Sci. plant nutrition</u> (<u>Tokyo</u>), 1964, <u>10</u>, 14.

42. Erner, Y. Revveni, O., Goldschmidt, E.E. <u>Plant Physiol.</u>, 1975, <u>36</u>, 279.

43. Bergel, F., Klein, R., Morrison, A.L., Rinderknecht, H., Moss, A.R., Ward, J.L. <u>Nature</u>, 1943, <u>152</u>, 750.

44. Norstadt, F.A., McCalla, T.M. <u>Science</u>, 1963, <u>140</u>, 410.

45. Owens, L.D. <u>Science</u>, 1969, <u>165</u>, 18.

46. Audus, L.J. <u>Plant Growth Substances</u>. "Chemistry and Physiology," Leonard Hill; London, Vol. 1, pg. 240.

47. Fisher, R. F. "Allelopathy" in "Plant Disease, an Advanced Treatise," Horsfall, F.G., and Cawling, E.B.,Ed., Academic Press, New York, 1979, p.313.

48. Einhellig, F.A. "Allelopathy – A Natural Protection, Allelochemicals," in "Handbook on Natural Pesticides: Methods, Mandava, N.B., Ed., CRC Press, Boca Raton, FL., 1984, Vol. 1 (in press).

49. Mandava, N.B. "Natural Products in Plant Growth Regulation" in "Plant Growth Substances," Mandava, N.B., Ed., ACS Symposium Series, Vol. 111, American Chemical Society, Washington, DC., 1979, p. 135.

50. Keeler, R.F. "Mammalian Teratogenicity of Steroidal Alkaloids" in "Isopentenoids in Plants: Biochemistry and Function," Nes, W.D., Fuller, G., Tsai, L.S., Ed., Marcel Dekker, New York, 1984, p.531.

51. Selber, J.N., Lee, S.M., Benson, J.M. "Chemical Characteristics
     and Ecological Significance of Cardenolides in Asclepias (Milk-
     weed) Species" in "Isopentenoids in Plants: Biochemistry and
     Function, " Nes, W.D., Fuller, G., Tsai, L.S., Ed., Marcel
     Dekker, New York, 1984, p.563.

52. Mandava, N.B., Thompson, M.J. "Chemistry and Functions of
     Brassinolide" in "Isopentenoids in Plants: Biochemistry and
     Function," Nes, W.D., Fuller, G., Tsai, L.S., Ed., Marcel
     Dekker, New York, 1984, p.401.

RECEIVED August 6, 1984

The Involvement of Allelochemicals in the Host Selection of Parasitic Angiosperms

DAVID G. LYNN

Department of Chemistry, The University of Chicago, Chicago, IL 60637

Several compounds have now been found which are capable of inducing the differentiation of the haustorium—a specialized organ in parasitic angiosperms which functions to attach the parasite to its host. A detailed description of the chemical nature of these compounds is presented. Measurements of the levels of these compounds exuded by a host have been made and correlated with levels required for haustorial induction. The data support a strict structural requirement for the factors involved in the induction of the haustorium. Although more than one structural class may be involved, the compounds appear to be linked with the host's allelochemicals and constitutive antibiotics. The evidence for synergistic activity among several different components in host exudate suggests the possibility of a more highly sophisticated mechanism of host selection through haustorial differentiation than previously anticipated.

As so many natural products chemists before us, we have become intrigued with the seemingly endless array of organic structures which are produced in nature. Over the past thirty years, questions directed primarily at the structures of these compounds and at their effects on man's physiology have led to profound technological advances. Man can now synthesize virtually any organic compound which exists in nature and has used these talents to develop new methods to study and combat human disease. We are now in a position to explore a fundamentally different frontier—the physiological and ecological significance of these compounds to the organisms which produce them. The greater our understanding of the chemical communication and chemical defenses inherent in all living organisms, the better we will be able to understand and maintain the intricate chemical nature of the world which surrounds us (1). A need for this understanding in the plant kingdom, the area on which so much of our food and oxygen supply is dependent, is becoming particularly apparent. We have focused our interest on the mechanisms of host selection in parasitic angiosperms—plants which have evolved the capability of parasitizing other plants. An earlier proposal (2) that

0097–6156/85/0268–0055$07.75/0

parasitic plants, like herbivorous insects, may key on host defense
chemicals as recognition cues suggested that a study of this system
may not only provide some access into the control of attachment in
some of these agriculturally devastating parasites, but also may un-
cover aspects of the development and specificity of the host's alle-
lochemistry.

The heterogenous group of plants that are classified as parasitic angiosperms are found across eight different families (for recent reviews see ref. 3). Most, but not all, are photosynthetic and
capable of maturing to seed set without a host; however, it is rare
if ever that field collections reveal plants that are devoid of attachments. The implication is that these attachments are uniquely
important to the viability of all of these parasites, and it should
be anticipated that specific early developmental events have evolved
to facilitate rapid and efficient host attachment. Other papers in
this Symposium will deal with strigol and derivatives of strigol
which function as specific and highly sensitive stimulants of germination in Striga asiatica (Witchweed). Germination appears to function as one viable level of chemical recognition in host selection.
However, the developmental feature that is uniquely common to all the
parasitic angiosperms is the haustorium (Figure 1) – the organ which
forms the physiological and morphological attachment between host
and parasite. These organs have been shown not to form when Agalinis
purpurea (4) or Striga (5) are grown axenically. The development of
these organs can, however, be very rapidly induced in the presence of
host roots or host root exudate. Laboratory cultures of both Agalinis purpurea and Striga asiatica have been developed and maintained. (4) When induced, the haustoria are fully formed within 6 to
12 hours. This system has been developed for both quantitative and
qualitative bioassays. Using these bioassays to direct the isolation, several haustorial inducing principles have now been characterized. This report details the chemical characterization of these
haustorial inducing factors and the present knowledge about the role
that these factors play in host selection and host allelopathy. More
specifically, I have focused on the chemical, spectroscopic and biological methods which we have developed during the course of this
work with the hope that these methods will be of some general use to
other scientists.

Plant Materials and Bioassays

Agalinis purpurea (L.) Raf. (Scrophulariaceae) seeds were collected
by Professor Lytton Musselman, Old Dominion University, and grown in
sterile culture in 60 x 15 mm Petri plates on Murashige-Skoog medium
as previously reported (4). Lespedeza sericea (Leguminosae) was
grown in vermiculite under greenhouse conditions. Bioassays for
haustorial induction were carried out by either spotting fractions on
0.5 cm filter paper discs and inserting the discs into agar in which
2 to 3 week old Agalinis plants were growing or removing the plants
individually and placing 4 to 5 each into depression slides containing distilled H_2O and known concentrations of inducer. The haustoria
in each case were read after 48 hours. The transfer of the plants
from agar to water introduces some stress, and the sensitivity of the
assay is diminished although the specificity seems unaltered.

Haustorial Inducing Factors from Gum Tragacanth

After finding that primary metabolites and hormones fail to initiate haustorial development in <u>Agalinis</u> (<u>4</u>) and realizing that these parasites must be cuing on substances exuded from the potential host, several commercially available plant exudates were screened for activity. Gum tragacanth, an exudate of <u>Astragalus</u> <u>gummifer</u> (Leguminosae), showed potent activity in eliciting haustorial development and was far more readily available than host exudate. Hoping that the tragacanth elicitors would provide some insight into the kinds of molecular species involved in haustorial induction, John Steffens and Dr. Mike Thompson focused their efforts on the factors present in gum tragacanth (<u>6</u>).

Soxhlet extraction of the dry gum (250 g) serially with hexanes, ether, and methanol resulted in an activity rich ether fraction (600 mg, dried <u>in vacuo</u>) that was partitioned between CCl$_4$ and 50% aqueous MeOH. The dried aqueous layer (300 mg) was applied directly to droplet countercurrent chromatography (DCC), C$_6$H$_6$/MeOH/CHCl$_3$/H$_2$O (2:3:2:1, upper phase mobile). The application of large quantities of this crude fraction directly to DCC proved critical to the purification and resolved two separate bands of activity. These bands, further purified by normal and reverse phase chromatography, existed as 8x10^{-5}% and 3.2x10^{-5}% of the dry gum and were named xenognosin A,<u>1</u> and B,<u>2</u>, respectively.

The UV (λmax 260, MeOH) of xenognosin A showed an hydroxystyrene chromophore and ^{1}H-NMR (100 MHz, acetone-d$_6$) double resonance experiments identified four isolated proton spin systems; a characteristic A$_2$B$_2$ (δ 7.20, 2H, d, J=8.5 Hz; δ 6.74, 2H, d, J=8.5 Hz) system, a 1,2,4-trisubstituted aromatic system (H-6, δ 6.90, d, J=8.4 Hz; H-5, δ 6.35, dd, J=8.4, 2.4 Hz; H-2, δ 6.44, d, J=2.4 Hz) with oxygen functionality at positions 2 and 4, a 1,3-substituted <u>trans</u> propene (δ 6.315, d, J=15 Hz; δ 6.16, dt, J=15, 5.6 Hz; δ 3.34, 2H, d, J=5.6 Hz) and a methoxy singlet (δ 3.77). The small sample size complicated these proton assignments and they were confirmed by a synthetic model system prepared by David Graden [2-(4'-methoxy)-3-(2',4'-dimethoxy)-diphenylpropene] (<u>7</u>) and spectral simulations. For example, the double triplet (δ 6.16) was obscured due to the poor signal-to-noise and the severe second order coupling at 100 MHz. Simulations (<u>7,8</u>) of the olefinic protons were critical to these assignments (Figure 3).

Two structural questions remained in the characterization of xenognosin A. The propene unsaturation could be conjugated with either of two possible aromatic phenolic chromaphores. Answers to related problems have been addressed through Overhauser experiments (<u>9</u>), but in this case the sample was too small to obtain sufficient signal-to-noise to detect the enhancements. Partially relaxed ^{1}H-NMR proved to be a far more sensitive method of detecting regiochemical proximities in this system. Figure 4 details the downfield region of the ^{1}H-NMR spectrum and compares it to a selected partially relaxed spectrum, [180°-t-90°]n. The must faster relaxation rate of the doublet at δ 7.2, relative to the other half of the A$_2$B$_2$ system or the protons of the other aromatic ring, confirms its proximity to the olefinic protons. This same distinctive difference was observed in several model systems including 4-methoxy styrene, which under similar

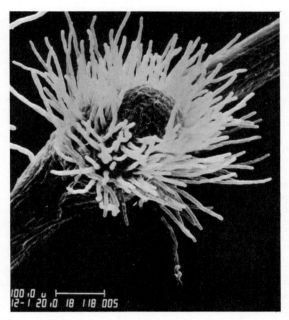

Figure 1. Haustorium on <u>Agalinis purpurea</u> 24 hours after induction.
Reprinted with permission from Dr. Vance Baird.)

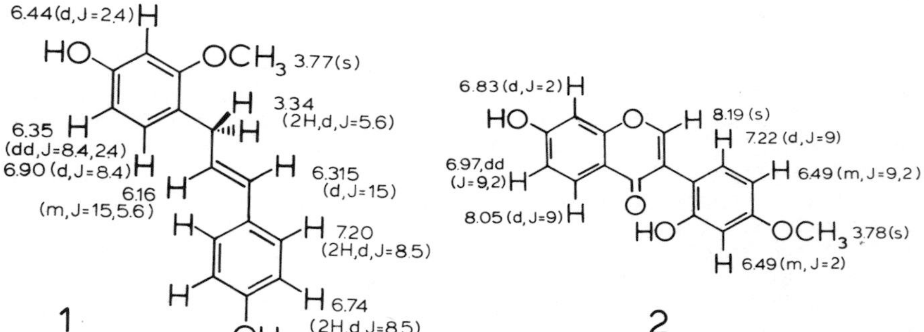

Figure 2. Structures of xenognosin A, 1, and B 2. ^{1}H–NMR
(100 MHz, acetone-d$_6$ signals are assigned.)

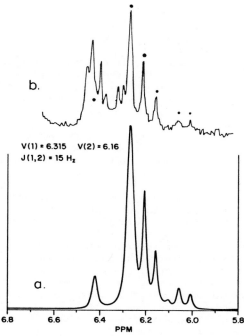

Figure 3. (a) The computer–simulated spectrum of the olefinic protons using chemical shifts of 6.3 and 6.1 ppm and a coupling constant (J) of 15 Hz. (b) The olefinic region of the 100 MHz ^{1}H–NMR spectrum of the originally isolated xenognosin A. The marked resonances correspond exactly with the resonances of the simulated spectrum.

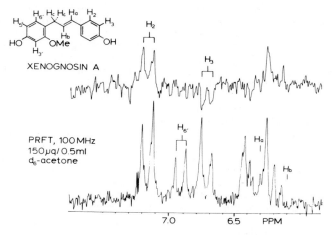

Figure 4.   Partially relaxed Fourier transform ^{1}H–NMR spectrum on 150 μg of xenognosin A shown above the normal ID–spectrum (downfield region only).

conditions gave a difference in T_1's of 0.5 sec between the doublets
of the A_2B_2 system. This was deemed sufficient support to assign
the styrene chromophore as in 1. Later work by David Graden with a
synthetic sample of 1 allowed for a quantitation of both the partially relaxed data (Figure 5) and nOe difference data (Figure 6).
     The second structural problem involved the regiochemical as-
signment of the single methoxyl group. This assignment could not be
verified by differences in T_1 times, so advantage was taken of specificity of the association between pyridine and the acidic phenolic
protons.   Titration of the acetone-d_6 solution to 20% pyridine-d_5
(v:v) gave clear shifts of the highfield and lowfield doublets of
the A_2B_2 of 3 Hz and 1 Hz respectively. Two additional aromatic
signals, δ 6.35 and δ 6.44, were shifted downfield by 4 Hz and 3 Hz
respectively, suggesting that the free hydroxyl group on the other
aromatic ring was flanked by two ortho protons as shown in 1. Mass
spectral confirmation of that assignment was obtained from CAD analyses of the m/z 137 fragment ion generated from chemical ionization
(CH_4) of 1 (Figure 7) (10). The m/z 137 ion comes from α-cleavage
of the olefin to generate the benzyl cation shown in Figure 7. It
was reasoned that the further fragmentation of that ion would be dependent on the regiochemistry of the methoxy group. The regioisome-
ric ions needed for mass spectral comparisons were accessible
through Dr. V. Kamat's synthesis of the isomeric benzyl alcohols.
Under chemical ionization conditions, these alcohols protonate and
lose H_2O to generate the same ions at m/z 137. CAD analyses confirm the suspected difference in fragmentation and establish the methoxy ortho to the alkyl substituent in 1 (Figure 8).
     The skeletal assignment of xenognosin B proved to be more
straightforward than did xenognosin A. The UV (MeOH, λmax 248 nm,
sh 300) and the characteristic one proton singlet at δ 8.19 (100 MHz,
acetone-d_6) are very diagnostic of isoflavanoids. Two trisubstituted
aromatic rings and an aromatic methoxy group (δ 3.78) completed the
^{1}H-NMR assignment and revealed the basic structure 2. The remaining
regiochemical assignment involved the placement of the methoxy methyl
group on one of three possible oxygen atoms - very similar to the
problem faced with xenognosin A. The chemical ionization MS(CH_4)
gave little fragmentation, but CAD (N_2) analyses of the M+1 molecular ion gave two ions at m/z 136 and m/z 148 (Figure 9) arising from
a retro-Diels-Alder fragmentation of the heterocyclic ring. The
masses of these ions established the association of the methoxy group
with the m/z 148 fragment, but did not locate it at either the 2' or
4' position of the ring. Capitalizing on the higher energy fragmentation afforded by electron impact mass spectrometry (70 ev, 150°C),
the retro-Diels-Alder fragmentation was completely suppressed and
the major fragment ion appeared at $(M-17)^+$. Through comparison with
other 2'-substituted isoflavanes (11), the M-17 fragment can be assigned to result from the loss of an hydroxyl radical to give the
stabilized oxonium ion (Figure 10).
     These data established the methoxy group at 4' and the struc-
ture of xenognosin B as 2. This sample also proved to be identical
by NMR, MS, and biological activity to a synthetic sample of 2',7-
dihydroxy-4'-methoxy isoflavone (generously provided by Professor
Paul Dewick, University of Nottingham, U.K.).
     Two different uses of mass spectrometry have proven critical to

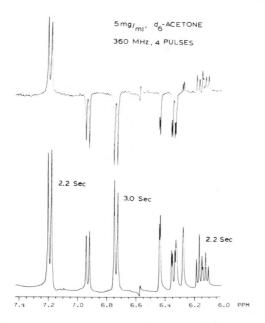

Figure 5. ^{1}H–NMR T_1 relaxation study of xenognosin A at 360 MHz.

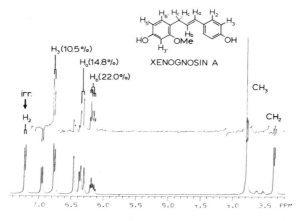

Figure 6.   Xenognosin A nOe difference spectrum.   (360 MHZ, acetone-d_6).   Irradiation of H_2 gives the enhancements shown in the upper trace.

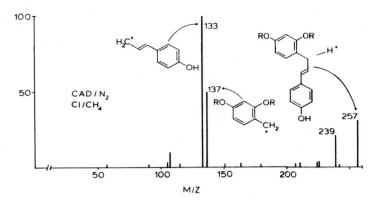

Figure 7. CI MS (CH$_4$, 150 ^{o}C) generated (M+1)$^+$ ion was quadrapole selected and subjected to collision activated decomposition
(N$_2$) to give the mass spectrum (6).

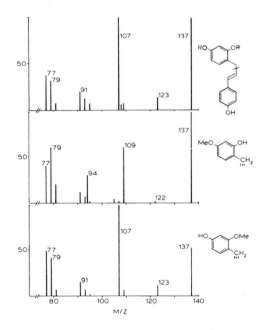

Figure 8. The m/z 137 ion was selected from the source and colli-
sionally decomposed (N$_2$) to give the CAD spectra. The top spectrum
is the fragment arising from the isolated xenognosin and the bottom
two spectra arise from the isomeric benzyl alcohols (6).

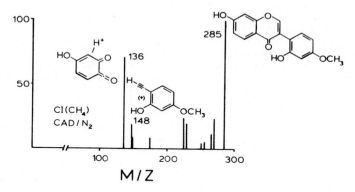

Figure 9.   Xenognosin B was chemically ionized (CH$_4$, 150 °C) and the (M+1)$^+$ ion quadrapole selected and subjected to collision activation decomposition (N$_2$) to give the mass spectrum.

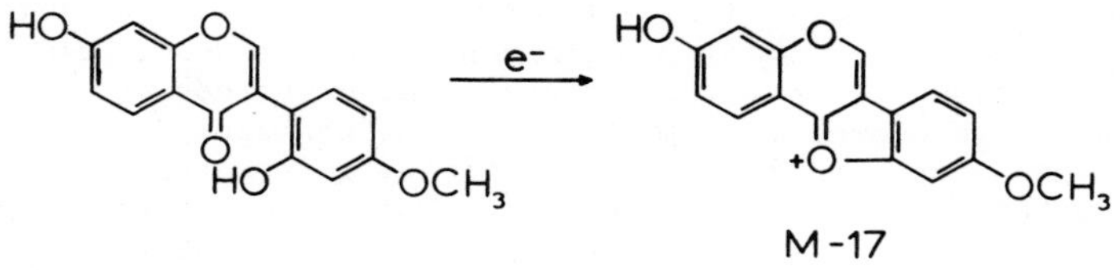

Figure 10   Postulated structure for the major fragment ion of xenognosin B under electron impact mass spectrometry.

the microgram scale structure assignment of regiochemistry of methoxyl substituents on these haustorial inducing factors. In fact,
the similarity of the dihydroxy and methoxy substituents on these
phenylpropenoids suggested that the biological activity may be linked to this functionality. Formononetin (7-hydroxy-4'-methoxy iso-
flavane), 3, isolated from the same source, was completely devoid
of biological activity and added further support for the necessity
of the functionality.

Dr. Kamat developed an efficient total synthesis of xenognosin
A (Scheme 1) which possessed sufficient flexibility for structure/
activity relationships to be explored (12). The several analogues
which he was able to prepare uncovered a strict structural specificity associated with certain parts of the xenognosin A skeleton.
Modification to the styrene system either through removing (4f) or
methylating (4g) the para-hydroxyl group had little effect on the
activity, but a simple change in the regiochemistry of the methoxy
substituent on the trisubstituted ring (4h) severely reduced the
activity (Figure 11). Further reduction (6) or oxidation (7) of the
propene bridge greatly reduced the biological activity, and removal
of either the methoxyl or hydroxyl groups from the trisubstituted
ring leaves the compound completely inactive.

These structure/activity studies point toward a precise structural dependence on certain functionality of the xenognosin A molecule (13). The meta-methoxyphenol functionality seems most significant since 5 shows slight activity in the agar assay. Even though
these studies are hampered by the differing solubility of the analogues and the stress implicit in removing the organisms from agar
and placing them in depression slides with distilled H_2O (the half
maximal response to xenognosin A in depression slides is $10^{-5}M$ and
in the agar assay is at least two orders of magnitude more sensitive), the development of some insight into the specificity of this
response has been gained. This specificity is remarkable in light
of the rather broad host range that Agalinis is capable of parasitizing. Nonetheless, flavanoids are well known as stable taxonomic
characters and several flavanoids and phenolics are often involved
in host allelopathy (14). Xenognosin B was suspected of being a
biosynthetic precursor of the red clover phytoalexin ± medicarpin,
and in fact was synthesized and shown to be biosynthetically incorporated into the phytoalexin before it was ever found to occur naturally (15). Xenognosin A has now been found in varieties of Pisum
as a stress metabolite associated with host defense (16). It seems
perfectly reasonable that these taxonomically characteristic and
physiologically active components which are produced by root tissue
would serve as cues for host selection in a root parasite.

Haustorial Inducers in a Natural Host

While the activity and the structural specificity of the xenognosins
were very encouraging, Astragalus gummifer is native to the Middle
East and would not be expected to have co-evolved as a host for the
southeastern U.S. native, Agalinis purpurea. The questions of whether molecules of this type were actually exuded from host roots
in sufficient quantities to constitute host selection still remained.
For that reason, John Steffens switched his attention to Lespedeza

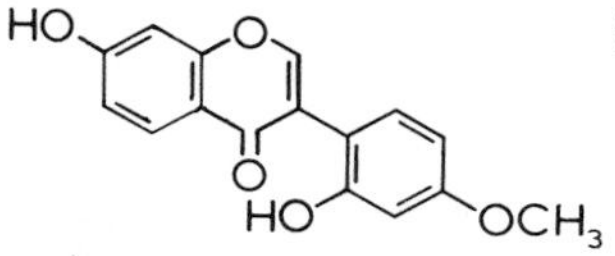

xenognosin B

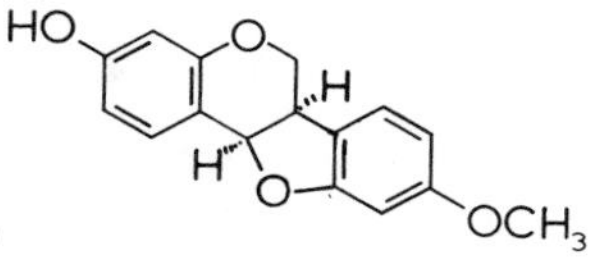

± medicarpin

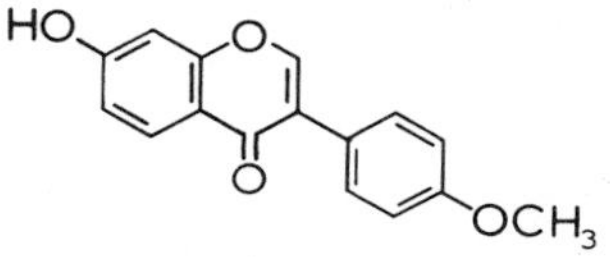

formononetin

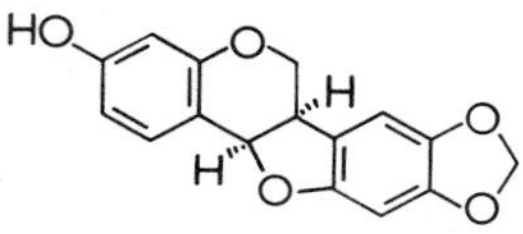

±maackiain

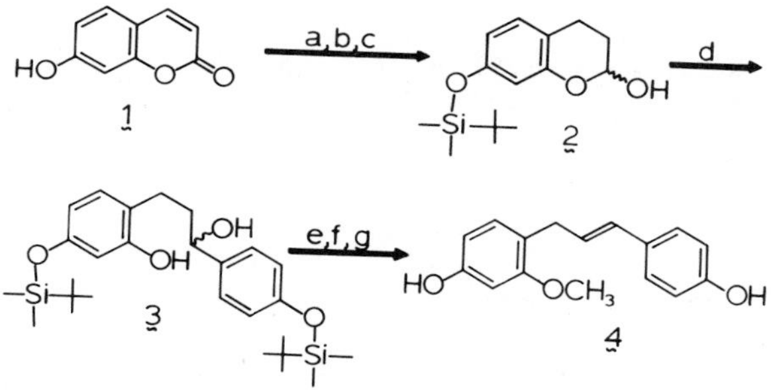

Scheme 1. Synthesis of xenognosin A (12).

a) H$_2$, Pd/C (10%), EtOH, 40 PSI, 24 hr, 95%.

b) TBDMS-Cl, (Et)$_3$N, DMAP, CH$_2$Cl$_2$, 25°C, 10 hr, quantitative.

c) 1 eq. DIBAL, toluene, -78°C, 2 hr, 95%.

d) p-TBDMS-O-C$_6$H$_4$Br, Mg, THF, 25°C, 1 hr, 85%.

e) CH$_2$N$_2$, MeOH, 4°C, 8 hr, 95%.

f) CH$_3$SO$_2$Cl, (Et)$_3$N, THF, 25°C, 2 hr, 60%.

g) (C$_4$H$_9$)$_4$NF, THF, 25°C, 0.5 hr, 90%.

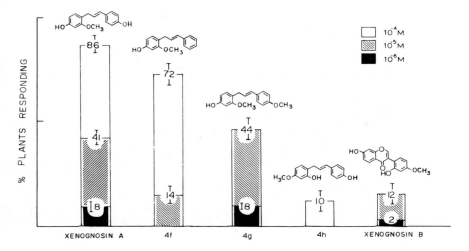

Figure 11. Relative activities of haustorial inducers when presented in solution to 2-to-3 week-old plants of <u>Agalinis</u> <u>purpurea</u>. Plants developed an average of two haustoria each when presented with xenognosin A at 10^{-4}M. (<u>13</u>).

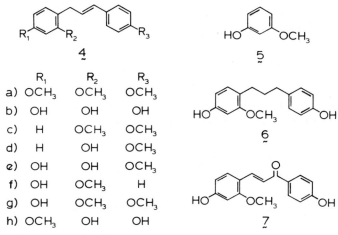

	R_1	R_2	R_3
a)	OCH$_3$	OCH$_3$	OCH$_3$
b)	OH	OH	OH
c)	H	OCH$_3$	OCH$_3$
d)	H	OH	OCH$_3$
e)	OH	OH	OCH$_3$
f)	OH	OCH$_3$	H
g)	OH	OCH$_3$	OCH$_3$
h)	OCH$_3$	OH	OH

Analogues of xenognosin A

sericea (Leguminosae), a forage legume which could be grown in the laboratory for the production of exudate and was readily parasitized by Agalinis (17). Four hundred grams (fresh weight) of 3-month old vermiculite-grown Lespedeza roots were extracted with 50% aqueous MeOH, dried in vacuo and lyophilized. Twelve grams of this material were extracted with acetone and concentrated to yield 2.5 g of a dark brown oil. This oil was applied directly to droplet counter-current chromatography ($CHCl_3$/MeOH/H_2O, 7:13:8, descending mode) and the early eluting active fraction was repeatedly flash chromatographed (SiO_2, CH_2Cl_2/MeOH) and finally purified by HPLC (reverse phase C_{18}, 83% MeOH/H_2O) giving 2 mg of a crystalline solid, 8.

Mass spectrometry analyses (18) (EI, 70 eV) gave a molecular ion at m/z 458.3757, $C_{30}H_{50}O_3$ (calcd 458.3760) and major fragment ions at m/z 234 and 224, suggesting a retro-Diels-Alder fragmentation of an olean-12-ene triterpene bearing two oxygen atoms on the A-B ring fragment (19). Seven methyl singlets in the highfield region of the ^{1}H-NMR (360 MHz, acetone-d_6) and a single olefinic proton (δ 5.5, t, J=7.5 Hz) supported the oleanene assignment (Figure 12). Three hydroxyl substituents were identified with deuterium-exchange negative ion CI MS (EtOD, N_2O) (20) by the appearance of a molecular ion at m/z 459 [(M-H)$^-$ with 2 exchanges]. The 3β- and 24-hydroxyl functionalities were suggested by ^{1}H-NMR (δ 3.45(H-3, dd, J=5.4, 11 Hz), δ 4.2 (H-24, d, J=11 Hz), δ 3.5 (H'-24, bd, J=11 Hz) (21). Disruption of the intramolecular hydrogen bonding with Me_2SO-d_6 and by acetylation, taken together with a 2.9% nOe of the C-25 methyl upon irradiation of H'-24 (22), confirmed this assignment.

Placement of the third hydroxyl group was more difficult. Electron impact fragmentation had restricted its location to the D-E ring fragment, and the proton residing on the same carbon (δ 3.42, t, J=4.5 Hz) was best assigned as an equitorial proton flanked by a single methylene. This restricted the hydroxyl group to an axial position at C-15, C-16, C-21 or C-22.

CAD-MS proved invaluable in establishing substituent regio-chemistry in the xenognosins and had proved useful in triterpenes (23). Comparison of the mass spectrum of the triacetate of 8 with 3-oxo-15-acetoxyolean-12-ene showed that both compounds gave an m/z 276 fragment from the retro-Diels-Alder rearrangement and both of these ions lost acetic acid to an m/z 216 ion. CAD analyses (1 μT Ar as collision gas) of these two fragment ions gave considerably different spectra, ruling out the C-15 hydroxylation of 8.

NOe difference experiments (22) to aid in methyl group assignment were conducted with the 3β,24-phenylborate of 8 (phenylboronic acid, benzene, reflux w/Dean Stark trap) since the methyl signals were more easily resolved following this modification. Key enhancements obtained from the nOe difference experiments and their assignments are listed in Table I. Additionally, CH_3-27 and CH_3-30 could be assigned from correlation with other triterpenoids (24). The remaining two methyl groups, CH_3-29 and CH_3-26, were tentatively assigned to the signals at δ 0.87 and δ 0.97 respectively. It became clear during the course of the work that derivatization of the 1,3-diol of 8 introduced small unexpected chemical shift changes in methyl signals distant from the A-ring, therefore making the predictive value of the empirically derived correlation tables unreliable.

Table I. Nuclear Overhauser Enhancement Difference Value for
Soyaspogenol B

Signal irradiated		Signal enhanced		
assignment	δ	assignment	δ	%
H-18	2.2(dd)	CH$_3$-28	1.05	1.3
H-24	3.17(bd)	CH$_3$-23	1.02	0.6
H-24'	4.1(d)	CH$_3$-25	0.93	1.9
CH$_3$-25	0.93(s)	H-24'	4.1(d)	4.9

Preirradiation (selective 180°) of each signal was followed by a
90° observed pulse delayed by 0.7s. This spectrum (550-1000 transients) was acquired simultaneously with a spectrum in which one
selective pulse was 3 ppm upfield of tetramethylsilane, and the
two spectra were computer subtracted to observe the enhancements.

Nevertheless, the use of the methyl groups as diagnostic signals for the localization of ring functionality is invaluable. Titration of the phenylborate ester of 8 with Eu(fod)$_3$ gave the data
shown in Figure 13. The invariance of the methyls CH$_3$-23 and CH$_3$-25
verified phenyl borate protection of the diol against shift reagent
complexation and restricts complexation to the third hydroxyl group
(Figure 13). Three methyl groups, CH$_3$-29, CH$_3$-28, and CH$_3$-30, displayed the greatest induced chemical shift with added shift reagent,
and this localized the hydroxyl group to the E-ring. Closer examination suggested that C-21 hydroxylation (β, axial) would predict
CH$_3$-29 and 30 to be most affected by the shift reagent, and that
placement at C-22 (α, axial) would predict CH$_3$-28 and 30 to experience the greatest downfield shift. Neither of these expectations
were realized, probably due to the angular dependence of the shift
reagent induced chemical shifts (25).

For that reason, attention was turned to experiments that would
increase the resolution obtainable in the ^{1}H-NMR spectrum. Professor
James Roark, visiting from Kearney State College, was experimenting
with methods for selectively observing the transmission of coupling
information over large numbers of bonds. The monoterpene dl-camphor
best represents what he found. Figure 14 shows a contour plot (26)
of the homonuclear proton correlation map of camphor. The corresponding 1D spectrum is shown along the ordinate. Connection of the
off-diagonal cross peaks with their corresponding diagonal partners
establishes the spin coupling between adjacent protons and allows
for virtually the complete proton assignment of camphor. Close in-
spection of the camphor map shows that the "W-type" 4-bond coupling
between H-2x and H-4x visible in 1D spectrum (J=3.8 Hz) has no cross
peak in the 2D spectrum of Figure 14. Protons with J couplins of
4 Hz or less have proven to give cross peaks that are weak or not
present in low resolution COSY experiments (27), but the resolution
of smaller couplings is possible by the insertion of fixed delays
following each of the two pulses (28). For example, an 0.2 sec
delay (Δ = 0.2 sec) following each of the two 90° pulses results in
the appearance of intense cross peaks between H-2x and H-4x.

Extending the delay to 400 msec results in the appearance of

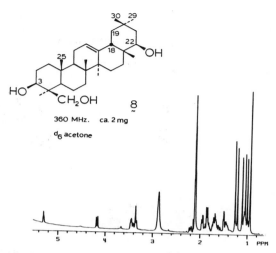

Figure 12. ^{1}H–NMR spectrum of the haustorial inducer from Lespedeza sericea.

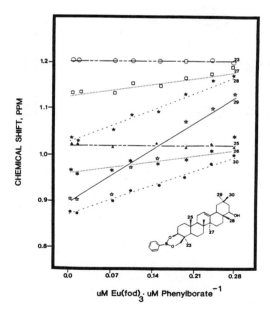

Figure 13. Effect of varying concentrations of the NMR shift reagent Eu(fod)$_3$ on methyl resonances of soyasapogenol B phenyl borate. Eu(fod)$_3$ was dissolved in a minimum amount of acetone-d$_6$ and added to a 2 ml solution of soyasapogenol B phenyl borate in the same solvent. Spectra were obtained at 360 MHz.

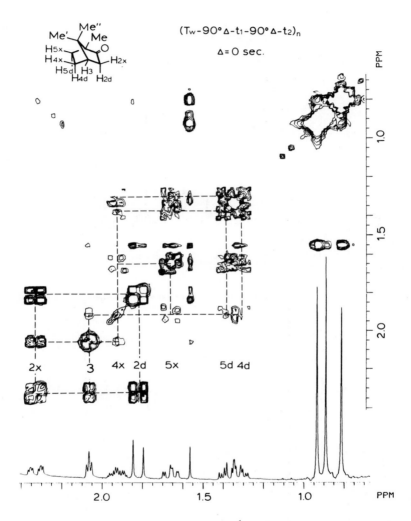

Figure 14. Contour plot of the 360 MHz ^{1}H–NMR correlation spectrum of dl–camphor. A 64 x256 data set was accumulated with quadrature phase detection in both dimensions and the data set was zero filled once in the F_2 dimension and symmetrized. T_w was 5 sec and t_1 was incremented by 1.63 msec. Total accumulation time was 24 minutes and data workup and plotting took 15 min.

several additional cross peaks. The higher contour slice in Figure 15a shows only the intense quaternary methyl diagonal signals, but with the addition of the delay (Figure 15b), cross peaks are observed between the high and the low field methyl singlets. The observation of these cross peaks arising from the ^{4}J coupling between the gem-dimethyls of camphor allows for the assignment of the bridgehead methyl to the central singlet at δ 0.89. This assignment is further supported by a lower contour slice of the same data, Figure 16, where cross peaks appear for the ^{5}J coupling between H-3 and the central methyl. Differentiation of the gem-dimethyls is readily apparent from the cross peaks appearing between H-4d and 5d and the high field methyl (δ = 0.81). The analogous coupling is also observed between H-2d and the downfield methyl (δ 0.94). Thus, in just two experiments that facilitated the observation of proximities over two to five bonds, all the proton assignments of camphor have been made. The assignments made here with delayed COSY agree with previous assignments of camphor made by rigorous comparisons with model systems (29).

Since the use of the ring protons of camphor had proven so effective in assigning the three methyl singlets through long range coupling interactions, the reverse usage of the methyl groups to assign functionality along the triterpene skeleton should be equally effective. Protection of the A-ring of 8 as an acetonide (2,2'-dimethoxypropane, acetone, pTsOH) and oxidation (PDC, pyridinium trifluoroacetate, rt, 4 h) generated the ketoacetonide 9. This oxidation shifted the protons α to the carbonyl downfield from the bulk of the backbone methylenes and greatly simplified their assignment (Figure 17). The cross peaks connecting H-18 (δ 2.3) to H-19e (δ 1.35) and H-19a (δ 2.07) are also pointed out in Figure 17. The insertion of a 0.25 s delay in the COSY sequence significantly changes the overall appearance of the 2D-map and the observed cross peaks (Figure 18). A ^{4}J coupling cross peak connects H-19a with the highfield methyl singlet (CH$_3$30). This same methyl singlet shows coupling to one of the protons adjacent to the carbonyl allowing its assignment as H-21a. Another cross peak verifying a ^{4}J coupling between H-19e and the other methylene proton adjacent to the carbonyl, H-21e, further confirms the site of oxidation as C-22 and so assigns the structure of the haustorial inducer as 3β,22β,24-trihydroxyolean-12-ene. During the course of this work Kitagawa et al. (30) revised the structures of soyasapogenol B to this structure on the basis of extensive chemical and X-ray analyses. Our work utilizing small scale NMR and MS spectra data is in agreement with that of Kitagawa.

Quantitation of Haustorial Inducers in Root Exudate

Now that an haustorial inducing factor has been characterized in a host plant that could be grown in the laboratory, the levels of the compounds actually exuded could be analyzed. John Steffens and Rody Spivey focused on developing methods that would allow for suitable quantitation. Efforts were made to quantitate not only the terpenoid components, but also the flavanoid, genistein (4',5,7-trihydroxyisoflavone), which was found to be a major isoflavone of Lespedeza. Genistein was analyzed to gain an estimate of levels of phenylpro-

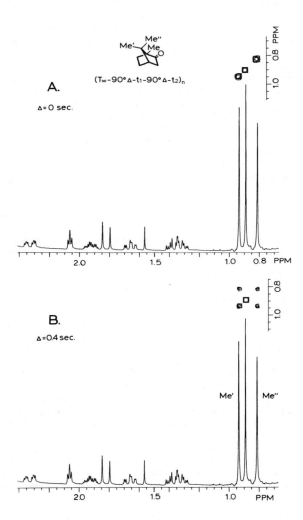

Figure 15. Higher contour slice of (A) the data in Figure 14 and (B) the data in Figure 17.

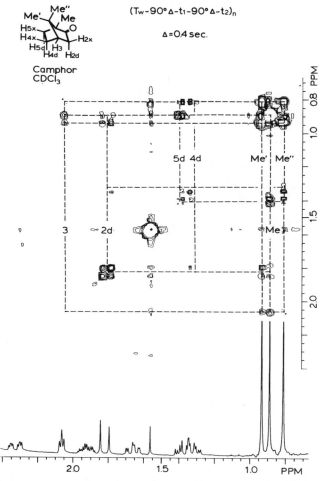

Figure 16. Same experiment as performed in Figure 14 with $\Delta = 0.4$ sec. Sixteen transients were collected rather than the 4 transients in Figure 14.

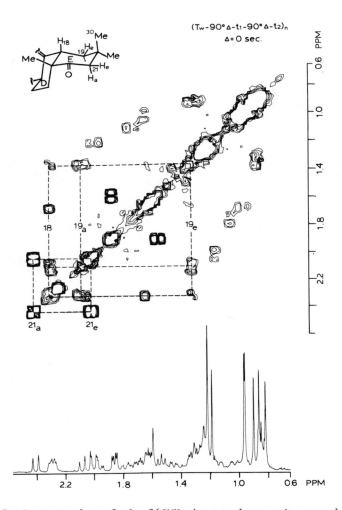

Figure 17. Contour plot of the 360MHz homonuclear spin correlation
mpa of 10 (2 mg, CDCL₃, high-field expansion) with no delay
inserted in the pulse sequence shown at the top of the figure.
Assignments of cross peaks indicating coupled spins in the E-ring
are shown with the dotted lines. The corresponding region of the
one-dimensional ¹H NMR spectra is provided on the abscissa. The
2-D correlation map is composed of 128 x 512 data point spectra,
each composed of 16 transients. A 4-s delay was allowed between
each pulse sequence (T$_w$) and t$_1$ was incremented by 554s. Data was
acquired with quadrature phase detection in both dimensions,
zero filled in the t$_1$ dimension, and the final 256 x 256 data was
symmetrized. Total time of the experiment was 2.31 h (17).

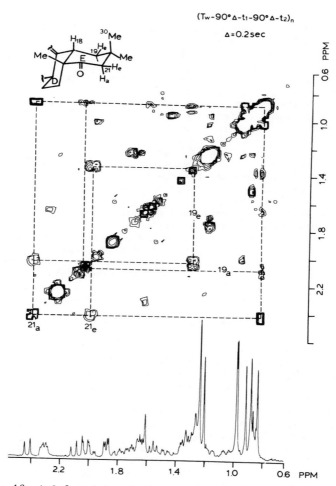

Figure 18. A 0.2-s delayed COSY spectrum of the aliphatic region of 10 (2mg, CDCL$_3$). Long-range "W-type" coupling of 19 and 21 axial protons to 30-CH$_3$ and coupling across the gem dimethyls from 19 eq to 21 eq establish the position of oxidation at C-22. The spectrum was obtained under conditions similar to those in Figure 1, except that 32 transients were acquired for each of 128 x 512 data point spectra (17).

panoids such as xenognosin A and B which might be exuded by plants
biosynthesizing molecules of this type.

Seeds of <u>Lespedeza sericea</u> were surface sterilized by soaking
in Captan (1 g/l) 1 h, 35% Clorox for 20 min, and washing with 1N
HCl and water. For quantitation of components in <u>Lespedeza</u> roots,
seeds were sown on moistened filter paper in plastic Petri dishes.
At 15d post-germination, roots (70 for isoflavone analysis; 210 for
triterpenes) were dissected off, frozen and lyophilized. The dried
tissue was then homogenized with 4 ml methanol and further extracted
with 3 additional volumes of the same solvent. After filtration
through glass wool and concentration <u>in vacuo</u> at 30°C, the extracts
underwent droplet counter-current chromatography in 7:13:8 chloro-
form/methanol/water (7:13:8, v/v) run in the descending mode with
the organic phase mobile. Fractions corresponding in retention
time to the soyasapogenols and genistein were collected, pooled and
concentrated <u>in vacuo</u>. Each fraction was then passed over a silica
gel column 0.5 x 2 cm eluted isocratically with 8% methanol dichloromethane (v/v). The eluate was then concentrated <u>in vacuo</u> and filtered prior to HPLC. HPLC analyses for genistein (65% MeOH:H_2O
(v/v), 260 nm detection) and for the terpenoids (84% MeOH:H_2O (v/v),
214 nm detection) were carried out on a Waters 5µ C_{18} reverse phase
column eluted at 1 ml/min.

For quantitation of root exudate components, surface sterilized
<u>Lespedeza</u> seeds were allowed to germinate and the seed coats were
separated from the young plants. At 3d post germination the root
of each plant was inserted through Nitex mesh suspended over water
in a sterile Petri dish. In this way, plants were supported by the
mesh so that only the roots were allowed to come into contact with
the water. Every 3d for the next 9d, exudate was removed and replaced with sterile distilled water; plates were discarded when signs
of contamination appeared. Exudate collected in this way from
3200 plants, filtered through an 0.8 Millipore filter and lyophilized, yielded 185 mg dry exudate. The lyophilized exudate was
extracted with methanol and the resulting extract, after filtration
and concentration <u>in vacuo</u>, was chromatographed under conditions
identical to those described above for root extracts.

Thin layer chromatograms and HPLC analyses showed a major terpenoid component that eluted with the synthesized 22-keto diol, <u>10</u>.
This compound, now known as soyasapogenol E, was present as a minor
component in the root extract, but in exudate appears to be present
in levels equal to or exceeding soyasapogenol B (Table II). Genistein is found in the roots at levels comparable to that of soyasapogenol B but is exuded at levels about two orders of magnitude lower
than the soyasapogenols. Also, the ratio of soyasapogenol B to E in
the root is 6 to 1, whereas greater levels of soyasapogenol E are
exuded than soyasapogenol B. This apparent selective exudation may
either represent an active aspect of root metabolism or reflect
cellular compartmentalization of secondary metabolites and their
passive leakage from the root. Bell and co-workers (31-32) have
shown a compartmentalization of plant chemical defenses in the cotton root, with triterpenoid synthesis occurring in the epidermis,
proanthocyanidins in the hypodermis, sesquiterpenoids in the cortex,
and proanthocyanidins in the endodermis. If a spatial distribution
of this kind is maintained in <u>Lespedeza sericea</u>, this could account

Table II. Root Exudate and Extract Quantitation Lespedeza sericea

	Soyasapogenol B	Soyasapogenol E	Genistein
Root Extract			
ρmol/root	291 ± 1	50.0 ± 0.2	304 ± 6
(% dry wt root)	(0.06)	(0.01)	(0.03)
Root Exudate (3245 plants)			
ρmol exuded/root/day	42.8 ± 0.1	56.7 ± 0.1	0.422 ± 0.01
(% dry wt exudate)	(0.34)	(0.45)	(0.002)

Each figure represents the mean and standard error of three repetitions. The exudation pattern from the first 5 days of collection
does not appear to differ from the second 5 days of collection. The
relative levels of soyasapogenols and genistein within the root are
not different at 15 days after germination, nor do they differ at
3 months of age. Soyasapogenols were not detected in shoot portions
of seedlings.

   Variance in root size of 5-day old plants was not taken into
account in either the calculation of ρmol/root or ρmold exuded/root.
The weight of filtered and lyophilized exudate of 375 plants was
used to extrapolate to the figure of 3245 individuals used for exudate collection in the analysis of exudation/root.

for the differential exudation of relatively large quantities of
epidermally localized soyasapogenol B and E, while the isoflavone
genistein, occurring in large quantities in cells distal to the epidermis, would be represented at much lower levels in the root exudate.

   Alternatively, the exudation of the soyasapogenols, which are
probably not differentially localized in the root, could represent
a more active process. According to the data in Table II, by day
15 approximately 94% of the soyasapogenol E synthesized in the root
has been released into the root exudate. In contrast, only 69% of
the soyasapogenol B synthesized in the root is found in root exudate. Thus, the differential secretion of soyasapogenol B and E
may account for the observed ratio of soyasapogenol B to E within
the root.

Conclusion

Several naturally occurring compounds have now been characterized
which are capable of inducing the differentiation of the haustorium
in Agalinis purpurea. They fall into two structural classes; the
phenylpropenoids, xenognosin A and B, and a terpenoid, soyasapogenol B. Both the biosynthetic connection of the xenognosins with
the taxononically useful flavanoids and the structural specificity
of their biological activity suggest that these compounds are ideal-
ly suited as host recognition cues for parasitic plants. The connection between flavanoids and root allelopathy (33) and the specific connections established by Dewick (15) and Carlson and Dolphin
(16) tying the xenognosins in with plant stress metabolites again

connects these compounds to specific, stable metabolites and therefore are likely cues of a suitable host.

The strengths of the arguments in favor of phenylpropenoid recognition cues identified in the nonhost gum tragacanth made the
isolation of the terpenoid from Lespedeza quite unexpected. However,
many of the arguments supporting the phenolics also support the
terpenoid. The haustorial inducing activity of soyasapogenol B is
dependent on certain structural features. Neither the 22-keto derivative, soyasapogenol E, nor a series of olean-12-ene triterpenes
from cactus (generously supplied by Professor Kirchner, University
of Arizona) possessed any detectable activity. The soyasapogenols
have only been found in the Leguminosae and therefore may have restricted occurrence, but their taxonomic usefulness is not as well
established as the flavanoids'. The soyasapogenols have also been
attributed with defensive roles. In their glycosylated form, they
are toxic and reduce herbivory in leguminous seeds (34). A group
of more highly oxidized oleanene triterpenes, the averacins, are
found in oat roots and are potent resistence factors to "take all"
disease caused by the fungus Gaeumannomyces graminis (35). If
Lespedeza and genistein exudation can be taken as representative,
then it seems unlikely that quantities of these phenolics sufficient
for the induction of haustoria would be exuded. Therefore the
structural specificity, the stable and specific metabolic production, and Lespedeza's metabolic control over the exudation of the
soyasapogenols suggest that the terpenoids are more appropriate
recognition cues for these parasites.

The haustorial inducing activity of soyasapogenol B is, however, much weaker than that of xenognosin A. In the filter paper
disk method, 20 nmol of soyasapogenol B is the lowest quantity
which will induce haustoria, whereas xenognosin A will induce large
numbers of haustoria at 1 nmol. Haustoria are not induced when
soyasapogenol B is presented as a 10 μM solution but xenognosin A
is quite active at the same concentration. Unlike the xenognosins
which constitute virtually all of the activity of gum tragacanth,
the soyasapogenol B activity represents only a portion of the
activity of Lespedeza. The continued reduction in biological activity during the fractionation greatly complicated the work with
Lespedeza until John Steffens was able to demonstrate a dramatic
increase in the activity of soyasapogenol B when it was combined
with another weakly active fraction. No such synergistic stimulation of activity has ever been detected with the xenognosins and
this demonstration with the terpenoids may indicate a considerably
more sophisticated, multi-component system of host recognition
than previously expected. At present, there is no data to suggest
that any of the synergistically acting substances are phenolic
although very little about the structure of these compounds is known.

The remarkably rapid process of cell division and differentiation which leads to the complete formation of the haustorium following stimulation suggests that a parasite cell or group of cells
are poised and ready for immediate response. It therefore may be
reasonable for a biologically active allelochemical, through some
fundamentally different pathway, to trip the initial physiological
events in the induction of the haustorium. In fact, xenognosin A
can induce haustorial development across a broad range of parasitic

angiosperms, including <u>Agalinis</u>, <u>Striga</u> <u>asiatica</u> (unpublished results), and <u>Sopubia</u> <u>delphinitolia</u> (Sahai and Shiranna, pers. comm.).
Soyasapogenol B only shows activity in the <u>Agalinis</u> system although
the synergistically active components have not been vigorously tested
on other parasites.

Only through more work on this system and the investigation of
other parasitic plants will the role of the haustorium in host selection be understood. The parasitic plants have provided a biological system to direct the isolation and identification of naturally occurring allelochemicals which have profound physiological effects on these plants. The cells that ultimately lead to the haustorium can then be chemically manipulated with the very real hope
of controlling haustorial expression. The specific chemical control
of this organ, an organ found only in the parasitic plants, provides
a fundamentally new and very specific way of combatting the agriculturally devastating parasites such as <u>Striga</u> and <u>Orobanche</u>. The
technologies for the further study of these systems and other allelochemical based systems are now in place in several laboratories
and the next ten years offer to provide rewarding insights into the
chemical basis of biological recognition phenomena.

<u>Acknowledgments</u>   The progress in these studies would never have
been possible without the collaborative effort between this laboratory and that of Professor James Riopel at the University of Virginia. The funding provided by USDA Competitive Research Grant
5901-0410-9-0257 and USDA Cooperation Agreement 58-7B30-0-196 made
this collaboration possible. We also gratefully acknowledge the
funding from Research Corporation and the Frasch Foundation. I
applaud the efforts of John Steffens, who has contributed in a major
way to every aspect of this work; Dr. Vinayak Kamat and David
Graden, whose synthetic and spectroscopic talents proved invaluable;
and Professor Jim Roark, who in a relatively short time made a
lasting contribution to the work.

<u>Literature Cited</u>

1. For an excellent although somewhat dated overview see "Chemical
    Ecology"; Sondheimer, E., Simeone, J. B., Eds., Academic Press:
    New York, 1970.
2. Atsatt, P. R.; Hearn, T. E.; Nelson, R. C.; Heineman, R. T.
    <u>Ann. Bot.</u> (London) 1978, 42, 1177-1184. Atsatt, P. R. <u>Am.</u>
    <u>Nat. III</u> 1977, 579-586.
3. Musselman, L. J. in "Biology and Ecology of Weeds", Holzner, W.,
    Numata, N., Eds. Junk Pub. The Hague, 1982. Musselman, L. J.
    <u>Ann. Rev. Phytopath.</u> 1980, 18, 463-489. Kuijt, J. <u>Symp. Bot.</u>
    <u>Upsal. XXII</u>, Uppsala, 1979, 4, 194-199. Knijt, J. <u>Ann. Rev.</u>
    <u>Phytopathol.</u> 1977, 15, 91-118. Knijt, J. "The Biology of
    Parasitic Flowering Plants", 1969, Berkeley, University of Cal-
    ifornia Press.
4. Riopel J. L.; Musselman, L. <u>J. Amer. J. Bot.</u> 1979, 66, 570-75.
    Riopel, J. L. in Proc. Second Symp. on Parasitic Weeds, 1979;
    North Carolina State University, Raleigh, NC, pp. 165-173.
5. Okonkwo, S. N. C. <u>Amer. J. Bot.</u> 1966, 53, 687-97.

6. Lynn, D. G.; Steffens, J. C.; Kamat, V. S.; Graden, D. W.;
    Shabanowitz, J.; Riopel, J. L. J. Am. Chem. Soc. 1981, 103,
    1868-70.
7. Graden, D. W. Ph.D. Thesis, University of Virginia, Virginia,
    1984.
8. Simulations were performed on an NTC 1280 computer with a
    program similar to the LAOCOON program. Castellano, S. A.;
    Bothner-By, A. A. J. Chem. Phys. 1964, 41, 3863.
9. Bachers, G. E.; Schaefer, T. Chem. Rev. 1971, 71, 617-26.
    Noggle, J. H.; Schirmer, R. E. "The Overhauser Effect: Chem-
    ical Applications"; Academic Press: New York, 1971.
10. All MS/MS data was obtained on a Finnigan 3200 spectrometer
    modified for triple quadrapole work.
11. Egushi, S.; Haze, M.; Nakayama, S.; Hayashi, S. Org. Mass
    Spec. 1977, 12, 51-2.
12. Kamat, V. S.; Graden, D. W.; Lynn, D. G.; Steffens, J. C.;
    Riopel, J. L. Tetrahedron Lett. 1982, 1541-44.
13. Steffens, J. C.; Lynn, D. G.; Kamat, V. S.; Riopel, J. L.
    Ann. Bot. 1982, 50, 1-7.
14. Bell, A. A. Ann. Rev. Pl. Physiol. 1981, 32, 21-81.
15. Dewick, P. M. J. Chem. Soc. Chem. Comm. 1975, 656-8.
16. Carlson, R. E.; Dolphin, D. H. Phytochem. 1982, 21, 1733-36.
    Carlson, R. E.; Dolphin, D. H. Phytochem. 1981, 20, 2281-84.
17. Steffens, J. C.; Roark, J. L.; Lynn, D. C.; Riopel, J. L.
    J. Am. Chem. Soc. 1983, 105, 1669-71.
18. High resolution EI MS was obtained through Harvey Laboratories,
    Inc. on a VG 7070.
19. Budzikiewicz, H.; Wilson, J. M.; Djerassi, C. J. Am. Chem.
    Soc. 1963, 85, 3688-3699.
20. Hunt, D. R.; Sethi, S. K. J. Am. Chem. Soc. 1980, 102, 6953-63.
21. Isuda, Y.; Sano, T.; Isobe, K.; Miyauchi, M. Chem. Pharm. Bull.
    1974, 22, 2396-2401.
22. Preirradiation (selective 180°) of 24-H' was followed after a
    delay (0.7 s) by a 90° observed pulse. This spectrum (550-1000
    transients) was acquired simultaneously with a spectrum in
    which one selective pulse was 3 ppm upfield of Me_4Si, and the
    two spectra were computer subtracted to observe the enhance-
    ments. Solomon, I. S. Phys. Rev. 1955, 99, 559-565. Hall, L.
    D.; Saunders, J. K. M. J. Chem. Soc., Chem. Commun. 1980,
    368-370.
23. Chen, M. T.; Barbalas, M. P.; Pegues, R. F.; McLafferty, F. W.
    J. Am. Chem. Soc. 1983, 105, 1510-1513.
24. Itô, S. in "Natural Products Chemistry"; Nakanishi, K.; Goto,
    T., Itô, S., Natori, S., Nozoe, S., Eds.; Kodansha Lts.:
    Tokyo, 1974, pp. 365-366.
25. McConnell, H. M.; Robertson, R. E. J. Chem. Phys. 1958, 29,
    1361-65.
26. Benn, R.; Günther, H. Angew Chem. Int. Ed. Engl. 1983, 22,
    350-380.
27. Lynn, D. G.; Phillips, N. J.; Hutton, W. C.; Shabanowitz, J.;
    Fennel, D. I.; Cole, R. J. J. Am. Chem. Soc. 1982, 104, 7319-
    22.
28. Bax, A.; Freeman, R.; Morris, G. J. Magn. Reson. 1981, 42, 164-
    8. Bax, A.; Freeman, R. J. Magn. Reson. 1981, 41, 542-61.

29. Baker, K. M.; Davis, B. R. Tetrahedron 1968, 24, 1663–72.
30. Kitagawa, I.; Yoshikawa, M.; Wang, H. K.; Saito, M.; Tosiri-
     suk, V.; Fujiwara, T.; Tomita, K.; Chem. Pharm. Bull. 1982,
     30, 2294–7.
31. Bell, A. A.; Stipanovic, R. D. Mycopathologia 1978, 65, 91–106.
32. Bell, A. A.; Mace, M. E. in "Fungal Wilt Diseases in Plants";
     Mace, M. E., Bell, A. A., Beckman, C. H., eds.; Academic Press,
     New York, 1980.
33. Bearder, J. R. in "Hormonal Regulation of Development" Encycl.
     Pl. Physiol., Vol. 9, Springer-Verlag: Berlin, 1980, pp. 9–112.
34. Langenheim, J. H. in "Advances in Legume Systematics"; Polhill,
     R. M., Raven, P. H., eds.; Royal Botanic Gardens New England,
     1981, pp. 627–55.
35. Crombie, L.; Crombie, W. M. L.; Whiting, D. A. J. Chem. Soc.
     Chem. Comm. 1984, 244–6. Crombie, L; Crombie, W. M. L.; Whit-
     ing, D. J. Chem. Soc. Chem. Comm. 1984, 246–8.

RECEIVED August 16, 1984

Sesquiterpene Lactones and Allelochemicals from *Centaurea* Species

K. L. STEVENS and G. B. MERRILL

Western Regional Research Center, Agricultural Research Service, U.S. Department of Agriculture, Berkeley, CA 94710

The allelopathic weeds, Russian knapweed (Centaurea repens) and yellow starthistle (C. solstitialis), have been examined and found to contain several sesquiterpene lactones and two chromenes. A detailed analysis of the ^{13}C-NMR spectra of the sesquiterpenes has been carried out, thus permitting structural analysis using ^{13}C-NMR alone. The sesquiterpene lactones acroptilin, repin, solstitiolide, and centaurepensin caused increased root elongation at 10ppm and inhibited it at higher concentrations. The chromenes had no significant biological activity.

From antiquity (1), the infestation of weeds into areas of agronomic importance has been of major concern to the farmer and later to the rancher. The loss in productivity, along with the cost and labor of eradication and control, has hampered full utilization of range and croplands as well as greatly diminished the monetary returns to the producer. Although many mechanisms, devices and procedures have been developed to contend with the weed problem, much work needs to be done to adequately control these pests. The use of herbicides has greatly changed American agriculture; however, it still is not the panacea one would hope for. A basic understanding of how weeds are able to effectively encroach into agricultural lands, seemingly without competition, would help to devise methods of control.

Many reports have appeared in the literature (2,3,4) concerning the allelopathic influence of weeds and how this mechanism is used to favor their propagation and survival. In 1962, Fletcher and Rennoy (5) investigated several Centaurea species and found growth inhibitory substances in Russian knapweed (C. repens) but did not identify them. They suggested that the mechanism of the Centaurea species "To form dense patches and to suppress the growth of other species" may well be the phytotoxins exuded by these weeds. Russian knapweed (Centaurea repens), referred to as Centaurea picris or Acroptilin repens in some literature (6) is a perennial herb which is rapidly becoming a major threat in many parts of the United States. Likewise yellow starthistle, another Centaurea species (C. solstitialis), a noxious annual that has become well established in many Western rangelands has, in many instances, left the rangeland not only nonproductive but counter-productive. Both Russian knapweed and yellow starthistle have been implicated (7) in the nervous disease in horses called equine nigropallidal encephalomalacia (ENE), a disorder caused by the necrosis and softening of specific brain tissue which often leads to death. An understanding of the mechanism of spread of these noxious weeds into useful agricultural lands might lead to a better and more economical method of control and thus not only return the land to productive use but eliminate the losses associated with ENE disease.

Composites have been reported (8) to contain a wide variety of sesquiterpene lactones and lactones of this type have been previously found in both Russian knapweed and yellow starthistle (9,10,11,12). The wide range of biological activity associated with sesquiterpene lactones, e.g., cytotoxicity, phytotoxicity, antineoplasticity, etc. (8) suggested the possibility that these compounds may play a role in the allelopathy of both Russian knapweed and yellow starthistle. Also, the presence of sesquiterpenes containing α-methylene-γ-butyrolactone groups suggested their implication in ENE disease since it has been shown that sesquiterpenes containing this particular functional group are often cytotoxic to mammalian cells (13).

In addition to the sesquiterpene lactones, two chromenes, Eupatoriochromene (I) and Encecalin (II) (14) have been isolated from yellow starthistle and have been shown to have some effects on germination and growth of lettuce.

Materials and Methods.

Plant Material. Russian knapweed and yellow starthistle were collected in June along Highway 4, Contra Costa County, CA, between Brentwood and Stockton. The aerial parts of the plants were air dried, ground in a hammermill with a 1/8" screen and extracted sequentially with Skellysolve-F and ether.

Isolation of Chromenes. The Skellysolve-F extract of yellow starthistle was partitioned between Skellysolve-F/benzene/methanol/water (1:1:4:1). The lower layer was repeatedly chromatographed on silicagel (column and preparative TLC) with Skellysolve-F/ether to give pure chromenes identified by comparison of physical properties with those reported (14) and by synthesis.

Isolation of Sesquiterpene Lactones. The ether extract was evaporated and dissolved in 95% ethanol. Then an equal volume of 4% aqueous lead acetate was added. After 1 hour the mixture was filtered to remove precipitated chlorophyll and phenolic products and the ethanol removed under vacuum. The aqueous layer was extracted with chloroform giving a dark colored oil from which the sesquiterpenes were isolated by a combination of chromatographic procedures, i.e., LH-20 gel permeation, silica gel using both packed columns and thin layer plates. A variety of solvents were also used to purify the individual sesquiterpene lactones, e.g., benzene-acetone (1:1), ethyl acetate, chloroform-methanol (9:1). On thin layer chromatographic plates, spots were visualized by spraying with 2% aqueous $KMnO_4$ solution.

<u>^{13}C-NMR Spectra of Sesquiterpene Lactones</u>. The ^{13}C spectra of the 14 guaianolide sesquiterpene lactones were obtained on a JEOL PFT-100 spectrometer operating at 25.03 MHz using a multiplicity pulse sequence (<u>15</u>) with a 90° flip angle at a 2 sec. repetition rate. Quarternary carbons were identified using a 30° flip angle at a 2 sec. repetition rate. Samples were run in pyridine-d_5 obtained from Aldrich Chemical Co, and used without further purification. Tetramethylsilane was used as an internal standard.

<u>Biological Assay</u>. Individual compounds, I–XVI, were tested for germination inhibition, and radicle and hypocotyl length alteration. All tests were on lettuce seeds (Black Simpson) or lettuce seedlings. Germination was carried out at 24°C in 9-cm petri dishes filled with 0.5% Agar (40 ml) each containing the required amount of test compound. Results were recorded after 24 hr. Root and shoot length studies were done on pregerminated seeds obtained in 24 hr. in full light on 0.5% Agar. Seedlings were placed on 0.5% Agar (40 ml) in 9-cm petri dishes containing the compound to be tested and kept in the dark at 24°C. At the end of 48 hr., both the shoot (hypocotyl) and root (radicle) lengths were recorded.

<u>Results and Discussion</u>

<u>^{13}C-NMR Spectra of Sesquiterpene Lactones</u>. Even though a great many sesquiterpene lactones (well over 1000 structures) have been isolated and their structures elucidated, the chemist is still faced with the problem of discerning the structure of new compounds because of the diversity of experimental techniques used by previous workers. Heavy reliance has been put on nuclear magnetic resonance and in particular ^{1}H-NMR. Also ^{13}C-NMR has been extensively used for sesquiterpene lactones; however, the data is often fragmented and obtained with a variety of solvents making it difficult to correlate newly isolated compounds with those reported in the literature. During the course of our work on <u>Centaurea</u> species, <u>viz</u>., Russian knapweed and yellow starthistle, we have isolated 14 sesquiterpene lactones all belonging to the guaianolide series and determined their ^{13}C-NMR spectra. Pyridine-d_5 was selected as the solvent for this study because of its ability to dissolve a wide range of compounds with differing polarities. The ^{13}C shielding data for the guaianolides (III–XVI) are shown in Table II.

For the sake of clarity, the resonances belonging to the individual groups have been isolated from the data and presented in tables III and IV. Table III shows the resonances associated with the side chain (carbons 16–19). Examination of the data reveals that each side chain presents a unique set of resonances (Table III). For example, resonances at δ 166.5, 136.8, 126.5 and 18.3 uniquely describe the side chain associated with elegin (XII), repdiolide (XIV) and epoxyrepdiolide (VII), i.e., a C_4 side chain ester with a double bond between C-17 and C-18. Likewise for each of the other three side chains one can completely describe them on the basis of their ^{13}C-NMR spectra alone.

Table IV shows the resonances associated with carbons 2–5, 9, 10 and 15, i.e., those carbons comprising the 5-membered carbocyclic ring. Again by examining these seven resonances one can uniquely describe the 5-membered carbocyclic ring structure. For example,

Table I. Structures and Origin of Sesquiterpene Lactones

Compound	Structure	Source*
Acroptilin (III)	A: R_1=H; R_2=OCO—Cl/OH	R, Y
Repin (IV)	A: R_1=H; R_2=OCO—O	R, Y
Subluteolide (V)	A: R_1=H; R_2=OCO—O	Y
Janerin (VI)	A: R_1=H; R_2=OCO—OH	R, Y
Epoxyrepdiolide (VII)	A: R_1=OH; R_2=OCO	R
Solstitiolide (VIII)	B: R_1=H; R_2=OCO—O	R, Y
epi-Solstitiolide (IX)	B: R_1=H; R_2=OCO—O	Y
Centaurepensin (X)	B: R_1=H; R_2=OCO—Cl/OH	R, Y
epi-Centaurepensin (XI)	B: R_1=H; R_2=OCO—Cl/OH	Y
Elegin (XII)	B: R_1=H; R_2=OCO	R, Y

Table I. Continued

Cynaropicrin   (XIII)   C: R_1=H; R_2=**OCO**—⟨   R, Y
                                          —**OH**

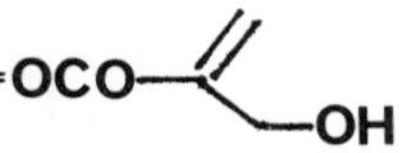

Repdiolide   (XIV)   C: R_1=OH; R_2=**OCO**—⟨   R

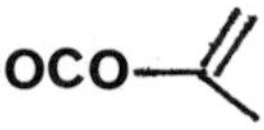

Chlorohydrin of   B: R_1=H; R_2=**OCO**—⟨   R, Y
   Cynaropicrin   (XV)   —**OH**

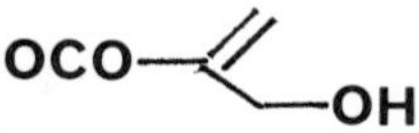

Solstitialin   (XVI)   Y

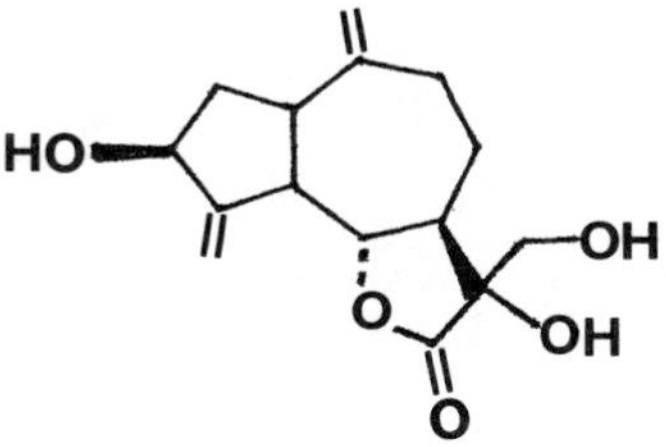

(*) R = Russian knapweed; Y = Yellow starthistle

Table II. Carbon Shifts of Guaianolides III–XVI

Compound

Carbon Number	III	IV	V	VI	VII	VIII	IX	X	XI	XII	XIII	XIV	XVI	XV
1	46.1	46.0	45.8	46.2	47.2	46.5	46.3	46.6	46.4	46.6	45.5	46.8	43.5	46.6
2	39.0	38.9	38.8	39.1	78.3	40.3	40.3	40.4	40.4	40.3	40.0	78.9	39.4	40.3
3	75.4	75.5	75.5	75.5	79.5	76.4	76.4	76.4	76.4	76.5	73.5	79.8	73.2	76.5
4	69.2	69.2	69.2	69.5	66.6	85.4	85.3	85.4	85.4	85.4	154.6	151.3	155.4	85.4
5	53.3	53.1	52.7	53.1	53.1	59.5	59.5	59.7	59.4	59.6	51.3	52.9	53.4	59.6
6	77.3	77.5	77.4	77.9	78.0	77.5	77.4	77.6	77.5	77.6	79.3	78.5	83.2	77.6
7	47.7	47.8	47.6	48.2	49.4	48.6	48.6	48.7	48.6	48.6	47.7	48.4	50.7	48.7
8	75.3	75.3	75.1	74.7	74.4	75.6	75.6	75.5	75.5	74.6	74.7	74.5	27.7	74.5
9	36.2	36.4	36.9	37.2	37.0	35.0	35.1	35.2	35.2	35.3	37.9	36.8	37.6	35.4
10	142.9	142.7	143.1	143.3	141.0	144.4	144.8	144.6	144.6	144.9	143.5	141.5	150.5	144.8

11	138.6	138.6	138.2	138.8	138.4	138.9	138.7	138.9	138.9	139.1	138.9	138.9	78.6	139.0
12	169.0	169.1	169.2	169.6	169.0	169.0	169.1	169.1	169.1	169.2	169.5	169.2	180.5	169.2
13	121.4	121.1	122.0	122.0	121.4	120.8	121.6	121.3	121.2	120.8	121.8	121.3	64.9	121.1
14	118.1	118.0	117.9	118.2	119.4	117.0	116.9	117.2	117.2	117.0	117.5	119.2	112.9	116.9
15	48.5	48.6	48.5	49.0	47.8	51.2	51.2	51.3	51.3	51.3	111.6	111.9	108.7	51.3
16	173.3	170.2	170.4	166.0	166.5	170.3	170.5	173.5	173.5	166.6	165.9	166.5	-----	165.9
17	75.4	54.3	54.2	142.1	136.6	54.4	54.3	75.5	75.5	136.9	142.3	136.7	-----	142.4
18	52.0	52.9	53.0	124.4	126.5	53.0	53.1	52.3	52.3	126.4	125.0	126.4	-----	124.9
19	24.3	17.5	17.4	61.3	18.3	17.6	17.5	24.4	24.4	18.4	61.2	18.3	-----	61.1

Table III. 13C-NMR Shifts of C-16, C-17, C-18 and C-19

Compound	group	C-16	C-17	C-18	C-19
elegin (XII)		166.6	136.9	126.4	18.4
repdiolide (XIV)		166.5	136.7	126.4	18.3
epoxyrepdiolide (VII)		166.5	136.6	126.5	18.3
cynaropicrin (XIII)		165.9	142.3	125.0	61.2
janerin (VI)		166.0	142.1	125.4	61.3
4,15-epichloro-hydrin of cynaro-picrin (XV)		166.0	142.4	124.9	61.1

repin (IV)	170.2	54.3	52.9	17.5
subluteolide (V)	170.4	54.2	53.0	17.4
solstitiolide (VIII)	170.3	54.4	53.0	17.6
epi-solstitiolide (IX)	170.5	54.3	53.1	17.5
acroptilin (III)	173.3	75.4	52.0	24.3
centaurepensin (X)	173.5	75.5	52.0	24.4
epi-centaurepensin(XI)	173.5	75.5	52.3	24.4

Table IV. 13C-NMR Shifts for C-2, C-3, C-4, C-5, C-9, C-10 and C-15

Compound	group	C-2	C-3	C-4	C-5	C-9	C-10	C-15
cynaropicrin (XIII)		40.0	73.5	154.2	51.3	37.9	143.5	111.6
solstitialin (XIII)		39.4	73.2	155.4	53.4	37.6	150.5	108.7
repdiolide (XIV)		78.9	79.8	151.3	52.9	36.8	141.5	111.9
solstitiolide (VIII)		40.3	76.4	85.4	59.5	35.0	144.4	51.2
epi-solstitiolide (IX)		40.3	76.4	85.3	59.5	35.1	144.8	51.2
centaurepensin (X)		40.4	76.4	85.4	59.7	35.2	144.6	51.3
epicentaurepensin (XI)		40.4	76.4	85.4	59.4	35.2	144.6	51.3
elegin (XII)		40.3	76.5	85.4	59.6	35.3	144.9	51.3
epichlorohydrin of cynaropicrin (XV)		40.3	76.5	85.4	59.6	35.4	144.8	51.3

acroptilin (III)	39.0	75.4	69.2	53.3	36.2	142.9	48.5
repin (IV)	38.9	75.5	69.2	53.1	36.4	142.7	48.6
subluteolide (V)	38.8	75.5	69.2	52.7	36.9	143.1	48.5
janerin (VI)	39.1	75.5	69.5	53.1	37.2	143.3	49.0
epoxyrepdiolide (VII)	78.3	79.5	66.6	53.1	37.0	141.0	47.8

resonances at δ 38.9, 69.2, 53.0, 36.5, 143 and 48.6 can be readily
associated with the group as found in acroptilin (III), repin (IV),
subluteolide (V) and janerin (VI). Each of the other four structures
can likewise be delineated on the basis of the ^{13}C-NMR spectra alone.

At the time these data were collected, the structure of compound
XV had not been elucidated. Its ^{13}C-NMR spectrum gave resonances at
δ166, 142.4, 124.9 and 61.1 associated with the side chain. Also
resonances were found at δ40.3, 76.5, 85.4, 59.6, 35.4, 144.8 and
51.3. Using Tables III and IV, it is readily apparent that the compound has the following side chain and 5-membered ring:

Therefore on the basis of the ^{13}C-NMR spectrum alone, it can be
assigned structure XV.

The biological evaluation of the compounds isolated from Russian
knapweed and yellow starthistle has led to some interesting features.
Encecalin (I) causes a dwarfing of lettuce shoots at concentrations
from 10ppm and up. On the other hand, eupatoriochromene (II) causes
very little dwarfing of the seedling but does inhibit germination of
lettuce at 100ppm and above. The concentration of these two chromenes in yellow starthistle may indeed allow for some effect to be
exerted on the surrounding plant community.

I   II

Extensive testing of the sesquiterpene lactones has shown that
they neither inhibit germination nor retard the growth of the hypocotyl in lettuce seedling after 48 hours. However, the growth of
the root is markedly altered. The data for acroptilin is shown in
Figure 1. The control has an average root length of 18mm whereas
at 20ppm of acroptilin (III) in the agar medium, root elongation is
increased to 22mm. At higher concentrations (80ppm), some inhibition takes place. This effect is even more dramatic for centaurepensin (X) (Figure 2) which gives an increase of 32% in root length
elongation at 10ppm. Also at 80ppm, a 47% reduction in root elongation is noted. Likewise repin (IV), Figure 3, and solstitiolide
(VIII), Figure 4, shows similar patterns giving a 46% and a 50%
increase in growth respectively. This growth pattern is characteristic of auxins which have been defined as substances which cause

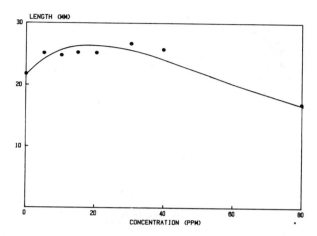

Figure 1.   Mean root length in mm of lettuce seedlings at various concentrations of Acroptilin.

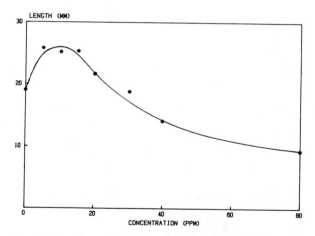

Figure 2.   Mean root length in mm of lettuce seedlings at various concentrations of Centaurepensin.

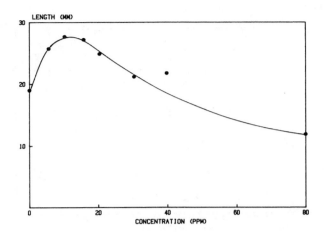

Figure 3.   Mean root length in mm of lettuce seedlings at
            various concentrations of Repin.

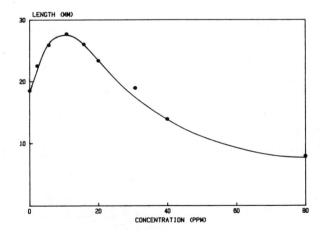

Figure 4.   Mean root length in mm of lettuce seedlings at
            various concentrations of Solstitiolide.

enlargement of cells. The most notable and the first to be characterized is indole acetic acid (IAA) which exhibits a similar growth
curve albeit at much lower concentrations (16). However, IAA causes
a slight increase in growth, on the order of 10% whereas these sesquiterpene lactones have increased growth by 50%! Other tests routinely used to determine auxin activity has led to inconclusive results. Namely, the lanolin paste method on young bean plants showed
little or no cell elongation. It appears that these lactones affect
only the roots.

The function of sesquiterpene lactones in plants has been the
subject of discussion for a long time and they are usually relegated
to the role of secondary metabolites which act as a "garbage dump"
for the plant. However, these compounds may indeed act as plant
growth regulators, both for themselves and the surrounding plant community, acting specifically within the root system of the plant.
Although the concentrations of the lactones within the plant are
often higher than 10ppm, the compartmentalization of components
within plant cells could well regulate the amount secreted into the
tissues of the plant and thus be controlled at the proper levels.
Further work on the role sesquiterpene lactones play in plant growth
regulation certainly deserves more detailed and rigorous study.

One final comment on the biological activity of the sesquiterpene lactones in relation to ENE disease in horses. Unfortunately,
horses are the only animal which contract this disease. Conse-
quently, acquiring sufficient pure material to be fed to horses be-
comes a formidable task. At the present time, trial studies are
being initiated with solstitialin-A to observe its effect on horses.
Insufficient material is available to permit the testing of the other
sesquiterpene lactones.

Literature Cited

1. Bible, Gen. 3:17,18
2. Rice, E. L. "Allelopathy"; Academic Press, N.Y.; 1974.
3. Rice, E. L. The Botanical Review 1979, 45, 15.
4. Silverstein, R. M.; Simeone, J. B. eds. J. Chem. Ecol 1983,
   9, 935-1281.
5. Fletcher, R. A.; Renney, A. J. Can. J. Plant Sci 1963, 43, 475.
6. Evstratova, R. I.; Rybalko, K. S.; Sheichenko, V. I. Khim. Prir.
   Soedin. 1972, 8, 451.
7. Cordy, D. R. In "Effects of Poisonous Plants on Livestock";
   Keeler, R. F.; Van Kampen, K. R.; James, L. F. eds.; Academic
   Press: New York, 1978; pp. 327-336.
8. Rodriguez, E.; Towers, G. H. N.; Mitchell, J. C. Phytochemistry
   1976, 15, 1573.
9. Thiessen, W. E.; Hope, H.; Zarghami, N.; Heinz, D. E.; Deuel,
   P.; Hahn, E. A. Chem. and Ind. 1969, 14, 460.
10. Harley-Mason, J.; Hewson, A. T.; Kennard, O.; Pettersen, R. C.
    J. Chem. Soc. Chem. Commun. 1972, 460.
11. Evstratova, R. I.; Scheichenko, V. I.; Rybalko, K. S. Khim.
    Prir. Soedin. 1973, 9, 161.
12. Gonzalez, A. G.; Bermejo, J.; Breton, J. L.; Massanet, G. M.;
    Dominguez, B.; Amaro, J. M. J. Chem. Soc. Perkin Trans. I. 1976,
    1663.
13. Kupchan, S. M. In "Recent Advances in Phytochemistry";

Runeckles, V. C., Ed.; Plenum: New York, 1975; Vol. 9,
p. 167.
14. Stoelink, C.; Marshall, G. P. *J. Org. Chem.* 1979, 44, 1429.
15. LeCocq, C.; Lallemand, J.-Y. *J. Chem. Soc. Chem. Commum.*
1981, 150.
16. Leopold, A. C.; Kriedemann, P. E. "Plant Growth and Development"; McGraw-Hill Book Co: New York, 1975; 111.

RECEIVED June 12, 1984

Fractionation of Allelochemicals from Oilseed Sunflowers and Jerusalem Artichokes

E. J. SAGGESE, T. A. FOGLIA, G. LEATHER[1], M. P. THOMPSON, D. D. BILLS, and P. D. HOAGLAND

Agricultural Research Service, Eastern Regional Research Center, U.S. Department of Agriculture, Philadelphia, PA 19118

The phenolic and related components present in stems and leaves of sunflower, Helianthus annuus L., and Jerusalem artichoke, Helianthus tuberosus L., were extracted sequentially and their activity as phytotoxic agents evaluated. Total acids and neutral compounds were isolated by extraction with methanol, acetone, and water. The free acids and neutral compounds were partitioned into the organic phase, whereas the acids, present as esters and aglycones, were liberated by subsequent alkaline hydrolysis of the aqueous phase. This procedure was compared with sequential extractive techniques employing alkaline hydrolysis of dried plant tissue followed by extraction of the acidified mixture with ethyl acetate. Fractions were individually evaluated for phytotoxic properties. Selected fractions from those showing a positive response were analyzed by gas-liquid chromatography. Structural identification and characterization of the individual components in these selected fractions were accomplished by gas chromatography-mass spectrometry.

The term allelopathy, when first proposed by Molisch (1), referred to either the beneficial or detrimental interaction between all types of plants and microorganisms. As presently used, this definition is generally accepted. Since 1970 a concerted effort has been made to understand the phenomenon of allelopathic interaction. The many interpretations resulting from these studies are well documented in the literature (2-4). An area currently receiving considerable attention is the allelopathic effect resulting from weed-crop and weed-weed interactions (2, 5-7). One study conducted by Wilson and Rice (7) showed that the common sunflower, Helianthus annuus L., possessed allelopathic properties. Realizing the inherent potential

[1]Current address: Weed Science Research, Agricultural Research Service, U.S. Department of Agriculture, Frederick, MD 21701

that these findings had for natural weed control, Leather (8)
undertook a systematic study to determine if the allelopathic
properties noted for the "wild sunflower" were exhibited by cultivated varieties. He found that cultivated sunflowers also contained
weed-suppressing allelochemicals. Earlier, several researchers
reported on the various structures of some of the secondary natural
compounds that are responsible for the inhibitory effect of the
sunflower on the growth rate of other plants (9-11). Inhibitory
effects noted for other plant species were attributed to the presence
of compounds which were subsequently identified as belonging to
several classes which included simple phenolic acids, coumarins,
terpenoids, flavanoids, alkaloids, cyanogenics, glycosides, and
glucosinolates (12-15). Although there are many reports in the
literature that describe the isolation and identification of these
major classes of compounds from a variety of plant species and
plant parts, such as seeds and roots, there is comparatively little
information concerning their isolation from the leaves and stems of
the sunflower and related species. Furthermore, many of these
studies have concentrated primarily on the major components present,
with little attention to the minor components. The phytotoxic
properties of some of the minor components have been reported only
recently (16-18). The identification and characterization of these
compounds also was described. The investigation reported here was
prompted by the findings of Leather (8). The purpose of this
initial investigation was to extract and fractionate components
from the leaves and stems of the sunflower and Jerusalem artichoke,
to evaluate the phytotoxic activity of the crude fraction by
bioassay, and to separate and identify major components comprising
the active fractions.

Experimental

Plant Material. Dried and fresh tissue from the leaf and stem of
sunflower (H. annuus L.) and the dried ground tissue from the leaf
and stem of the Jerusalem artichoke (H. tuberosus L.) were used as
source materials for the investigation reported here. The fresh
tissues were harvested from plants grown in pots under illumination
provided by 1000-watt, metal halide lamps for a photoperiod of
12 hr in a greenhouse maintained at 75°F and 80-85% RH. The plants
were approximately 4 months old at time of harvesting. Immediately
after collection, the fresh material was stored at -60°C. Sufficient
material was removed for the extractions and either lyophilized or
used directly, depending on the extraction procedure. Dried material
was ground in a Wiley mill to pass through a No. 40 mesh screen.
Except for the fresh leaves and stems of the sunflower grown in our
greenhouse, all of the other dried tissues were obtained from plants
grown in a greenhouse with supplemental light from full-spectrum
metal halide lamps at the USDA Weed Science Research Laboratory,
Frederick, MD. Leather (8) found no difference in allelopathic
potential between sunflower plants grown under these conditions and
field-grown plants. (Reference to brand or firm name does not
constitute endorsement by the U.S. Department of Agriculture over
others of a similar nature not mentioned.)

Extraction. A variety of extraction procedures were evaluated from
a simple leaching of the macerated tissue with warm water or dilute
alcohol to the more harsh procedures employing alkaline hydrolysis.
Except for a small number of the hydroxylated benzoic acid derivatives, leaching under mild conditions did not effectively extract
many of the potentially allelopathic chemicals which were covalently
bound as esters and in other forms. Many of these higher molecular
weight compounds remained insoluble but contained the allelopathic
constituent acids which are released slowly during natural biodegradation of the plant debris in the soil (3). In order to isolate
these bound acids within the tissue, procedures were employed which
would effectively hydrolyze these chemical entities and thereby
release many of the acids from their bound form. Solvents used in
the extraction procedures described below were all HPLC grade and
residue free.

Extraction Procedure A. Based on a method reported by Krygier et
al. (19), fractions containing the free, esterified, and insoluble-
bound organic acids were obtained. One or two grams of either the
fresh or dried sunflower plant material was homogenized in a Polytron
with 20 to 40 ml of 70% methanol:70% acetone (1:1, v:v) for 5 min
and then centrifuged. This step was repeated five times. The
supernatants were combined and reduced to one-fifth of the original
volume with a rotary evaporator at 40°C and 20 Torr vacuum. The
resulting solution was acidified to pH 2 with 6N HCl and filtered
to remove a small amount of precipitate. The filtrate was extracted five times with 25 ml hexane to remove the lipids, and then
extracted five additional times with 20 to 40-ml portions of ethyl
acetate-ethyl ether (1:1, v:v). The extracts were combined, dehydrated with anhydrous sodium sulfate, and filtered. The filtrate
was evaporated to dryness to give a residue containing the free
uncombined acids. The aqueous solution remaining after the above
extraction was hydrolyzed with 20 to 40 ml of 4N NaOH for 4 hr
under nitrogen at room temperature. The hydrolysate was acidified
to pH 2 with 6N HCl and extracted as above—hexane followed by
ethyl ether-ethyl acetate (1:1, v:v). This gave a residue containing the byproducts from the hydrolysate from the ester-bound
compounds. The insoluble-bound compounds, which were contained in
the residue remaining from the original Polytron extraction with
70% methanol/ H_2O:70% acetone/H_2O (v:v), also were hydrolyzed with
20-40 ml of 4N NaOH at room temperature under a nitrogen atmosphere
to release the acids from the compounds in which they were bound.
After 4 hr the mixture was acidified as above and centrifuged. The
supernatant was extracted as above with hexane, followed by ethyl
acetate-ethyl ether (1:1, v:v). The organic extracts were combined
and dehydrated over sodium sulfate. After filtering, the solvent
was removed and the residue dried to constant weight at 40°C and
20 Torr vacuum. Each of these residues was used for the bioassay
described below and in subsequent analysis.

Extraction Procedure B. Figure 1 gives a flow diagram for this
fractionation procedure, which was based on a modification of the
simplified methods described by Serve et al. (20) and Hartley and
Buchan (21). Two grams of ground dried sunflower leaves were added

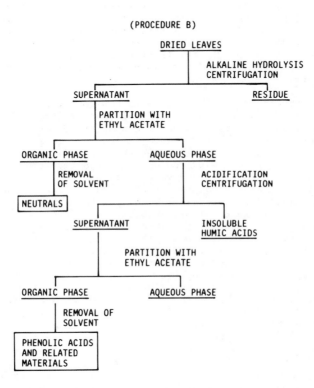

Figure 1. Fractionation of plant material.

to 200 ml of 1N NaOH. The mixture was stirred at room temperature for 24 hr under a nitrogen atmosphere and then centrifuged. The supernatant was decanted and the residue was washed with two 100-ml portions of distilled water. The combined wash water and the supernatant were extracted four times with 100-ml portions of ethyl acetate. The extracts were combined and dried with anhydrous sodium sulfate, filtered, and the solvent removed by rotary evaporation at 40°C and 20 Torr vacuum. The residue, containing neutral components, was dried under vacuum at 20 Torr to constant weight. The aqueous fraction was acidified to pH 1 with 6N HCl, and the small amount of humic acids which precipitated was removed by filtration. The filtrate was extracted three times with 100-ml portions of ethyl acetate. The organic extracts were combined, dried over anhydrous sodium sulfate, and filtered. The solvent was removed by rotary evaporation and the residue contained the freed byproducts from the hydrolyzed esterified and insoluble-bound compounds.

In addition to the above, a variety of other extractive methods were evaluated (21, 22), but those described here were judged most satisfactory for our purposes.

Analytical Methods

The extracted fractions were esterified with either BF_3-MeOH reagent or diazomethane and analyzed by GLC. Gas liquid chromatography (GLC) was conducted with a Perkin-Elmer Sigma 3 equipped with flame ionization detector. Separations were obtained on a Hewlett Packard 12 m x 0.2 mm i.d. capillary column coated with methyl silicon fluid (OV-101). The temperature was maintained at 80°C for 2 min then programmed from 80 to 220°C at 8°C/min. The injector temperature was 250°C. Mass spectra were obtained on a Hewlett Packard model 5995 GC-MS mass spectrometer, equipped with a 15 m fused silica capillary column coated with 5% phenyl methyl silicone fluid. Spectra were obtained for major peaks in the sample and compared with a library of spectra of authentic compounds.

Bioassay for Phytotoxic Properties

The phytotoxicity of the crude residues obtained in the various fractions was assayed with the aquatic macrophyte plant, Lemna minor L. (23). The plants, each a rosette of three fronds (a mother and two daughters), were placed in 24-well tissue culture cluster plates with 1.5 ml of medium containing mineral additives (24, 25). Except where noted, all test samples were dissolved in 50 µl of absolute ethanol. Einhellig et al. (23) have shown that ethanol amendments amounting to 0.3% (v/v) did not adversely effect the growth of L. minor in the bioassay. The test was replicated 6X with one control for each three test treatments. A 5-µl aliquot of test sample was added to each well and the system cultured for 5 to 7 days under constant light at 28°C after which the effect on growth rate of L. minor was noted by a count of the final number of fronds, and the weight of the dried fronds was determined and compared with the dried weight of the controls. The data were analyzed by analysis of variance with Duncan's multiple-range test.

Discussion

Tables I and II show the response noted for the samples obtained
from the various sources of sunflower and Jerusalem artichoke. The
two extractive procedures, the sources, and the samples are compared.
The tables show that all fractions, regardless of extractive method
or source, were phytotoxic to various degrees. The bioassay was
designed to evaluate the response of L. minor at the highest concentrations that the solubility of the extracted material in the
substrate would permit. In the case of the sunflower, the amount
of extracted material per assay varied from 28 ppm to 620 ppm; for
the Jerusalem artichoke, 286 ppm to 686 ppm. Where an effect was
noted, a second assay was performed with one-third the original
concentration of extracted material. This is shown in the last
column of the tables. Reducing the concentration did not, in all
instances, proportionately reduce the phytotoxic effect. As
shown in Table I, fractions C and G, and Table II, fractions B, I,
and K, the inhibitory effect was equal to or more pronounced at the
diluted concentration than at the original higher concentration,
but this was only true for the fractions where the initial concentration was the highest of the fractions tested. This observation is
not uncommon for crude plant extracts which may sometimes stimulate
growth at higher concentrations. As a rule, however, growth is
inhibited at the higher concentrations and stimulated at the lower.
In this study, except for instances noted, the effect was concentration dependent and the degree of growth inhibition varied with the
concentration. Generally, there is little difference in phytotoxicity
between the fractions obtained from the sunflower when compared
with Jerusalem artichoke. Further, little difference is shown in
phytotoxicity among the fractions regardless of source or procedure
used. What was surprising was that all fractions were phytotoxic,
and most to a great degree when compared to control. Still more
surprising, as shown later, was the finding that even fractions
which did not appear to contain any phenolic acid were equally
phytotoxic. Since the acidified hydrolysate fractions containing
the ester and insoluble-bound components gave phytotoxic responses
comparable to those noted for the unhydrolyzed fraction (neutral
compounds and free acids), the question arises, does hydrolysis
release from the complex the compounds found in the unhydrolyzed
fraction, and are these the same or related compounds or are they
completely different? Preliminary analysis using high performance
liquid and gas chromatography indicates that the fractions all
contain, among other things, similar and related chemical species.
The major components have been identified tentatively as phenolic
and fatty acids. At this time, seven phenolics have been identified
in only four of the fractions. These are shown in Table III. A
measure of the magnitude of the confidence level (cc) with a
spectrum of standards is given. The first three entries are from
the sunflower; the last, from the Jerusalem artichoke. In all
fractions isolated, both from the sunflower and the Jerusalem
artichoke, a homologous series of fatty acids ranging from C_{10} to
C_{18} have been identified also by GC-MS. Even-chain, C_{16} to C_{18}
saturated and C_{18} mono- and di-unsaturated, predominated. This is
not surprising, since fatty acids are major constituents of plant

Table I. Helianthus annuus. Growth Inhibition of L. minor. The Effect of the Fractions of Fronds as Percentage of Control after 5 Days Exposure

Sample	Source	Procedure	Fraction	Application Conc. (ppm)	% of Control at Original Conc.	% of Control at 1/3 Dilution
A	Lyophilized leaves	B	1[a]	66	22*	87
B			2[b]	98	8*	68*
C	Frozen fresh leaves	A	1[c]	620	12*	8*
D			2[d]	520	7*	22*
E			3[e]	120	12*	38*
F	Lyophilized leaves	A	1[c]	28	88*	79*
G			2[d]	780	10*	6*
H			3[e]	514	9*	21*
I	Lyophilized stems	A	1[c]	392	8*	29*
J			2[d]	256	7*	15*
K			3[e]	100	13*	79*

a Unbound neutral components and lipids.
b Hydrolyzed compounds from ester and insoluble-bound compounds.
c Free acids and unbound simple compounds.
d Hydrolyzed products from ester-bound compounds.
e Hydrolyzed products from insoluble-bound compounds.
* Significantly different from controls, $P \leq 0.05$.

Table II. Helianthus tuberosus. The Effect of the Fractions on Growth Inhibition of L. minor. Dry Weight as a Percentage of the Control after 7 Days Exposure

Sample	Source	Procedure	Fraction	Application Conc. (ppm)	% of Control at Original Conc.	% of Control at 1/3 Dilution
A	Dried leaves	A	1[c]	392	4*	109*
B			2[d]	-	6*	3*
C			3[e]	200	6*	-
D	Dried leaves	A	1[c]	484	4*	6*
E			2[d]	-	-	-
F			3[e]	350	4*	6*
G	Dried leaves	B	1[a]	340	6*	10*
H			2[b]	200	122	-
I	Dried stems	A	1[c]	664	4*	3*
J			2[d]	286	5*	15*
K			3[e]	686	4*	3*

a Unbound neutral components and lipids.
b Hydrolyzed compounds from ester and insoluble-bound compounds.
c Free acids and unbound simple compounds.
d Hydrolyzed products from ester-bound compounds.
e Hydrolyzed products from insoluble-bound compounds.
* Significantly different from controls, $P \leq 0.05$.

Table III. Phenolics Identified by GC-MS

Phenolic	Sample			
	D[a]	K[a]	J[a]	I[b]
Gallic acid		+++[c]		
Protocatechuic acid		+++[c]		++[d]
P-hydroxybenzoic acid		+++[c]	++[d]	
Benzoic acid	+++[c]		++[d]	
Vanillic acid		++[d]		
Syringic acid		+++[c]		
Salicyclic acid	+[e]	++[d]	+[e]	++[d]

[a] Sunflower.
[b] Jerusalem artichoke.
[c] +++ (cc^f > 0.95).
[d] ++ (cc^f > 0.85).
[e] + (cc^f < 0.85).
[f] Confidence level for match of mass spectrum of known phenolic acids.

membranes as phospholipids, glycolipids, waxes, and triglycerides and are readily released during the hydrolysis and extraction procedures.

Tables IV and VII give the percentage of the final frond number noted compared to the control for the same fractions as given in Tables I and II. The magnitude of the response in all cases is proportional to that noted on the basis of dried weight of fronds. The difference, where present, may be due to the size of the fronds since the effect of the phytochemical may be to limit frond size but not necessarily the number of fronds. Tables V, VI, and VIII give a description of the visual appearance of the fronds treated. Tables V and VI give the observations after 5 and 7 days, respectively, for the fraction from the sunflower, and Table VIII, for those from Jerusalem artichoke after 7 days.

The tables show that the fronds treated with extracted materials exhibited bleaching, chlorosis, and other morphological changes. The phytotoxic effect may be due to interference with chlorophyll production or other metabolic processes. The observations noted in Table VI suggest that there is some recovery at 7 days from the conditions described in Table V for 5 days. However, this is not the case, since the same chlorotic effect is noted for the new fronds which were the first to emerge early in the observation period. This indicates that the phenomenon is due to the growth of new fronds which have not yet been exposed to the test media.

Those fractions showing activity were equal to or greater in toxicity to that noted for the crude water extracts of sunflower. Although the techniques employed to extract the fractions described are not the same as those which prevail in nature, the purpose of these investigations was to isolate and test the compounds indigenous

Table IV. <u>Helianthus annuus</u>. Growth Inhibition of <u>L. minor</u>. The Effect of the Fractions on Final Frond Number as Percentage of Control after 5 Days Exposure

Sample	Source	Procedure	Fraction	Application Conc. (ppm)	% of Control at Original Conc.	% of Control at 1/3 Dilution
A	Lyophilized leaves	B	1[a]	66	35*	88*
B			2[b]	98	8*	78*
C	Frozen fresh leaves	A	1[c]	620	10*	5*
D			2[d]	520	8*	19*
E			3[e]	120	9*	36*
F	Lyophilized leaves	A	1[c]	28	78	79*
G			2[d]	780	10*	5*
H			3[e]	514	10*	13*
I	Lyophilized stems	A	1[c]	392	9*	32*
J			2[d]	256	9*	20*
K			3[e]	100	21*	81*

[a] Unbound neutral components and lipids.
[b] Hydrolyzed compounds from ester and insoluble-bound compounds.
[c] Free acids and unbound simple compounds.
[d] Hydrolyzed products from ester-bound compounds.
[e] Hydrolyzed products from insoluble-bound compounds.
* Significantly different from controls, $P \leqq 0.05$.

Table V. Helianthus annuus. Description of Fronds after
5 Days Exposure

Sample	Procedure	Observation
A[a] B[b]	B	Fronds light green; some white, small in size Fronds all pure white
C[c] D[d] E[e]	A	Fronds appear beige; roots beige Fronds off-white Fronds pure white
F[c] G[d] H[e]	A	One to two white fronds (total 3.7) Fronds brown; roots black; veins darkened Fronds off-white; roots brown

[a] Unbound neutral components and lipids.
[b] Hydrolyzed compounds from ester and insoluble-bound compounds.
[c] Free acids and unbound simple compounds.
[d] Hydrolyzed products from ester-bound compounds.
[e] Hydrolyzed products from insoluble-bound compounds.

Table VI. Helianthus annuus. Description of Fronds after
7 Days Exposure

Sample	Procedure	Observation
A[a] B[b]	B	Fronds and roots green Fronds and roots lighter shade of green
C[c] D[d] E[e]	A	Fronds and roots beige in color, veins visible Fronds part green, yellow, and white; daughter fronds green; fronds smaller than control; roots white Fronds green but light shade, some light markings; roots white
F[c] G[d] H[e]	A	Fronds green in color, some fronds with white markings; roots green Fronds brown with veins visible (brown); roots dark brown Fronds green, yellow, and white; daughter fronds same as adults; fronds clumped together, roots white

[a] Unbound neutral components and lipids.
[b] Hydrolyzed compounds from ester and insoluble-bound compounds.
[c] Free acids and unbound simple compounds.
[d] Hydrolyzed products from ester-bound compounds.
[e] Hydrolyzed products from insoluble-bound compounds.

Table VII. Helianthus tuberosus. The Effect of the Fractions on Growth Inhibition of L. minor. Final Frond Number as a Percentage of Control after 7 Days Exposure

Sample	Source	Procedure	Fraction	Application Conc. (ppm)	% of Control at Original Conc.	% of Control at 1/3 Dilution
A	Dried leaves	A	1[c]	392	6*	112*
B			2[d]	-	6*	6*
C			3[e]	200	10*	-
D	Dried leaves	A	1[c]	484	6*	9*
E			2[d]	-	-	-
F			3[e]	350	6*	19*
G	Dried leaves	B	1[a]	340	6*	15*
H			2[b]	200	114	-
I	Dried stems	A	1[c]	664	6*	6*
J			2[d]	286	6*	22*
K			3[e]	686	6*	7*

a Unbound neutral components and lipids.
b Hydrolyzed compounds from ester and insoluble-bound compounds.
c Free acids and unbound simple compounds.
d Hydrolyzed products from ester-bound compounds.
e Hydrolyzed products from insoluble-bound compounds.
* Significantly different from controls, P ≤ 0.05.

Table VIII. <u>Helianthus</u> <u>tuberosus</u>. Description of Fronds
after 7 Days Exposure

Fraction	Observation	
	Original Concentration	1/3 Dilution
A[c] B[d] C[e]	Fronds and roots white Fronds and roots beige Most fronds white; younger fronds green; roots beige	Fronds and roots green Fronds and roots beige -
D[c] E[d] F[e]	Most of frond white; daugh- ter frond part green; roots white - Fronds and roots white	Larger frond mostly white; daughter fronds green and yellow - Fronds pale green and roots white
G[a] H[b]	Fronds and roots white Fronds and roots green	Fronds are whitening (larger fronds); daugh- ter fronds green -
I[c] J[d] K[e]	Fronds and roots white Fronds and roots beige Fronds and roots white	Fronds and roots beige Some fronds pale green; some beige; roots beige Fronds and roots white

[a] Unbound neutral components and lipids.
[b] Hydrolyzed compounds from ester and insoluble-bound compounds.
[c] Free acids and unbound simple compounds.
[d] Hydrolyzed products from ester-bound compounds.
[e] Hydrolyzed products from insoluble-bound compounds.

to the plant tissues studied. Since some of them are conjugated
and released slowly in nature, the harsh alkaline hydrolysis was
employed to liberate the potential allelochemical moieties from the
compounds in which they were covalently bound and which would not
be readily extracted under milder conditions. All fractions showing
a phytotoxic effect are being further characterized and their
phytotoxicity evaluated.

Acknowledgment

We appreciate the assistance of Mr. Peter Vail, Food Science Laboratory, in providing the analytical services.

Literature Cited

1. Molisch, H. "Der Einfluss einer Pflanze and die andere—Allelo-pathic" 1937, Gustav Fisher Verlag, Jena.
2. Rice, E. L. "Allelopathy"; Academic Press, New York, 1974; 353 pp.
3. Rice, E. L. Bot. Rev. 1979, 45, 15-109.
4. Putman, A. R. Chemical and Engineering News 1983, 61(14), 34-45.
5. Colton, C. E.; Einhellig, F. A. Amer. J. Bot. 1980, 67(10), 1407-13.
6. Bell, D. T.; Koeppe, D. E. Corn. Agron. J. 1972, 64, 321-25.
7. Wilson, R. E.; Rice, E. L. Bull. Torrey Bot. Club 1968, 95, 423-48.
8. Leather, Gerald R. Weed Science 1983, 31, 37-42.
9. Koeppe, D. E.; Rohrbaugh, L. M.; Rice, E. L.; Winden, S. H. Phytochemistry 1970, 9, 297-301.
10. Koeppe, D. E.; Southwick, L. M.; Bittell, J. E. Can. J. Bot. 1976, 54, 593-99.
11. Lehman, R. H.; Rice, E. L. The American Midland Naturalist 1972, 87(1), 71-80.
12. Putnam, A. R.; Duke, W. B. Ann. Rev. Phytopathol. 1978, 16, 431-51.
13. Swain, T. Ann. Rev. Plant Physiol. 1977, 28, 479-501.
14. Moreland, D. E.; Egley, G. H.; Worsham, A. D.; Monaco, T. J. Adv. Chem. 1966, 53, 112-41.
15. Harborne, J. B. "Phytochemical Ecology"; London: Academic Press, 1972; 272 pp.
16. Spring, O.; Albert, K.; Gradmann, W. Phytochemistry 1981, 20(8), 1883-85.
17. Spring, O.; Albert K.; Hager, A. Phytochemistry 1982, 21(10), 2551-53.
18. Spring, O.; Hage, A. Planta 1982, 156, 533-40.
19. Krygier, L.; Sosulski, F.; Hogge, L. J. Agric. Food Chem. 1982, 30, 330-34.
20. Serve, L.; Provetti, L.; Longuemard, N. J. Chrom. 1983, 259, 319-28.
21. Hartley, R. H.; Buchan, H. J. Chrom. 1979, 180, 139-43.
22. Laird, W. M.; Mbadiive, E. I.; Synge, R. L. M. J. Sci. Food Agric. 1976, 27, 127-30.
23. Einhellig, F. A.; Leather, G. R.; Hobbs, L. L. J. Chem. Ecol. 1984, (in press).
24. Cleland, C. F.; Briggs, W. R. Plant Physiol. 1967, 1553-61.
25. Hillman, W. S. Amer. J. Bot. 1961B, 48, 413-19.

RECEIVED August 6, 1984

Biosynthesis of Phenolic Compounds
Chemical Manipulation in Higher Plants

STEPHEN O. DUKE

Southern Weed Science Laboratory, Agricultural Research Service, U.S. Department of Agriculture, Stoneville, MS 38776

Biosynthetic pathways of phenolic compounds, including many allelochemicals, in higher plants are influenced by various chemicals such as certain herbicides and growth regulators. Although most of these compounds have little or no direct effect on phenolic compound biosynthesis, others such as glyphosate, have profound effects that may seriously affect allelochemical interactions under field conditions. Two known primary sites of chemical manipulation of phenolic production exist: 1) 5-enolpyruvylshikimate-3-phosphate synthase (EPSP synthase) and 2) phenylalanine ammonia-lyase (PAL). Certain amino-phosphonic acids, such as glyphosate, strongly inhibit aromatic amino acid synthesis by inhibiting the activities of EPSP synthase and other enzymes of the shikimate pathway. PAL, a pivotal enzyme in the transformation of aromatic amino acids to secondary phenolic compounds is strongly and specifically inhibited by ∝-aminooxy-β-phenyl-propionic acid (AOPP) and its analogues. The herbicides chlorsulfuron and acifluorfen induce high levels of PAL and aromatic compounds. These compounds have potential in studies of the role of phenolic compounds in allelopathy.

Many phytotoxic compounds produced by higher plants are phenolic compounds. Several of these have been implicated in allelopathy. Based on the biosynthetic pathway from which they are derived, phenolic compounds produced by higher plants fall into two general categories: 1) terpenoid phenolic compounds derived from five carbon isoprene units (e.g. gossypol) and 2) phenolic compounds derived from the shikimic acid pathway (Figure 1). In the last decade, several compounds have been found to have profound and specific effects on certain enzymes involved in the synthesis of shikimic acid pathway-derived phenolic compounds. These compounds

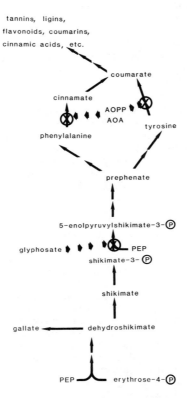

tannins, ligins,
flavonoids, coumarins,
cinnamic acids, etc.

Figure 1. Biosynthetic pathway for production of shikimic acid
pathway-derived phenolic compounds in higher plants.

have the potential to manipulate phenolic allelochemical
production, thereby allowing the determination of the role of
specific families of compounds in allelopathy at the whole plant
level. This chapter is intended to review the physiological and
biochemical literature on these compounds in order that
ecophysiologists might be encouraged to use them to probe the roles
of phenolic allelochemicals in plant interactions. Manipulation of
phenolic allelochemical production at the enzyme level will be
emphasized.

Inhibitors of Aromatic Amino Acid Synthesis

Until the discovery of the herbicide glyphosate (N-(phosphonomethyl)glycine) (Figure 2, I), no chemicals were known to directly
affect production of aromatic amino acids in situ. Jaworski (1)
first hypothesized that glyphosate inhibits aromatic amino acid
synthesis. His hypothesis was based on feeding studies in which he
reduced glyphosate-caused inhibition of growth in Rhizobium
japonicum and duckweed (Lemna gibba L.) by supplying the organisms
with aromatic amino acids in their growth medium. In later
studies, in which profiles of free pools of aromatic amino acids
from higher plants were examined, levels of aromatic amino acids
were found to be greatly reduced in comparison to other amino acids
(2-6). Studies by Amrhein et al. (7-9) established that reduced
levels of aromatic amino acids resulted from inhibition of
5-enolpyruvylshikimate-3-phosphate synthase (EPSP synthase, EC
2.5.1.19) (Figure 1). Glyphosate is competitive with respect to
phosphoenolpyruvate and uncompetitive with respect to
shikimate-3-phosphate (10), the two substrates of EPSP synthase.
Tissue cultures of higher plants with elevated levels of EPSP
synthase are tolerant of glyphosate (9); thus far, however, no
higher plants with an EPSP synthase resistant to glyphosate have
been reported, although microbial strains with glyphosate-
insensitive EPSP synthase have been found (11, 12).
    Glyphosate also inhibits the activity of a form of
3-deoxy-D-arabino-heptulosonate-7-phosphate synthase (13), however,
this inhibition is relatively weak. Effects of glyphosate on the
enzymes of aromatic amino acid synthesis have been described in
more detail in recent reviews (14-17).
    The effects of glyphosate on phenolic compound production are
two-fold: 1) accumulation of phenolic compounds that are
derivatives of aromatic amino acids is reduced and 2) pools of
phenolic compounds derived from constituents of the shikimate
pathway prior to 5-enolpyruvylshikimate-3-phosphate become larger.
Assays that do not distinguish between effects on these two groups,
such as that for hydroxyphenolics of Singleton and Rossi (18), can
lead to equivocal and difficult to interpret results (e.g. 3-5).
I will, therefore, discuss only those studies in which the nature
of the phenolic compounds being studied can be determined.

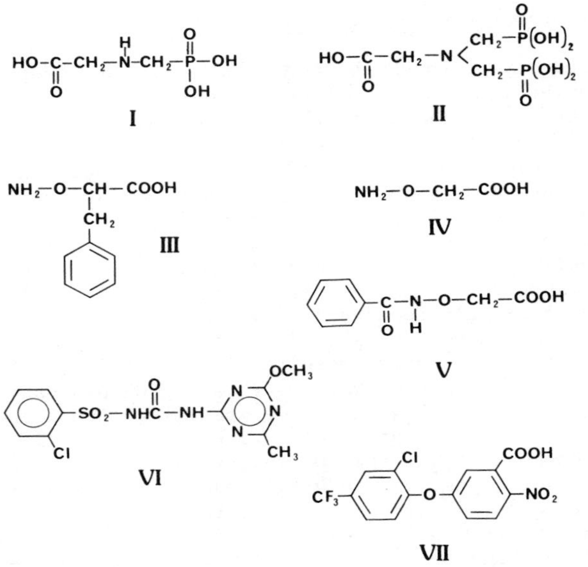

Figure 2. Chemical structures of several chemicals referred to in the text.

Anthocyanins are the most easily assayed and commonly studied
derivatives of aromatic amino acids (Figure 1). Glyphosate
drastically reduces accumulation of anthocyanin flavonoids in
treated tissues (6, 19) (Figure 3). Levels of rutin and
procyanidin, both flavonoids, are reduced in glyphosate-treated
buckwheat hypocotyls (6). Glyphosate would presumably similarly
affect levels of flavonoids and flavonoid derivatives that are
known to be allelochemicals.

Cinnamic acid-derived aromatic acids that have been reported
to be allelochemicals are reduced by glyphosate. Chlorogenic acid
in buckwheat hypocotyls (6) and caffeic acid production in Perilla
cell suspensions are greatly reduced by glyphosate (20).
Accumulation of other non-flavonoid aromatic amino acid
derivatives, such as cinnomyl putrescines, is also reduced by
glyphosate (21). Levels of glyceollin, a flavonoid-derived
phytoalexin, were greatly reduced in soybean (Glycine max) leaves
and hypocotyls by glyphosate (22). This reduction resulted in the
loss of resistance in soybean to certain microbial pathogens (23).
Unfortunately, similar studies have not been conducted to examine
the role of phenolic compounds in allelopathic interactions.

Glyphosate greatly enhances accumulation of shikimate which
ultimately accumulates in the vacuole (9, 24-27). Gallic acid also
accumulates in glyphosate-treated plants (25), apparently due to
accumulation of a pre-aromatic acid precursor, presumably
dehydroshikimic acid.   Thus, glyphosate increases the accumulation
of certain phenolic compounds derived from shikimic acid pathway
intermediates and decreases synthesis of aromatic amino
acid-derived secondary aromatic compounds. How such an alteration
in the qualitative pattern of phenolic compounds will affect
allelopathic interactions is unknown. This type of study can only
be conducted if sub-lethal glyphosate treatments can produce
significant and relatively long-lived changes in phenolic content.
If not, glyphosine (N,N-bis(phosphonomethyl)glycine) (Figure 2,
II), a growth regulator may be useful in such a study, because it
apparently has a mechanism of action that is similar to that of
glyphosate (6, 19).

Inhibitors of Phenylalanine Ammonia-Lyase

Phenylalanine ammonia-lyase (PAL; EC 4.3.1.5) is a pivotal enzyme
in controlling flow of carbon from aromatic amino acids to
secondary aromatic compounds (Figure 1) (28). PAL primarily
deaminates phenylalanine to form t-cinnamic acid, however, in many
species, it also less efficiently deaminates tyrosine to form
p-coumaric acid. Because PAL is restricted to plants and is an
important enzyme in plant development, Jangaard (29) suggested that
PAL inhibitors might make safe and effective herbicides, however,
in his screen of several herbicides, he found no compound to have a
specific effect on PAL. This was also the case in studies by
Hoagland and Duke (30, 31) in which 16 herbicides were screened.

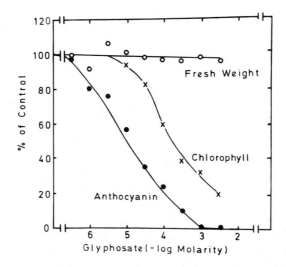

Figure 3. Effect of glyphosate on fresh weight (o), chlorophyll
content (x), and anthocyanin content (•) of excised buckwheat
cotyledons after 24 h of treatment in the light. Reproduced with
permission from Ref. 6. Copyright 1980, Plant Physiology.

Duke and Hoagland (32) suggested that compounds which greatly increase in vivo PAL activity would be phytotoxic because of depletion of aromatic amino acid pools needed for protein synthesis and overproduction of phytotoxic secondary compounds, leading to autoallelopathy. Their early results indicated that glyphosate might have such a mechanism of action, however, later work indicated that glyphosate causes increases in extractable PAL activity by decreasing production of PAL products that reduce PAL levels by a feedback inhibition process (33, 34). Glyphosate has no in vitro effect on PAL activity.

A large number of in vitro inhibitors of PAL are known, however, few of these effectively inhibit PAL activity in vivo without also causing phytotoxic effects unrelated to their effects on PAL. PAL inhibitors fall into four categories: 1) PAL product inhibitors such as t-cinnamic acid and p-coumaric acid (e.g. 35); 2) general carbonyl reagents, such as cyanide, bisulfite, or borohydride (e.g. 36); 3) L-phenylalanine or L-tyrosine analogues, such as D-phenylalanine or p-fluorophenylalanine (e.g. 37); and 4) the more specific carbonyl reagents, such as α-aminooxy-β-phenylpropionic acid (AOPP) (Figure 2, III) (38, 39), and its analogues, such as aminooxyacetic acid (AOA) (Figure 2, IV) (40) and D,L-α-hydrazino-β-phenylpropionic acid (41, 42). PAL products are very often allelochemicals with a great deal of non-specific phytotoxicity due to several physiological mechanisms (e.g. 43). Carbonyl reagents are also non-specific and phytotoxic because of their reaction with many different proteins. Aromatic amino acid analogues are generally toxic because they inhibit protein synthesis (e.g. 44).

AOPP and AOA inhibit transaminase enzymes (39, 44) and other pyridoxylphosphate-dependent enzymes, presumably by interference with the carbonyl group of pyridoxyl phosphate (45). They apparently inhibit PAL by interaction with the carbonyl-like group involved in catalysis by PAL (36). AOA is not an effective PAL inhibitor for in vivo studies because of its lack of specificity that results in a relatively high degree of phytotoxicity (e.g. 40). It has been patented as a herbicide (46), and the herbicide benzadox (Figure 2, V) is apparently metabolized by plants to AOA, its active in vivo component (47).

The high toxicity of AOA is due to its very high efficiency as a transaminase inhibitor (K_i = 0.45 μM) as compared to its efficacy as a PAL inhibitor (K_i = 120 μM) (48), making it impossible to effectively inhibit PAL in vivo without also greatly inhibiting amino acid metabolism. Other pyridoxyl phosphate-requiring enzymes, such as ACC synthase (an enzyme involved in ethylene production) (49), are also more sensitive to AOA than to AOPP. With AOPP, the selectivity is reversed with the K_i for PAL being 1.4 nM and that for phenylalanine transaminase being 3 μM (48). Thus, AOPP can effectively block PAL activity in vivo without being strongly phytotoxic (e.g. 20, 32, 38).

AOPP has been used in many studies to examine the role of PAL
in the synthesis of secondary aromatic compounds. The results
summarized in Table I indicate that levels of AOPP that have little
or no effect on growth can strongly affect production of secondary
aromatic products. Other studies have shown rapid cessation of
isoflavone synthesis in Cicer ariethinum by 0.3 mM AOPP (61).
Intact plants can be treated with little effect on growth (33, 50,
51, 54-59), yet few studies have been done on the effects of this
compound on plant interactions with other organisms. The only
successful published work in this area is that from the laboratory
of Grisebach (50, 51), in which AOPP reduction of glyceollin
production in Glycine max led to loss of resistance to Phytophthora
megasperma (Figure 4). In a similar study, Keen and Holliday were
unsuccessful in reducing glyceollin levels with AOPP (22, 23).
Because many allelochemicals are cinnamic acid derivatives, use of
AOPP to manipulate and modulate their synthesis should lead to a
better understanding of the role of these compounds in plant
interactions.

Influences on Polyphenol Oxidase (PPO)

Because polyphenol oxidase (PPO, EC 1.10.3.1) can orthohydroxylate
certain secondary phenolic compounds in vitro (e.g. 62), it has
been assumed by many that the enzyme is involved in synthesis of
orthohydroxylated phenolic compounds such as caffeic acid and
certain flavonoids. The fungal toxin, tentoxin, totally eliminates
the development of PPO activity in certain higher plants (63-65)
without affecting the content of ortho-hydroxylated phenolic
compounds (66). Tentoxin has no effect on in vitro PPO activity
(63). Furthermore, in healthy plant tissues, PPO is usually latent
as a phenol oxidase and is intracellularly separated from most of
its potential substrate (67). Thus, it is unlikely that PPO is
involved in synthesis of phenolic compounds in healthy tissues.
However, during senescence or cellular damage latent PPO is
activated and compartmentalization between PPO and phenolic
substrates is broken down. There is good evidence that the
quinones produced from phenols by PPO under such conditions are
involved in resistance to pathogens (e.g. 60), however, whether
these quinones are involved in plant interactions with each other
is unknown. The most likely situations in which this might be the
case are in the cases of decaying vegetation (much of the browning
of senescing and decaying plant tissues is due to condensation of
PPO-produced quinones) and of actual physical damage caused by
contact between plants (most likely in subterranean environments).
       Although many in vitro inhibitors and inducers of latent PPO
activity are known (69), tentoxin is the only compound known to
strongly inhibit potential in vivo PPO activity without being
extremely phytotoxic. Tentoxin causes extreme chlorosis without
directly affecting growth or morphogenesis (70). Testing the
allelopathic potential of tentoxin-affected tissues might provide
insight into the role of PPO in production of allelochemicals.

Table I. Effects of AOPP on Secondary Aromatic Compounds
in a Variety of Tissues

Tissue	AOPP Conc. (mM)	Growth reduction (%)	Secondary compound and % reduction	Reference
Perilla frutescens callus	0.1	20	caffeic acid-80	(20)
Glycine max seedlings	1.0	--	glyceollin-90	(50, 51)
Coleus blumei cell suspension cultures	0.0005	--	rosmarinic acid-50	(52)
Daucus carota suspension culture	0.1	0	anthocyanin-80	(53)
Fagopyrum esculentum hypocotyls	0.1	0	anthocyanin-85	(37)
seedlings	0.3	0	vanillin-liberated from lignin-65	(54)
Brassica oleracea hypocotyls	0.1	0	anthocyanin-80	(55)
Raphanus sativus cotyledons	0.5	0	a feruloyl derivative-70	(56)
			kaempferol-80	
			pelargonidin-85	
Sphagnum magellanicum plants	0.1	0	sphagnorubin-80-95	(57, 58)
Cucumis sativus hypocotyls	0.1	--	hydroxycinnamic acids-80	(59)
roots	0.1	--	hydroxycinnamic acids-60	(59)
Nicotiana tabacum suspension culture	0.02	0	secondary phenolics-90	(60)

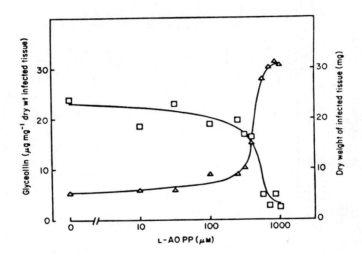

Figure 4. Dependence of glyceollin content (□) of soybean
seedlings and amount of Phytophthora affected tissue (△) on the
concentration of L-AOPP. Reproduced with permission from Ref. 51.
Copyright 1982, Physiological Plant Pathology.

Effects of Chemicals with Unknown Mechanisms

Many herbicides and other chemicals have been reported to influence levels of various phenolic compounds in higher plants by unknown mechanisms. It is unlikely that more than a few of these compounds have a primary influence on secondary phenolic compound synthesis. For instance, in our survey of the effects of 17 herbicides on anthocyanin accumulation, only glyphosate appeared to directly influence accumulation (31). The effects of several compounds on secondary phenolic compound production for which the mechanism of influence is unknown are summarized in Table II. A much longer list could be derived from the literature. Unfortunately, many of these compounds are phytotoxic or are known to have effects other than on secondary aromatic compound production. In most cases the effects on these compounds correlate well with extractable PAL activity (31, 71, 72, 73, 74) (Figure 5), even though they do not directly affect the enzyme.

Of the compounds listed in Table II, chlorsulfuron and acifluorfen (Figure 2, VI and VII, respectively) look particularly interesting with regard to manipulation of allelochemical production. Both of these compounds can cause large increases in the phenolic titre of plant tissues (e.g. Figures 6 & 7). Suttle et al. (75) suggested that chlorsulfuron would be an excellent chemical for manipulation of the quality and quantity of secondary phenolic compounds in studies of the role of these potential allelochemics in ecological interactions.

The concept of using chemicals for regulation of secondary compound production in order to give a crop plant an advantage in an agricultural ecosystem is well illustrated by results of Langcake et al. (84, 86). Their studies indicated that the systemic fungicide WL 28325 (2,2-dichloro-3,3-dimethylcyclopropane carboxylic acid) acts in rice plants by enhancing production of two diterpene phytoalexins, momilactones A and B. Rapid accumulation of these diterpenes coincided with inhibition of post-penetrative fungal growth.

Conclusions and Prospects

Chemical manipulation of phenolic allelochemical production in plants has two potential values: 1) for study of the role of phenolic allelochemicals in plant interactions with other organisms and 2) to alter such interactions for agricultural purposes. The first of these uses has already been accomplished on a limited scale (21, 22, 50, 51, 84, 86), however, there is no published evidence of the latter. This does not mean that herbicide and growth regulator-influences on plant secondary metabolism do not affect agricultural ecosystems by changing allelochemic compositions of plants. It is likely that this is the case, but it

Table II. Effects of Various Xenobiotics on Production of Various Secondary Aromatic Compounds

Compound - conc.	Tissue	Aromatic compound	% Reduction (-) or increase (+)	Reference
metribuzin 0.05 µM	Glycine max hypocotyls	anthocyanin	-75	(31)
propham 0.10 µM	Glycine max hypocotyls	anthocyanin	-90	(31)
propanil 0.2 mM	Glycine max hypocotyls	anthocyanin	-70	(31)
DSMA 0.5 mM	Glycine max hypocotyls	anthocyanin	+20	(31)
acifluorfen - 1.6 ppm	Spinacea oleracea leaves	phenolic amine derived from ferulic acid	+2500	(71)
5.0 ppm	Glycine max leaves	glyceollins	>+7500	(72)
		glyceofuran	>+1600	(72)
5.0 ppm	Phaseolus vulgaris leaves	phaseollin	>+4700	(72)
5.0 ppm	Pisum sativum seedlings	pisatin	+1900	(72)
5.0 ppm	Vicia faba seedlings	medicarpin	>+4500	(72)
		wyerone	>+1500	(72)
50.0 ppm	Apium graveolens	xanthotoxin	+400	(72)
5.0 ppm	Gossypium hirsutum	hemigossypol	+5900	(73)
alachlor 20-50 µM	Sorghum bicolor mesocotyls	anthocyanin	-90	(73)
		lignin	-50	(73)
chlorosulfuron 10 µg/seedling	Glycine max hypocotyls	anthocyanin	+400	(74)
10 µg/seedling	Helianthus annuus hypocotyls	p-coumaric acid	+3200	(75)
		caffeic acid	+3200	
		ferulic + sinapic acid	+1300	

sethoxydim	0.28 kg/ha	Zea mays leaves	anthocyanin	+20	(76)
salicylhy-droxamic acid	1 mM	Solanum tuberosum tuber discs	rishitin	-90	(77)
ammonium nitrate	1%	Fagopyrum esculentum hypocotyls	anthocyanin	-26	(78)
		cotyledons	anthocyanin	-45	(78)
silver nitrate	1 µM	Glycine max hypocotyls	glyceollin	>+1300	(79)
2,4-D	1000 ppm	Nicotiana tobacum plants	scopolin	>+3000	(80)
	1000 ppm	Helianthus annuus plants	scopolin	>+3000	(81)
ethylene	10 µl/1	Sorghum vulgare seedlings	anthocyanin	-32	(82)
		Brassica rapa seedlings	anthocyanin	-47	(82)
		Vaccinium macrocarpon fruit	anthocyanin	+400	(82)
		Euphorbia pulcherrima bracts	anthocyanin	+30	(82)
kinetin	10 µg/ml	Amaranthus tricolor seedlings	amaranthin	+2500	(83)
WL 28325	3 mg/12.5 cm pot	Oryza sativa leaves	momilactone A	+725	(84)
			momilactone B	>+1400	(84)
p-chloromer-curibenzene sulfonic acid	10 mM	Glycine max hypocotyls	glyceolin	>+1000	(85)

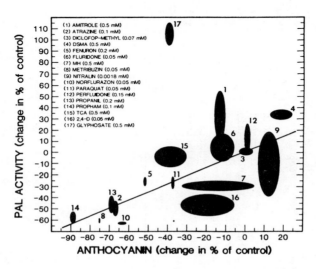

Figure 5.   Relationship between extractable PAL activity and
anthocyanin content of soybean hypocotyls of light-grown seedlings
48 h after treatment with various herbicides.   Data points are
mean values bounded by ellipses with axes ± 1 SE for each mean.
Reproduced with permission from Ref. 31, Copyright 1983, Weed
Science.

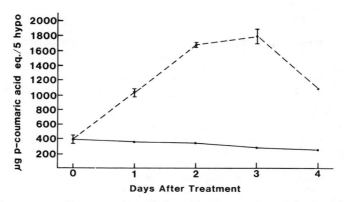

Figure 6.   Time course of effects of 10 µg chlorsulfuron per seedling on total phenolic content of sunflower hypocotyls expressed as p-coumaric acid equivalents.   Reproduced with permission from Ref. 75.   Copyright 1983, J. Plant Growth Regul.

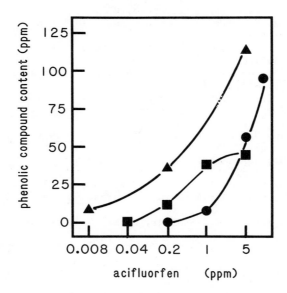

Figure 7.   Leaf content of pisatin in pea (▲) glyceollin in soybean (●), and phaseollin in bean (■) after 48 h after treatment with various levels of acifluorfen.   Redrawn from (72).

has not been documented because the effect was overlooked or was
too subtle to detect. A likely situation in which there may be
such an effect is under no-tillage conditions in which glyphosate
is used as a dessicant (e.g. 87) or with chlorsulfuron in
no-tillage wheat. Using a slow-acting level of glyphosate as a
dessicant should change the levels and ratios of phytotoxic
phenolic constituents of crop residues (e.g. 88), thus influencing
weed seedling growth and development.
    Chemical manipulation of secondary compound composition of
crop plants offers several advantages over genetic control of their
production. Chemical manipulation allows for timing the
manipulation as well as possibly determining the quality and
quantity of the desired response.
    Several compounds with excellent potential for manipulation of
phenolic allelochemic production have been discussed. These
include glyphosate, AOPP, chlorsulfuron, and acifluorfen.
Sublethal doses of these compounds can drastically alter the
phenolic compound content of higher plants. This type of
manipulation is a direct method for determining the role of these
compounds in plant interactions with their biological environment
that should be exploited.

Literature Cited

1. Jaworski, E. G. J. Agric. Food Chem. 1972, 20, 1195-8.
2. Nilsson, G. Swed. J. Agric. Res. 1977, 7, 153-7.
3. Hoagland, R. E.; Duke, S. O.; Elmore, C. D. Plant Sci. Lett.
    1978, 13, 291-9.
4. Hoagland, R. E.; Duke, S. O.; Elmore, C.D. Physiol. Plant.
    1979, 46, 357-66.
5. Duke, S. O.; Hoagland, R. E.; Elmore, C. D. Physiol. Plant.
    1979, 46, 307-17.
6. Holländer, H.; Amrhein, N. Plant Physiol. 1980, 66, 823-9.
7. Amrhein, N.; Shab, J.; Steinrücken, H. C. Naturwiss. 1980,
    67, 356-7.
8. Steinrücken, H. C.; Amrhein, N. Biochem. Biophys. Res. Comm.
    1980, 94, 1207-12.
9. Amrhein, N.; Johanning, D.; Shab, J.; Schulz, A. FEBS Lett.
    1983, 157, 191-6.
10. Boocock, M. R.; Coggins, J. R. FEBS Lett. 1983, 154, 127-33.
11. Schulz, A.; Sost, D.; Amrhein, N. Arch. Microbiol. 1984, 137,
    121-123.
12. Comai, L.; Sen, L. C.; Stalker, D. M. Science 1983, 221,
    370-1.
13. Rubin, J. L.; Gaines, C. G.; Jensen, R. A. Plant Physiol.
    1983, 70, 833-9.
14. Hoagland, R. E.; Duke, S. O. In "Biochemical Responses
    Induced by Herbicides"; Moreland, D. E.; St. John, J. B.;
    Hess, D., Eds.; ACS SYMPOSIUM SERIES No. 181, American
    Chemical Society: Washington, D.C., 1982; pp. 175-205.

15. Duke, S. O.; Hoagland, R. E. In "The Herbicide Glyphosate";
    Grossbard, E.; Atkinson, D., Eds.; Butterworths: London,
    1984; Chap. 6.
16. Cole, D. J. In "The Herbicide Glyphosate"; Grossbard, E.;
    Atkinson, D., Eds.; Butterworths: London, 1984; Chap. 5.
17. Duke, S. O. In "Weed Physiology"; Duke, S. O., Ed.; CRC
    Press: Boca Raton, 1985; Vol. II, In Press.
18. Singleton, Y. L.; Rossi, J. A. Am. J. Enol. Viticult. 1965,
    16, 144-58.
19. Hoagland, R. E. Weed Sci. 1980, 28, 393-400.
20. Ishikura, N.; Iwata, M.; Mitsui, S. Bot. Mag. Tokyo 1983, 96,
    111-20.
21. Berlin, J. and Witte, L. Z. Naturforsch. 1980, 36c, 210-4.
22. Holliday, M. J.; Keen, N. T. Phytopath. 1982, 72, 1470-4.
23. Keen, N. T.; Holiday, M. J.; Yoshikawa, M. Phytopath. 1982,
    72, 1467-70.
24. Amrhein, N.; Deus, B.; Gehrke; P.; Holländer, H.; Schab, J.;
    Schulz, A.; Steinrücken Proc. Plant Growth Regulator Soc. Am.,
    1981, 8, 99-106.
25. Amrhein, N.; Holländer-Czytko, H.; Leifeld, J.; Schulz, A.;
    Steinrücken, H.-C.; Topp, H. Proc. internationales d'etudes
    et assemblies du Groupe Polyphenols, 1982, pp. 21-30.
26. Amrhein, N.; Deus, B.; Gehrke, P.; Steinrücken, H. C. Plant
    Physiol. 1980, 66, 830-4.
27. Holländer-Czytko, H.; Amrhein, N. Plant Sci. Lett. 1983, 29,
    89-96.
28. Camm, E. L.; Towers, G. H. N. Prog. Phytochemistry 1977, 4,
    169-88.
29. Jangaard, N. O. Phytochemistry 1974, 13, 1769-75.
30. Hoagland, R. E.; Duke, S. O. Weed Sci. 1981, 29, 433-9.
31. Hoagland, R. E.; Duke, S. O. Weed Sci. 1983, 31, 845-52.
32. Duke, S. O.; Hoagland, R. E. Plant Sci. Lett. 1978, 11,
    185-90.
33. Duke, S. O.; Hoagland, R. E.; Elmore, C. D. Plant Physiol.
    1980, 65, 17-21.
34. Duke, S. O.; Hoagland, R. E. Weed Sci. 1981, 29, 297-302.
35. Engelsma, G. In "Regulation of Secondary Product and Plant
    Hormone Metabolism"; Luckner, M.; Schreiber, K., Eds.;
    Pergamon: New York, 1979; pp. 163-72.
36. Hanson, K. R.; Havir, E. A. Recent Adv. Phytochemistry 1972,
    4, 46-85.
37. Szkutnicka, K.; Lewak, S. Plant Sci. Lett. 1975, 5, 147-156.
38. Amrhein, N.; Gödeke, K. H. Plant Sci. Lett. 1977, 8, 313-17.
39. Amrhein, N. In "Regulation of Secondary Product and Plant
    Hormone Metabolism"; Luckner, M.; Schreiber, K., Eds.;
    Pergamon: New York, 1979; pp. 173-82.
40. Hoagland, R. E.; Duke, S. O. Plant Cell Physiol. 1982, 23,
    1081-8.
41. Brand, L. M.; Harper, A. E. Biochemistry, 1976, 15, 1814-21.

42. Holländer, H.; Kiltz, H.-H.; Amrhein, N. Z.Naturforsch. 1979,
     34c, 1162-73.
43. Duke, S. O.; Williams, R. D.; Markhart, A. H. Ann. Bot. 1983,
     52, 923-6.
44. Lea, P. J.; Norris, R. D. Phytochemistry 1976, 15, 585-95.
45. John, R. A.; Charteris, A.; Fowler, L. J. Biochem. J. 1978,
     171, 771-9.
46. Dow Chem. Co., U.S. Patent 3 162 525, 1964.
47. Nakamoto, H.; Ku, M. S. B.; Edwards, G. E. Photosynthesis
     Res. 1982, 3, 293-303.
48. Amrhein, N.; Gödeke, K.-H.; Kefeli, V. I. Ber Deutsch. Bot.
     Ges. 1976, 89, 247-59.
49. Amrhein, N.; Wenker, D. Plant Cell Physiol. 1979, 20,
     1635-42.
50. Grisebach, H.; Börner, H.; Moesta, P. Ber. Deutsch. Bot. Ges.
     1982, 95, 619-42.
51. Moesta, P.; Grisebach, H. Physiol. Plant Pathol. 1982, 21,
     65-70.
52. Ellis, B.E.; Remmen, S.; Goeree, G. Planta 1979, 147, 163-7.
53. Noe, W.; Langebartels, C.; Seitz, H.V. Planta 1980, 149,
     183-7.
54. Amrhein, N.; Frank, G.; Lemm, G.; Luhmann, H.-B. Eur. J. Cell
     Biol. 1983, 29, 139-44.
55. Amrhein, N.; Holländer, H. Planta 1979, 144, 385-89.
56. Strack, D.; Tkotz, N.; Klug, M. Z. Pflanzenphysiol. 1978, 89,
     343-53.
57. Tutschek, R. Planta 1982, 155, 301-6.
58. Tutschek, R. Planta 1982, 155, 307-9.
59. Amrhein, N.; Gerhardt, J. Biochim. Biophys. Acta 1979, 583,
     434-42.
60. Berlin, J.; Vollmer, B. Z. Naturforsch. 1979, 34c, 770-5.
61. Amrhein, N.; Diederich, E. Naturwiss. 1980, 67, 40.
62. Schill, L.; Grisebach, H. Hoppe Seyler's Z. Physiol. Chem.
     1973, 354, 1555-62.
63. Vaughn, K. C.; Duke, S. O. Physiol. Plant. 1981, 53, 421-8.
64. Vaughn, K. C.; Duke, S. O. Protoplasma 1982, 110, 48-53.
65. Vaughn, K. C.; Duke, S. O. Physiol. Plant. 1984, 60:257-62.
66. Duke, S. O.; Vaughn, K. C. Physiol. Plant. 1982, 54, 381-5.
67. Vaughn, K. C.; Duke, S. O. Physiol. Plant. 1984, 60, 106-12.
68. Montalbini, P.; Buchanan, B. B.; Hutcheson, S. W. Physiol.
     Plant. Pathol. 1981, 18, 51-7.
69. Mayer, A. M.; Harel, E. Phytochemistry 1981, 20, 955-9.
70. Duke, S. O.; Lane, A. D. Physiol. Plant. 1984, 60, 341-6.
71. Komives, T.; Casida, J. E. Pest. Biochem. Physiol. 1982, 18,
     191-6.
72. Komives, T.; Casida, J. E. J. Agric. Food Chem. 1983, 31,
     751-5.
73. Molin, W. T.; Anderson, E.; Porter, C. A. Abstr. Weed Sci.
     Soc. Am. 1984, 81.
74. Suttle, J. C.; Schreiner, D. R. Can. J. Bot. 1982, 60, 741-5.

75. Suttle, J. C.; Swanson, H. C.; Schreiner, D. R. J. Plant
    Growth Regul. 1983, 2, 137-49.
76. Asare-Boamah, N. K.; Fletcher, R. A. Weed Sci. 1983, 31,
    49-55.
77. Stelzig, D. A.; Allen, R. D.; Bhatia, S. K. Plant Physiol.
    1983, 72, 746-9.
78. Margna, V.; Laanest, L.; Margna, E.; Otter, M.; Vainjarv, T.
    EESTI NSV. Tenduste Akad. Toimetised 1974, 23, 298-304.
79. Stossel, P. Phytopath Z. 1982, 105, 109-19.
80. Dieterman, L. J.; Lin, C.-Y.; Rohrbaugh, L. M.; Thiesfeld, V.;
    Wender, S. H. Anal. Biochem. 1964, 9, 139-45.
81. Dieterman, L. J.; Lin, C.-Y.; Rohrbaugh, L. M.; Wender, S. H.
    Arch. Biochem. Biophys. 1964, 106, 275-79.
82. Craker, L. E.; Wetherbee, P. J. Plant Physiol. 1973, 52,
    177-9.
83. Piattelli, M.; de Nicola, M. G.; Castrogiovanni, V.
    Phytochemistry 1971, 10, 289-93.
84. Cartwright, D. W.; Langcake, P.; Ride, J. P. Physiol. Plant
    Path. 1980, 17, 259-67.
85. Stössel, P. Planta 1984, 160, 314-9.
86. Langcake, P.; Wickins, S. G. A. Physiol. Plant Path. 1975, 7
    113-26.
87. Putnam, A. R.; DeFrank, J. Crop Protection 1983, 2, 173-81.
88. Liebl, R. A.; Worsham, A. D. J. Chem. Ecol. 1983, 9, 1027-57.

RECEIVED August 6, 1984

Allelopathic Agents from Common Weeds

Amaranthus palmeri, Ambrosia artemisiifolia, and Related Weeds

NIKOLAUS H. FISCHER and LEOVIGILDO QUIJANO

Department of Chemistry, Louisiana State University, Baton Rouge, LA 70803

In a search for allelopathic agents from common weeds, Amaranthus palmeri S. Wats (Palmer amaranth) and Ambrosia artemisiifolia L. (Louisiana annual ragweed) have been analysed for their organic natural products. From A. palmeri phytol, chondrillasterol, vanillin, 3-methoxy-4-hydroxynitrobenzene and 2,6-dimethoxy- benzoquinone were isolated. From the roots of Ambrosia artemisiifolia four polyacetylenes, a mixture of sesquiterpene hydrocarbons, methyl caffeate, and a mixture of β-sitosterol and stigmasterol were obtained.

The germination regulation effects of ten sesquiterpene lactones, the flavonoid artemetin and the diterpene lactone, 17-acetoxyacanthoaustralide, on 12 crop and weed seeds are presented.

A mechanism is proposed by which water-insoluble plant lipids (sterols etc.) may act as allelochemicals via micelle formation with long-chain fatty acids. By this process plant lipid solubility and transport in the aqueous medium are enhanced. This might suggest a reevaluation of water-insoluble plant constituents such as sterols as potential allelopathic agents.

Amaranthus palmeri S. Wats., commonly referred to as Palmer amaranth, is a common weed belonging to the family Amaranthaceae. The observation that A. palmeri inhibits the growth of onion and carrot and also exhibits self growth-inhibition (communicated by Dr. R.M. Menges, USDA, Wesleco, TX) has led to an interdisciplinary search for the possible allelopathic activity of A. palmeri. The study involved Dr. R.M. Menges who also provided the plant material for our natural products investigation. Dr. Judith M. Bradow, USDA, Southern Regional Research Center, New Orleans, LA, performed the bioassays of all plant extracts and the pure constituents described in this paper. Her findings related to Palmer amaranth and the

0097-6156/85/0268-0133$06.00/0

common ragweed, <u>Ambrosia artemisiifolia</u> L., are described in another
paper of this book. Since field data indicated that the residues of
<u>A. palmeri</u> as well as <u>Ambrosia artemisiifolia</u> seemed to cause
significant growth inhibitions, it was our initial goal to perform
an extensive phytochemical study of <u>Amaranthus palmeri</u>, <u>Ambrosia
artemisiifolia</u> and related weeds in search for possible growth
inhibitors and/or promoters.

Materials and Methods

1. <u>Extractions and Chemical Analyses of Amaranthus palmeri S. Wats
(Amaranthaceae)</u>. The plant material for the chemical studies was
supplied by Dr. R. M. Menges, USDA, Weslaco, TX. The air-dried
aerial parts (518 g) were extracted at ambient temperature with
petroleum ether (PE), dichloromethane (CH_2Cl_2) and methanol (MeOH)
and the solvents were evaporated <u>in vacuo</u> providing 4.5 g, 4.7 g and
12.1g crude extracts, respectively. Samples of the crude extracts
were tested by Dr. J. M. Bradow for germination regulation activity
and the biological data are described in her paper.
     The PE extract (4.5 g) was pre-adsorbed on silica gel (10 g)
and chromatographed over 50 g silica gel (70-230 mesh) using as
eluants PE followed by mixtures of PE/CH_2Cl_2 (1:1) and
CH_2Cl_2/acetone (10,20,40,80%), 100 ml fractions being taken. The PE
fractions contained glycerides (^{1}H NMR, MS). Fraction 5 (PE/CH_2Cl_2)
was further purified by preparative TLC (PE/ether, 3:1, 2 x). The
low polarity band contained glycerides and a more polar band
provided phytol (<u>1a</u>) which was shown to be identical with authentic
material by ^{1}H NMR and MS comparison. Fractions 6-8 (PE/CH_2Cl_2)
gave after further purification by preparative TLC (PE/ether, 19:1,
5 x) 116 mg chondrillasterol (<u>2</u>), mp 154-6° [lit. (<u>2</u>) 168-9°].
     Extraction of 650g of air-dried, ground <u>A. palmeri</u> roots with
CH_2Cl_2 at ambient temperature provided after solvent evaporation <u>in
vacuo</u> 4.7g of a crude syrup which was chromatographed over 100g of
silica gel (70-230 mesh). Solvents were chloroform ($CHCl_3$) and
mixtures of $CHCl_3$ and acetone (5, 10, 20, 40%) with 100 ml fractions
being taken. Early fractions eluted with $CHCl_3$ contained fats ([1]H
NMR, MS). Fractions 7-10 were further purified by prep. TLC
(PE/ether, 2:1, 2 x) to give a mixture of fatty acids and chondrillasterol (<u>2</u>). Fraction 20, after purification by prep. TLC
(PE/ether, 2:1, 2 x), provided a mixture of chondrillasterol (<u>2</u>) and
an aromatic compound. Further separation by preparative TLC
(CH_2Cl_2/acetone, 9:1) gave 2mg pure 3-methoxy-4-hydroxynitrobenzene
(<u>6</u>). Since an authentic sample of 6 was not available for
comparison the 200 MHz ^{1}H NMR and MS data are given: ^{1}H NMR, δ
($CDCl_3$): 4.02s (OMe), 6.99d (J=9.0 Hz) (H-6), 7.77d (J=2.5 Hz) (H-
2), 7.89dd (J=9.0 Hz, J=2.5 Hz) (H-5). EIMS (probe) m/z (rel.
int.): 169 (M^+) (100.0), 139 (M^+-OCH_2) (21.5), 123 (M^+-NO_2) (19.9),
108 (22.9).
     Fraction 22 was further separated by repeated preparative TLC
(PE/ether, 2:1, 6 x) and CH_2Cl_2/acetone, 9:1) to give,
besides <u>2</u> and <u>3</u>, vanillin (<u>7</u>) which was identified by ^{1}H NMR and MS
comparison with authentic material. The low polarity TLC bands
contained traces of a mixture of aromatic compounds (^{1}H NMR, MS)
which were not further investigated due to the lack of material.

Trituration of fraction 26 with $CHCl_3$ provided 45mg of yellow needles which were spectroscopically shown to be 2,6-dimethoxy-benzoquinone (8), mp 250-2° [lit. (3) 254°].   200 MHz ^1H NMR, δ ($CDCl_3$): 3.85s (2 OMe), 5.86s (H-3, H-5); EIMS (probe) m/z (rel. int.): 168 (M^+) (37.3), 138 (10.3), 80 (23.9), 69 (100.0).

The CH_2Cl_2 extract (4.7 g) was mainly composed of the more polar constituents found in the PE extract. The MeOH extract (12.1g) remains to be chemically investigated.

2. <u>Extraction and Chemical Analysis of Ambrosia artemisiifolia L.</u>
For preliminary germination tests on lettuce and carrot seeds, extracts of <u>Ambrosia artemisiifolia</u> L. were used which had been collected on 13 May, 1979 by Mr. Axel Ohmstede in the East Baton Rouge Parish near the Louisiana State University Campus.   Air-dried ground aerial parts (495g) were stirred with P.E. for 90 min. and filtered by suction. The extraction process of the marc was repeated with CH_2Cl_2, MeOH and H_2O. Evaporation of the solvents <u>in vacuo</u> provided the four crude extracts used for the bioassays. Samples of the crude extracts were prepared such that the ratio of the weight of the sample to the total weight of the crude extract was known. The concentrations of the test solutions are described in terms of % of dry weight of the plant material.

A bulk collection of <u>Ambrosia artemisiifolia</u> L. was made on 1 June, 1983 in Louisiana, East Baton Rouge Parish on Bluebonnet Road behind Bluebonnet Ridge Apartments. Fresh roots (332 g) were extracted at ambient temperature twice with 1 L each of $CHCl_3$. The combined extracts were evaporated <u>in vacuo</u> to provide 800 mg of a red oil which was triturated with PE and filtered. The P.E.-soluble material (630 mg) was chromatographed over 15 g of silica gel to give eleven fractions of 100 mL each. Fraction 1 was re-chromato-graphed over a small silica gel column using P.E. as a solvent, three fractions being taken. Further purification by preparative TLC (P.E./CH_2Cl_2,4:1) provided, besides triglycerides (^1H NMR), a mixture of sesquiterpene hydrocarbons. GC-MS analysis of this mixture using a 30m SE-30 capillary column (injection temp.: 200°; column temp.: 60° for 30 sec., then increased by 15°/min. to 250°) showed in the GC-trace eight peaks which gave mass spectra with parent ions at m/z 204 ($C_{15}H_{24}$). The five major constituents were: α-guaiene, β-patchoulin, α-bulnesene, bergamotene and β-bisabolene (14), the latter being the major constituent.

Fraction 2 and 3 (P.E.) provided, besides the sesquiterpene hydrocarbon mixture, a compound which on the basis of ^1H NMR and MS spectral evidence was shown to be the acetylenic hydrocarbon pentayneene (9) $C_{13}H_6$, previously isolated from <u>A. artemisiifolia</u> (4).

Fraction 4 (P.E./CH_2Cl_2, 1:1) was a bright red solution the ^1H NMR spectrum of which indicated a mixture of two acetylenic constituents (A and B) which were separated by preparative TLC on 1 mm silica gel plates. The red band (A), when left in soln for 15 hr at r.t., lost its color. The ^1H NMR spectrum of the decolored soln was identical with the spectrum of compound B. Comparison of the physical data (^1H NMR and MS) of the two compounds with the spectral data reported for the dithiophene (A) (10) and the thiophene (B) (11) (5,6) established their identity. Both compounds had been

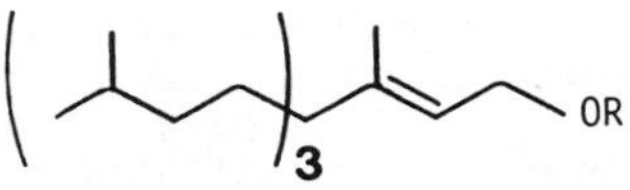

1a, R = H

1b, R = Ac

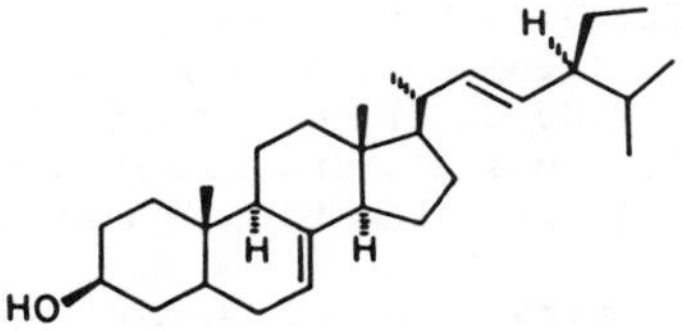

2

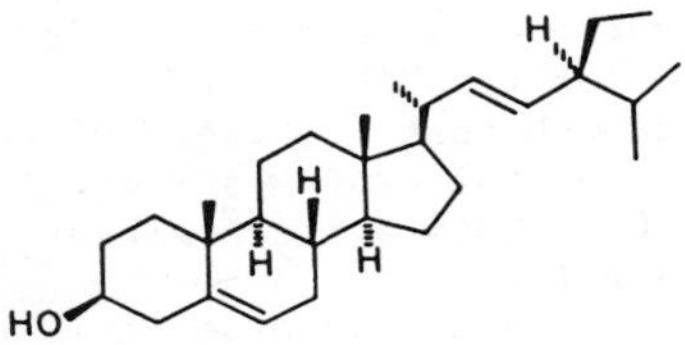

3

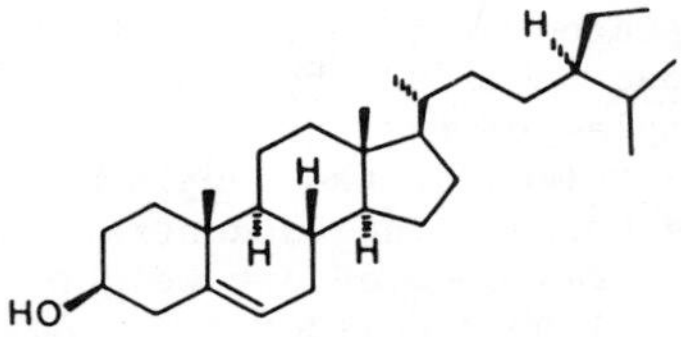

4

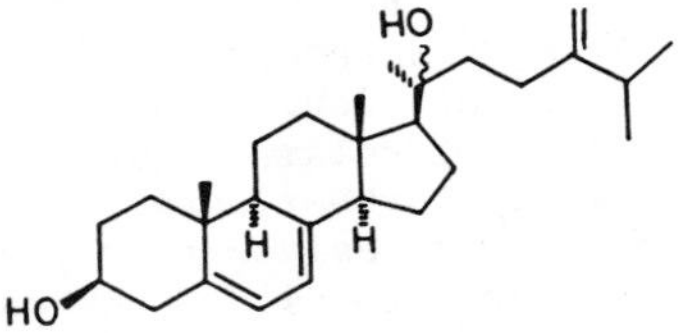

5

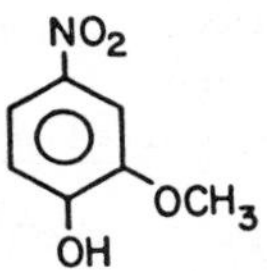

6

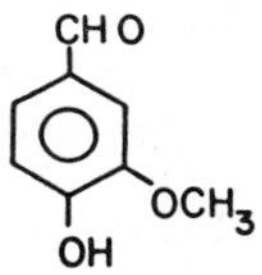

7

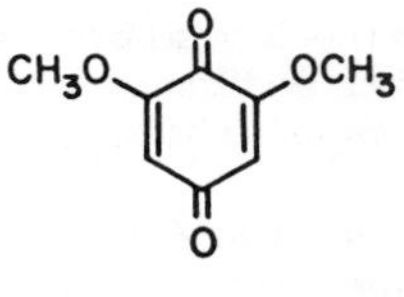

8

previously isolated from A. artemisiifolia L. (4). Fraction 5,
(P.E./Ch$_2$Cl$_2$, 1:1), after further preparative TLC separation,
contained a minor amount of an acetylenic compound which on the
basis of it ^{1}H NMR spectrum was tentatively assigned structure
(12). The acetate of a triterpene with ^{1}H NMR spectral features
very similar to those of β-amyrin acetate was obtained from a more
polar TLC band. Re-isolation of more material is necessary for an
unambiguous structural assignment of this triterpene.

From fractions 6 and 7 (CHCl$_3$), after further separations by
preparative TLC, a mixture of β-sitosterol (4) and stigmasterol
(3) was obtained. The identity was established by comparison of [1]H
NMR and MS data with those of authentic samples. Another TLC band
provided minor amounts of a phenolic compound which on the basis of

^{1}H NMR and MS data was tentatively assigned methyl caffeate (13).
The alternative structure of ferulic acid was excluded by spectral
comparison with an authentic sample of ferulic acid.

3. Germination Regulation by Sesquiterpene Lactones (15-24), 17-Acetoxyacanthoaustralide (25) and Artemetin (26).

Pure sesquiterpene
lactones were tested by Dr. Bradow. Since their isolation and
physical data have been previously reported, plant sources and
literature references are given in Table I. Table II contains the
bioassay results of the sesquiterpene lactones (15-24). The
flavonoid artemetin (26) was isolated from Melampodium cinereum DC,
(16) and the diterpene 17-acetoxyacanthoaustralide (25) was obtained
from Melampodium longipilum Robins. (18). The effect of artemetin
(26) and 17-acetoxyacanthoaustralide (25) in seed germination tests
are given in Table II.

Results and Discussion

1. Chemical Studies on Amaranthus palmeri S. Wats. Knowledge
related to natural products of the genus Amaranthus (Amaranthaceae)
is limited. Therefore, a systematic search for secondary
metabolites in A. palmeri was performed. The air-dried aerial parts
and the roots were separately investigated since the composition and
type of constituents can vary dramatically within different plant
parts. The ground aerial parts were extracted with organic solvents
of increasing polarity (PE, CH$_2$Cl$_2$ and MeOH). Chromatography of the
low polarity extract (PE) provided fats as well as phytol (1a). The
structure of the latter diterpene and its acetate (1b) was
established by comparison of the ^{1}H NMR and mass spectra with the
data of authentic samples of phytol (1a) and the acetate (1b).
Intermediate fractions provided chondrillasterol (2) (0.02 of dried
leave weight) which was identical with authentic material as shown
by spectral methods (^{1}H NMR, MS).

Members of the family Amaranthaceae are known to produce
ecdysteroids such as β-ecdysone (19) and inocosterone (20). From
petroleum ether extracts of Amaranthus spinosus Linn. Behari and
Andhiwal (21) obtained β-sitosterol (4), stigmasterol (3),
campesterol and cholesterol. From the roots of the same species,
Banerji (22) isolated two new saponins, a diglucoside and a
triglucoside of α-spinasterol. More recently, Roy et al. (23)

$$CH_3-(C \equiv C)_5-CH=CH_2$$

9

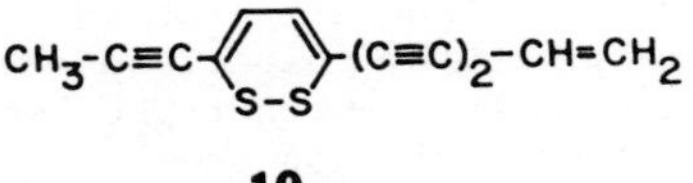

$$CH_3-C \equiv C \underset{S-S}{\swarrow} (C \equiv C)_2-CH=CH_2$$

10

$$CH_3-C \equiv C \underset{S}{\swarrow} (C \equiv C)_2 - \overset{2}{C}H = \overset{1}{C}H_2$$

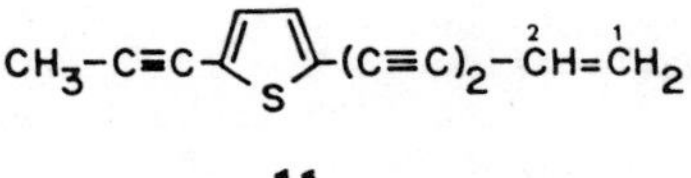

11
12; 1,2-epoxide

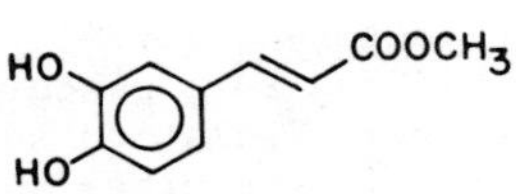

13

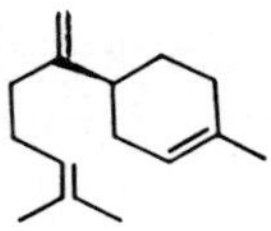

14

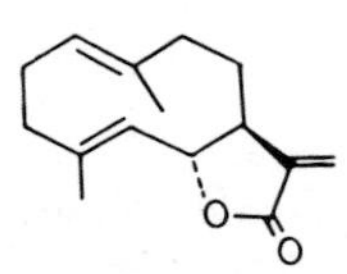

15

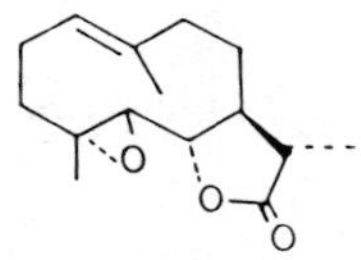

16

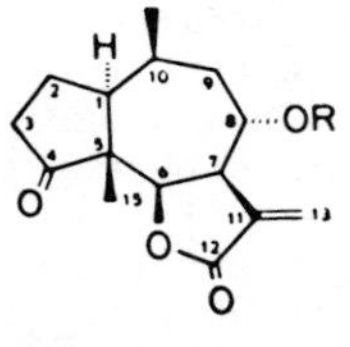

17, R=H
18, R=Ac

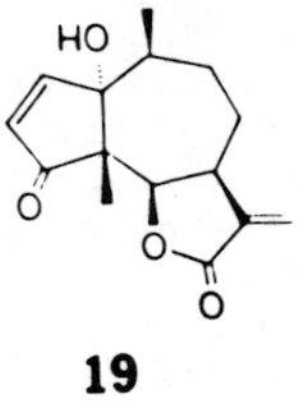

19

Table I.   Sesquiterpene Lactones and their Plant Sources used for Germination Regulation Tests.

Compound	Plant Source	Reference
Costunolide (15)	Saussurea lappa Clark	7
11,13-Dihydro-parthenolide (16)	Ambrosia artemisiifolia L.	8
Confertiflorin (17)	Ambrosia confertiflora DC.	9
Desacetylconfertiflorin (18)	Ambrosia confertiflora DC.	9
Parthenin (19)	Parthenium hysterorophorus Linn.	10
	Ambrosia psilostachya	11
Enhydrin (20)	Enhydra fluctuans Lour.	12
	Melampodium longipilum Robins.	13
Melampodinin A (21)	Melampodium americanum L.	14,15
Melampodin B (22)	Melampodium cinereum DC.	16
Cinerenin (23)	Melampodium cinereum DC.	16
Calein (24)	Calea ternifolia var. calyculata	17

Table II. EFFECTS OF SESQUITERPENE LACTONES (15-24), 17-ACETOXYACANTHOAUSTRALIDE (25) and ARTEMETIN (26) ON SEED GERMINATION[1,2]

Species	Compound											
	15 [10^{-4}]	16 [10^{-4}]	17 [7×10^{-5}]	18 [7×10^{-5}]	19 [10^{-4}]	20 [10^{-4}]	21 [10^{-4}]	22 [10^{-4}]	23 [6×10^{-5}]	24 [5×10^{-5}]	25 [10^{-4}]	26 [5×10^{-5}]
Onion	103	102	98	94	109	101	109	101	96	99	103	93
Sorghum	111	92	87	97	95	110	99	87	93	90	95	98
Oats	108	100	104	98	95	86	105	96	98	99	104	91
Ryegrass	75	102	86	114	97	93	104	119	88	101	101	96
Wheat	91	116	111	93	114	105	77	106	86	105	96	105
Lettuce	99	100	99	100	107	105	99	101	97	102	99	98
Clover	92	111	112	96	89	109	96	101	97	93	112	100
Carrot	121	113	97	104	92	99	96	98	120	101	91	116
Cress	99	100	99	99	100	100	99	99	100	101	100	100
Cucumber	114	111	96	102	111	99	104	96	106	93	109	101
P. amaranth	82	93	111	101	115	97	83	97	109	75	111	113
Tomato	97	104	110	100	101	104	100	100	104	99	100	95

[1] The bioassay germinations were performed by Dr. Judy Bradow at 25° for 72 hr in the dark in 3ml solutions containing 0.1% dimethylsulfoxide (DMSO) on a sheet of Whatman paper No.1 in 9cm petri dishes. Comparison was made with a parallel 0.1% DMSO control solution. Paired Student's analysis indicated that only wheat was significantly affected by the DMSO solution (-6%).

[2] The detailed botanical names for the seed species used in the germination test are summarized in Table I of Dr. J. Bradow's paper.

[3] The numbers in brackets below the structure numbers represent the molar concentrations of the compounds used in the tests.

described a new sterol, amasterol (5), from roots of <u>Amaranthus</u>
<u>viridis</u> L., a cultivated crop in India. The authors reported
inhibition of seed germination as well as growth inhibition of
lettuce seeds (<u>Lactuca sativa</u> cv White). Amasterol was also shown
to be a potent growth inhibitor of <u>Helmintho sporium</u> oryzi, a
pathogenic fungus (23).

Dichloromethane extracts of roots of <u>A. palmeri</u>, after
chromatography over silica gel, gave chondrillasterol in non-polar
fractions followed by 3-methoxy-4-hydroxynitrobenzene (6). Since
nitrobenzene derivatives of type 6 might have been previously used
in field experiments, it is possible that this compound is an
artifact. Later chromatographic fractions contained vanillin (7)
which was identified by ^{1}H NMR and mass spectral comparison with an
authentic sample. Subsequent fractions afforded 2,6-dimethoxy
benzoquinone (8), a constituent in members of a number of plant
families (24). To our best knowledge this is the first report
of the presence of 8 in the family Amaranthaceae. It is noteworthy
that the quinone 8 exhibits cytotoxicity in the KB cell culture
system <u>in vitro</u> (ED$_{50}$ = 2.8 μg/ml) (25,26). In P-388 lymphocytic
leukemia <u>in vitro</u> tests 2,6-dimethoxybenzoquinone displayed a highly
cytotoxic activity of 0.0015 μg/ml (24). The ecological function of
this widespread natural product is not known.

The effects of the above constituents of <u>A. palmeri</u> on the
germination and growth of a number of plant seeds are discussed by
Dr. J. M. Bradow in her chapter of this book.

2. The Chemical Constituents of Ambrosia artemisiifolia L. Most
ragweeds are distinguished by their adjustment to different
ecological environments and a successful interspecific competition.
Jackson and Willemsen (1) were able to demonstrate that <u>A.</u>
<u>artemisiifolia</u> is a dominant species in the first year of old field
succession in the Piedmont of New Jersey. Root exudates and shoot
extracts of ragweed inhibited the germination and growth of other
early invaders of abandoned fields. The presence of caffeic acid
and chlorogenic acid in addition to unknown phenolic compounds could
be demonstrated. Rice had shown earlier that similar phenolic acids
were present in <u>A. psilostachya</u> DC. (27).

Anaya and del Amo (28) provided evidence that aqueous extracts
of leaves and roots of the semi-arid and subtropical ragweed, <u>A.</u>
<u>cumanensis</u> H.B.K., inhibited the germination and growth of several
plant species and was also auto-inhibitory. Water washes of the
leaves depressed the growth of some species and stimulated others.
Decomposition products of leaves and roots were highly inhibitory to
some seedlings, suggesting that in these cases microorganisms play a
major role in the allelopathic process. Allelopathic effects of
extracts of aerial parts and rhizomes of western ragweed, <u>Ambrosia</u>
<u>psilostachya</u> DC., on seed germination and seedling growth of a
number of plants was recently shown by Dalrymple and Rogers (29).

In preparation for the isolation of allelopathics from root
exudates of <u>A. artemisiifolia</u> using the method by Tang and Young
(30) we wished to learn first about the major root constituents and
possibly establish their structures. We describe here our
preliminary findings on the root natural products of Louisiana <u>A.</u>
<u>artemisiifolia</u> populations. Freshly obtained roots of the ragweed,

which had been collected in Baton Rouge, LA, were extracted with
chloroform. Extensive chromatography of the petroleum ether-soluble
extracts provided, besides triglycerides, an oil which was analysed
by gas chromatography-mass spectrometry (GC-MS). The first eight
peaks of the GC-trace showed in the mass spectrum parent ions at $\underline{m/z}$
204. The mass spectral data together with the 200 MHz ^{1}H NMR
spectrum of the mixture suggest a sesquiterpene hydrocarbon mixture
of the empirical formula $C_{15}H_{24}$. The spectral data were in good
agreement with absorptions typical of the β-bisabolene skeleton
(14) and other farnesol cyclization products. Structural
assignments of the five major constituents in the sesquiterpene
hydrocarbons mixture were based on MS data of a GC-MS analysis. The
major constituents with increasing retention time were: α-guaiene,
β-patchoulin, α-bulnesene, bergamotene and β-bisabolene (14), the
latter sesquiterpene being the major constituent. Pretreatment of
seeds with a CH_2Cl_2 solution (0.25 wt.%) of the sesquiterpene
hydrocarbon mixture caused significant seed germination inhibitions.
Onion, ryegrass and Palmer amaranth were particularly sensitive to
the hydrocarbon mixture, but cucumber germination was slightly
increased. Application of the sesquiterpene mixture in 0.1% DMSO
aqueous solution inhibited germination of oat, ryegrass and cucumber
and, in particular, Palmer amarath. Slightly more polar chromatographic fractions contained the three polyacetylenes 9-11 which had
been previously isolated from A. artemisiifolia and are common root
constituents of Ambrosia species (4). Compound 10 is thermally very
unstable and extrudes elemental sulfur under formation of the
thiophene derivative 11. The 1,2-epoxide derivative of 11 appears
to be present in very low concentration as shown by ^{1}H NMR.
Subsequent fractions provided very small amounts of a triterpene
with ^{1}H NMR absorptions similar to those of β-amyrin, followed by a
band which contained stigmasterol (3) and β-sitosterol (4). From
fraction 7 a phenolic compound was isolated which exhibited ^{1}H NMR
and MS data in agreement with the methyl caffeate structure (13).
Spectral comparison with ferulic acid together with polarity
considerations excluded the latter structure. Re-isolation of
larger quantities of the root constituents of A. artimisiifolia and
their separations will be necessary for detailed bioassays of the
sesquiterpene hydrocarbons and triterpenes. In a previous
investigation of the aerial parts of the common Louisiana ragweed,
A. artemisiifolia (8), we had isolated the germacranolide 11,13-
dihydroparthenolide (16) the germination activity effects of which
will be discussed in the following section together with the other
sesquiterpene lactones.

3. Effects of Sesquiterpene Lactones on Seed Germination.
Sesquiterpene lactones are common constituents of the Asteraceae but
are also found in other angiosperm families and in certain
liverworts (31,32). These highly bitter substances exhibit a wide
spectrum of biological activities (33) which include cytotoxicity,
anti-tumor, anti-microbial, insecticidal (34) and molluscicidal (35)
properties. Furthermore, they are known causes for livestock
poisoning and contact dermatitis in humans (33). Structure-activity
relationship studies on sesquiterpene lactones have demonstrated
that biological activity frequently depend on the presence of the α-

methylene-γ-lactone moiety A, which is a potent receptor for
biological nucleophiles,

in particular, thiol groups to form the Michael addition
product B which can result in the inhibition of key enzymes such as
phosphofructokinase (36).

The function of sesquiterpene lactones as growth regulators has
been reviewed by Gross (37). The lactones heliangine (27),
helianginol, pyrethrosin and cyclopyrethrosin acetate promote the
adventitious root formation on hypocotyls of cuttings taken from
Phaseolus mungo seedlings and inhibit elongation of Avena coleoptile
sections (38). Compounds with the methylene-γ-lactone group reduced
to the saturated lactone showed no effect on the root formation.
Since all active compounds reacted with cysteine which was not the
case with the reduced, saturated lactones, the Japanese authors
concluded that the root growth promoting activity must be related to
the reactivity of the methylene γ-lactones towards biological thiol
groups such as cysteine which deactivates heliangine in its root
formation activity (39). The sesquiterpene dilactone vernolepin
obtained from Vernonia hymenolepis A. Rich. inhibits extension
growth of wheat coleoptile sections at a 1.8×10^{-5} molar
concentration but simultaneous administration of increasing amounts
of auxin reduced the inhibitory effect of vernolepin (40).
Parthenin (19) has at a concentration of 50 ppm no effect on the
germination of the bean Phaseolus vulgaris but inhibits the
development of radicles and hypocotyls (41). Similar effects were
observed by Kanchan for Parthenium hysteropherus and Eleusine
coracana coleoptiles (42) and it was shown that besides parthenin
(19), caffeic acid, vanillic acid, ferulic acid, chlorogenic acid
and anisic acid were major constituents in P. hysteropherus (43).
The germination and growth of velvet leaf (Abutilon theophrasti
Medic.), a weed causing severe problems in corn and soybean fields,
is inhibited by tomentosin (28) and axivalin (29), two lactones
isolated from seeds of povertyweed (Iva axillaris Pursh) (44).

The availabity of a number of sesquiterpene lactones with
different skeletal types (31) from our previous structural studies
of this group of biologically active compounds suggested their
testing for possible germination and growth regulation activities.
Their plant sources are given in Table I and the biological actions
on a number of crop and weed seeds are summarized in Table II. A
0.1 mM solution of the simplest germacranolide, costunolide (15),
significantly promoted germination of sorghum (111%) carrot (121%),
and cucumber (114%) but inhibited ryegrass (75%), wheat (91%) and
Palmer amaranth (82%). In contrast, 11,13-dihydroparthenolide (16)
from Ambrosia artemisiifolia, which lacks the methylene moiety in
the γ-lactone group, inhibits sorghum (92%) and Palmer amaranth
(93%) but significantly promotes germination of wheat (116%), clover
(111%), carrot (113%) and cucumber (111%). This demonstrates that a
compound without a nucleophile receptor such as the methylene

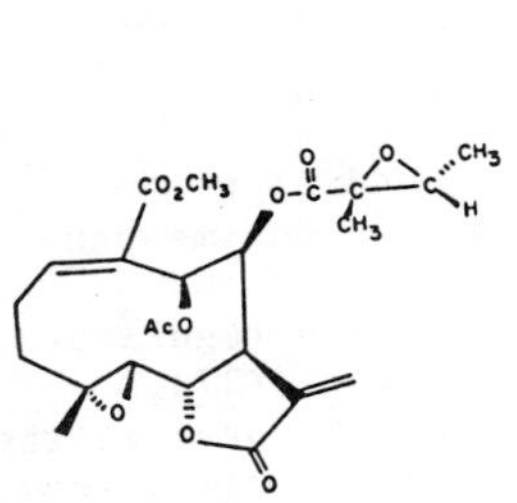

20

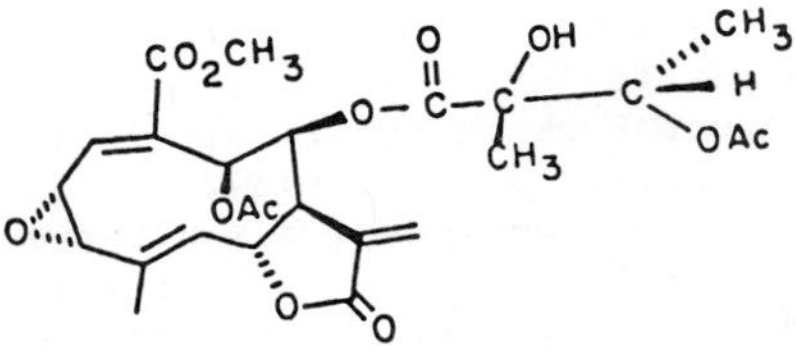

21

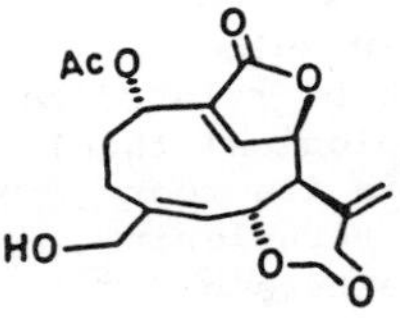

22

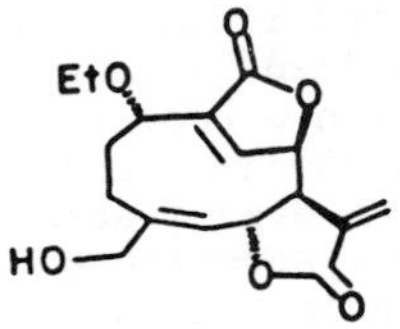

23

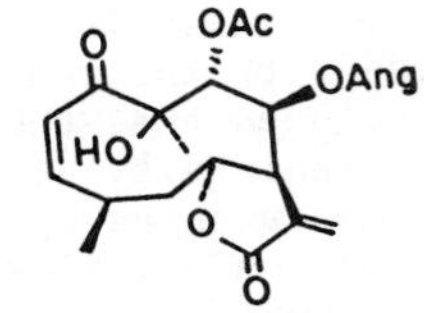

24

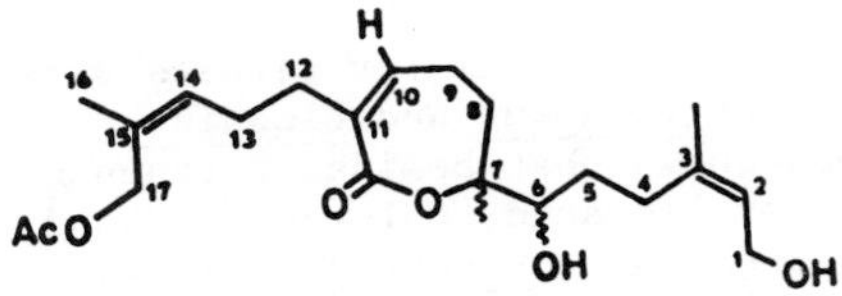

25

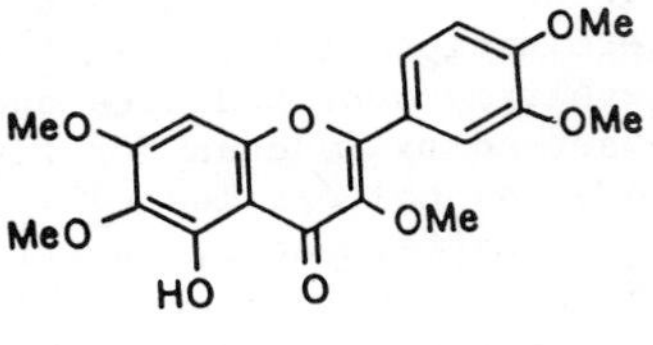

26

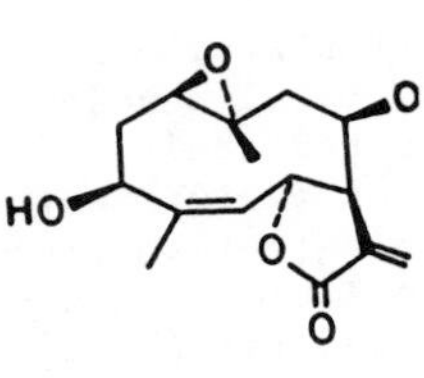

27, R = Tig

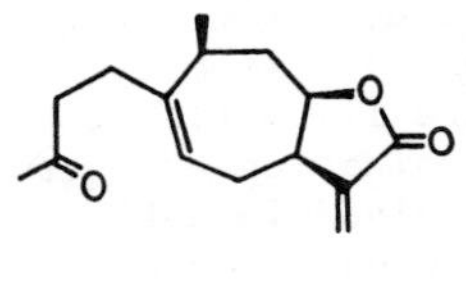

28

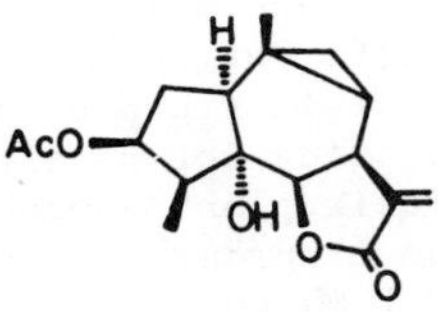

29

γ-lactone group is able to exhibit seed germination activities. The above data also support earlier observations (38) that the same substance can inhibit germination of one species and promote another.

The two pseudoguaianolides confertiflorin (17) and its desacetyl derivative (18), both isolated from A. confertiflora, show considerable differences in activity. Sorghum (87%) and ryegrass (86%) are inhibited and wheat (111%), clover (112%) and Palmer amaranth (111%) are promoted by 17. To the contrary, ryegrass (114%) is promoted by 18 while most other seeds show little effects. Future studies with a series of ester derivatives of 18 with increasing chain length will give clues about the influence of chain length and lipophilicity on the seed germinations. Parthenin (19) had been previously shown to inhibit the development of radicals and hypocotyls of beans (Phaseolus vulgaris) (41) and Eleusine coracana (42). In our tests, clover (89%) and carrot (92%) are also inhibited, but onion (109%), wheat (114%), cucumber (111%) and P. amaranth (115%) are promoted. The allelopathic effects of the two melampolides enhydrin (20) and melampodinin A (21) differ in a number of seed germination tests. Sorghum (110%) and clover (109%) are promoted and oat (86%) is inhibited by enhydrin (20). Compound (21) promotes onion (109%) and significantly inhibits wheat (77%) and Palmer amaranth (83%). Melampodinin A (21) is an anti-leukemic (P-388) compound in in vivo tests on laboratory animals (15) and also exhibits insect allelochemic effects with high larval mortality on the fall armyworm (Spodoptera frugiperda J. E. Smith) (34). Enhydrin (20) which lacks a 2,3-epoxide function shows no activity (34). Based on model reactions with sodium methoxide, nucleophilic attacks occur at the methylene lactone group but predominantly at C-2 under opening of the 2,3-epoxide (45). This suggests that biological nucleophiles (R-SH) should react at the same centers. The difference between lactones 20 and 21 with respect to the 2,3-epoxide function might explain their different bioactivities. The structural difference between the co-occuring melampodin B (22) and cinerenin (23) lie in the acetoxy- and ethoxy-attachment at C-2, respectively, causing major differences in their polarity, 22 being more polar. Significant seed germination effects are shown on ryegrass which is promoted by 22 (119%) and inhibited by 23 (88%). Sorghum is depressed by both whereas wheat (86%) is inhibited by 23 only. At the 6×10^{-5} molar level cinerenin promotes seed germination of carrot (120%) and to a lesser extent Palmer amaranth (109%).

Calein A (24) which contains one α, β-unsaturated ketone, ester and lactone moiety has only minor effects on most seed germinations but strongly inhibits Palmer amaranth (75%) at the 0.05 mM level. The diterpene lactone 17-acetoxyacanthoaustralide (25) has minor effects, showing promotion of clover (112%), cucumber (109%) and Palmer amaranth (111%) and inhibition of carrot germination (91%). The flavonoid artemetin (26) at 0.05 mM concentrations slightly affects onion (93%), oat (91%) and promotes seed germinations in carrot (116%) and Palmer amaranth (113%). In closing, it is noteworthy to point out that lettuce and cress are not significantly affected by all ten sesquiterpene lactones as well as the diterpene (25) and artemetin (26).

4. Possible Mechanism of Allelopathic Action of Water-Insoluble
Plant Lipids. Many non-polar natural products with germination and
growth regulation activities in laboratory tests are in pure form
not sufficiently water soluble to account for their allelopathic
activities observed in the field. For this reason the notion exists
that sterols and other non-polar plant constituents are not likely
to play a role in allelopathic actions, and it is generally
concluded that the bioactivity data observed in the laboratory are
therefore coincidental.

We have recently observed in our laboratory that water washes
of undamaged leaves in a number of plants contained sterols and
other lipids in sufficiently high concentration comparable with
concentrations used in typical laboratory bioassays. These aqueous
lipid solutions are frequently accompanied by long-chain (C-12 to C-
18) fatty acids. We therefore suggest that micelle formation
between the lipids and fatty acids may occur. By this mechanism the
lipid solubility in the aqueous medium is significantly enhanced,
thus allowing the release of otherwise water-insoluble plant
constituents into the environment. Presently, experiments are in
progress in our laboratory to provide further evidence for the
"micelle-mechanism" of allelopathic lipids.

Acknowledgments
We thank Mr. Kenneth R. Wilzer, Jr. Plant Protection Institute,
USDA, Beltsville, MD for a sample of chondrillasterol and Helga D.
Fischer and Joel Carpenter for technical assistance. Support for
this work by the USDA-ARS Cooperative Agreement No. 58-7B30-2-399 is
acknowledged.

Literature Cited

1. Jackson, J.R.; Willemsen, R.W. Amer. J. Bot. 1976, 63, 1015-
    23.
2. Bergmann, W.; Feeney, R.J. J. Org. Chem. 1950, 15, 812-4.
3. Richtzenhain, H. Chem. Ber. 1944, 77, 409-17.
4. Bohlmann, F.; Burkhardt, T.; Zdero, C. "Naturally Occuring
    Acetylenes"; Academic: London, 1973.
5. Bohlmann, F.; Kleine, K.M.; Arndt, C. Chem. Ber. 1964, 97,
    2125-34.
6. Bohlmann, F.; Kleine, K.M. Chem. Ber. 1965, 98, 3081-6.
7. Rao, A.S.; Kelkar, G.R.; Bhattacharyya, S.C. Tetrahedron 1960,
    9, 275-83.
8. Fischer, N.H.; Wu-Shih, Y.F.; Chiari, G.; Fronczek, F.R.;
    Watkins, S.F. J. Nat. Prod. 1981, 44, 104-10.
9. Fischer, N.H.; Mabry, T.J. Tetrahedron 1967, 23, 2529-38.
10. Herz, W.; Miyazaki, M.; Kishida, Y. Tetrahedron Letters 1961,
    1961, 82-7.
11. Geissman, T.A.; Griffin, S.; Waddell, T.G.; Chen, H.H.
    Phytochemistry 1969, 8, 145-50.
12. Ali, E.; Ghosh Dastidar, P.P.; Pakrashi, S.C.; Durham, L. J.;
    Duffield, A.M. Tetrahedron, 1972, 28, 2285-98.
13. Seaman, F.C.; Fischer, N.H. Phytochemistry 1978, 17, 2131-3.
14. Fischer, N.H.; Wiley, R.A.; Perry, D.L. J. Org. Chem 1976,
    1976, 41, 3956-9.

15. Malcolm, A.; DiFeo, D.; Fischer, N.H. Phytochemistry 1982, 21, 151-5.
16. Perry, D.L.; Fischer, N.H. J. Org. Chem. 1975, 40, 3480-6.
17. Lee, I.Y.; Olivier, E.J.; Urbatsch, L.E.; Fischer, N.H. Phytochemistry 1982, 21, 2313-6.
18. Quijano, L.; Fischer, N.H. Phytochemistry 1984, 23, 829-31.
19. Takemoto, T.; Ogawa, S.; Nishimoto, N.; Yen, K.Y.; Abe, K.; Sato, T.; Takashi, M. Yakugaku Zasshi 1967, 87, 1521-3.
20. Takemoto, T.; Hikino, Y.; Nishimoto, N. Yakugaku Zasshi 1967 87, 325-8.
21. Behari, M.; Andhiwal, C.K. Curr. Sci. 1976, 45, 481-2.
22. Banerji, N. J. Indian Chem. Soc. 1980, 57, 417-9.
23. Roy, S.; Dutta, A.K.; Chakraborty, D.P. Phytochemistry 1982, 21, 2417-20.
24. Handa, S.S.; Kinghorn, A.D.; Cordell, G.A.; Farnsworth, N.R. J. Nat. Prod. 1983, 46, 248-50.
25. Cordell, G.A.; Chang, P.T.O.; Fong, H.H.S.; Farnsworth, N.R. Lloydia 1977, 40, 340-3.
26. Jones, E.; Ekundayo, O.; Kingston, D.G.I. J. Nat. Prod. 1981, 44, 493-4.
27. Rice, E.L. Southwest Nat. 1965, 10, 248-51.
28. Anaya, A.L.; del Amo, S. J. Chem. Ecol. 1978, 4, 305-13.
29. Dalrymple, R.L.; Rogers, J.L. J. Chem. Ecol. 1983, 9, 1073-8.
30. Tang, C.S.; Young, C.C. Plant Physiol. 1982, 69, 155-60.
31. Fischer, N.H.; Olivier, E.J.; Fischer, H.D. In "Progress in the Chemistry of Organic Natural Products"; Herz, W.; Grisebach, H.; Kirby, G.W., Eds.; Springer: Vienna, 1979; Vol. 38, pp.47-390.
32. Seaman, F. C. Bot. Rev. 1982, 48, 121-595.
33. Rodriguez, E.; Towers, G.H.N.; Mitchell, J.C. Phytochemistry 1976, 15, 1573-80.
34. Smith, G.M.; Kester, K.M.; Fischer, N.H. Biochem. Syst. Ecol. 1983, 11, 377-80.
35. Fronczek, F.R.; Vargas, D.; Fischer, N.H.; Hostettmann, K. J. Nat. Prod., in press.
36. Kupchan, S.M. In "Recent Advances in Phytochemistry"; Runeckles, V.C., Ed.; Plenum: New York, 1975; Vol. 9.
37. Gross, D. Phytochemistry 1975, 14, 2105-12.
38. Shibaoka, H.; Shimokoriyama, M.; Iriuchijima, S.; Tamura, S. Plant & Cell Physiol. 1967, 8, 297-305.
39. Shibaoka, H.; Mitsuhashi, M.; Shimokoriyama, M. Plant & Cell Physiol. 1967, 8, 161-70.
40. Sequeira, L.; Hemingway, R.J.; Kupchan, S.M. Science 1968, 161, 789-90.
41. Garciduenas, M.R.; Dominguez, X.A.; Fernandez, J.; Alanis, G. Rev. Latinoamer. Quim. 1972, 3, 52-3.
42. Kanchan, S.D. Curr. Sci. 1975, 44, 358-60.
43. Kanchan, S.D. Plant Soil 1980, 55, 67-75.
44. Wolf, R.B.; Spencer, G.F. Book of Abstracts, Paper No. 275, Meeting of the Weed Science Society of America, Feb. 8-10, 1984, Miami, FL.
45. Malcolm, A. J. Ph.D. Dissertation, Louisiana State University, Baton Rouge Campus, 1983.

RECEIVED September 21, 1984

Allelopathic Agents from *Parthenium hysterophorus* and *Baccharis megapotamica*

BRUCE B. JARVIS[1], N. B. PENA[1], M. MADHUSUDANA RAO[1], NILGUN S. CÖMEZOGLU[1], TAHA F. CÖMEZOGLU[1], and N. B. MANDAVA[2,3]

[1]Department of Chemistry, University of Maryland, College Park, MD 20742
[2]Plant Hormone Laboratory, U.S. Department of Agriculture, Beltsville, MD 20705

Parthenium hysterophorus is a weed with a history of being an agricultural pest. A water extraction of P. hysterophorous has yielded parthenin, coronopilin, and two new sesquiterpene lactones closely related in structure to the former compounds. These compounds are believed to be responsible, at least in part, for the allelopathic properties of this plant. An alcohol extract of the Brazilian shrub, Baccharis megapotamica has yielded a large number of closely related, highly toxic sesquiterpenes belonging to a class of mycotoxins known as the macrocyclic trichothecenes. Evidence has been gathered to show that these compounds are acquired by B. megapotamica from a fungal source and oxygenated by the plants to yield the baccharinoids which are potent in vivo active antileukemic agents. A large scale extraction (1800 Kg) has yielded over twenty different macrocyclic trichothecenes.

In this chapter we are focusing on two different plants which appear to express allelopathy in quite different fashions. The first, Parthenium hysterophorus Linn. is recognized in many parts of the world as causing serious agricultural problems due principally to its invasion of crop lands and the subsequent lowering of crop yields (1). Although this plant is native to the North and Central Americas, it has now been spread to many other

[3]Current address: Environmental Protection Agency, Washington, D.C. 20406

0097–6156/85/0268–0149$06.00/0

areas including Australia, Africa, China, India, Vietnam
and the Pacific Islands (2). In contrast, the other
plant, Baccharis megapotamica Spreng which is found in
several parts of Brazil, is a plant of limited agricultural interest. Both B. megapotamica and P.
hysterophorus belong to the Asteraceae family (formerly
Compositae), a class of plants which has furnished the
natural products chemist with a rich array of interesting
chemicals (3).

The behavior of P. hysterophorus in the field would
appear to fall under the classic definition of allelopathy as defined by Rice (4), i.e., this plant adversely
affects other nearby vegetation by extruding chemicals
which inhibit the growth of these other plants (5-10).
In contrast, nothing is known about B. megapotamica and
its relationship with other plants in its habitat.
However, both P. hysterophorus and B. megapotamica pose
a serious hazard to animals who ingest these plants. P.
hysterophorus has been implicated in allergic contact
dermatitis common in many parts of India (11,12), and, as
we shall see, B. megapotamica contains a series of
sesquiterpenes which are probably the most potent skin
irritants known.

There are, of course, many plants toxic to animals
which also may cause dermatitis. Such properties in
principle are not necessarily related to allelopathy.
What makes B. megapotamica unusual is that the chemicals
responsible for the toxic properties of the plant are
produced not by the plant itself but by a fungus associated with the plants. It is this plant-fungus interaction at the chemical level which suggests that B.
megapotamica is involved in a most extraordinary form of
allelopathy (41).

Parthenium hysterophorus

Since our principal interest in P. hysterophorus has been
to identify the chemicals involved in the allelopathy, we
have employed water as the extraction media for the plant
material. Presumably, in the environment, chemicals
extruded by P. hysterophorus find their way to neighboring plants by transport through the water in the soil.

An aqueous extract of P. hysterophorus (collected in
Puerto Rico) was partitioned into methylene chloride at
pH 7, pH 10 and pH 2. Bioassays of the methylene
chloride soluble fractions, using the bean second internode bioassay (13), showed that the highest activity was
concentrated in the methylene chloride extract at pH 7.
Extensive chromatographic purification (flash chromatography, medium pressure LC, preparative TLC) monitored
by bioassay led to the isolation of the four sesquiter-

pene lactones 1-4. Parthenin (1), which was the major
congener isolated, and coronopilin (2) have been isolated
previously from P. hysterophorus (parthenin, 14-16;
coronopilin, 17). Compounds 3 and 4 have not been
reported previously to have been isolated from natural
sources, although 3, 11-epitetrahydroparthenin, has been
synthesized by the hydrogenation of parthenin (14).

A number of sesquiterpene lactones exhibit interesting biological activity (18), and several accounts of
the bioactivity of parthenin have been reported (18-21).
However, our experience suggests that these results
should be interpreted with some caution because in the
case of parthenin, we have been unable to obtain this
compound in >98% purity. Although we obtained beautiful
crystals of 1 whose melting point (163-166°C) and optical
rotation match those reported in the literature (14),
analytical HPLC showed our sample to be contaminated with
two other compounds. Repeated efforts to remove these
contaminating substances by recrystallization, TLC and
preparative HPLC (both normal and reversed phase) failed.
Although it is likely that parthenin is responsible at
least in part for the allelopathic properties of P.
hysterophorus, biological testing, to be really meaning-
ful, must be conducted with a pure sample of parthenin.
One way around the problem of contamination might be to
use a sample of parthenin obtained by total synthesis, a
feat which has recently been completed (22).

Baccharis megapotamica

Several years ago, a collection of B. megapotamica found
growing in a saltwater marsh near Curitiba, Brazil, was
shown to contain a series of compounds which exhibited
very high activity *in vivo* against mouse P388 leukemia

(23). The National Cancer Institute (NCI) has been
screening natural sources (higher plants, cultures of
microorganisms, and marine animals) for a number of years
in search of new and unusual chemotherapeutic agents to
be used in cancer treatment (24,25), and B. megapotamica
was screened as part of this program. There are over 400
species of Baccharis which are found mainly in Central
and South America; none appears to be of any significant
economic or agricultural interest to man (3), and of
nearly 100 species of Baccharis screened by the NCI (26),
only B. megapotamica exhibited significant in vivo P388
activity.
      What has made this initial finding (23) so unusual
is the nature of the compounds found in B. megapotamica
responsible for the anticancer activity; they belong to
the class of trichothecene mycotoxins. Heretofore,
trichothecenes have been found to be associated only with
cultures of various common soil fungi, e.g., Fusarium,
Trichothecium, Trichoderma, Myrothecium, Cephalosporium,
Stachybotrys, Verticimonosporium, and Cyclindrocarpon
(27). In recent years, trichothecenes have generated a
great deal of interest because of their extraordinary
range of bioactivity. They have been implicated as chemical warfare agents (28) [so-called "Yellow Rain", a
topic of some controversy (29)], and employed in clinical
trials in the treatment of cancer (30), but they are probably best known for their role as mycotoxins, fungal
metabolites which pose a serious hazard to man and farm
animals by their contamination of food and feeds
(27,31-36). The presence of trichothecenes in B. megapotamica is all the more surprising when one notes that
these toxins also are potent phytotoxins (37, 41).
      The trichothecenes isolated from B. megapotamica,
the baccharinoids, belong to the class known as the
macrocyclic trichothecenes (38-42). The baccharinoids
are closely related to the roridins whose structures were
elucidated about fifteen years ago (38). The principal
difference in structure between the roridins and
baccharinoids lies in the presence of an oxygen atom, in
the form of either an 8β-hydroxyl group or a 9β,10β-epoxy
group, in the A-ring of the baccharinoids (See Figure 2
for structures). The roridins lack oxygen functionality
in the A-ring, with roridin K (43), which possesses an
8α-hydroxyl group, being the only known exception. This
latter compound is a very minor metabolite isolated from
a culture of Myrothecium verrucaria (43). However trivial the presence of the A-ring oxygen functionality in
the baccharinoids may appear, this functionality is
clearly responsible for the difference in P388 in vivo
activity between the inactive roridins and the potently
active baccharinoids (41,44).
      We have grown B. megapotamica from seed and shown

that over a period of several years none of the macro-cyclic trichothecenes is produced (45). Furthermore, when B. megapotamica is fed roridin A (Figure 1) through the roots, roridin A is translocated to the leaves, rapidly metabolized to 8β-hydroxyroridin A which is then slowly epimerized to baccharinoid B7 (Figure 1), with no apparent harm to the plant (45). These data strongly suggest that B. megapotamica, as it is found in its native Brazilian habitat, is associated with a roridin-producing fungus, most likely a Myrothecium. The fungus produces the roridins which are taken up by B. megapotamica, metabolized and stored as baccharinoids in the plant tissue. It should be noted that this metabolism to the baccharinoids is not a detoxification since the baccharinoids and roridins exhibit essentially the same high degree of phytotoxicity toward a variety of plants including tomatoes, corn, tobacco, and beans (46). Recently, another Baccharis species, B. coridifolia, has been shown to contain appreciable quantities of roridins A and E (47).

In April of 1984, a team of scientists from the Universities of Maryland and Minnesota (48) returned from Brazil with a collection of Baccharis species gathered in two regions of southern Brazil, Curitaba and Santa Maria. The original collections of B. megapotamica were made in a marshy area several miles outside Curitaba. The latest collections from this same marsh contained not only B. megapotamica, but also four other Baccharis species, B. unicinella, B. semiverata, B. myriocephala, and B. camporum which were found growing along with the more predominant species, B. megapotamica. However, as shown by HPLC analysis, only B. megapotamica appears to contain any appreciable amounts of the macrocyclic trichothecenes. And, interestingly, it is the female plants, particularly the flowering portions, which have the highest concentration (ca. one part per thousand) of the baccharinoids (see Figure 1).

However, the most interesting of the Baccharis species is most certainly B. coridifolia. This plant is well-known in Brazil as being one of that country's most toxic plants (49) and is a serious hazard to livestock who graze in pastures populated by B. coridifolia (49). Four separate collections were made of B. coridifolia (48). Three of these collections were from several miles outside Curitaba and a fourth collection in a pasture near Santa Maria, the site of an earlier collection of B. coridifolia (47), and several hundred miles away from Curitaba. This latter collection and two of the three former collections contained appreciable quantities of roridins A and E. A third collection from near Curitaba appeared to contain no macrocyclic trichothecenes. From a 20 g sample of one of the collections of B. coridifolia

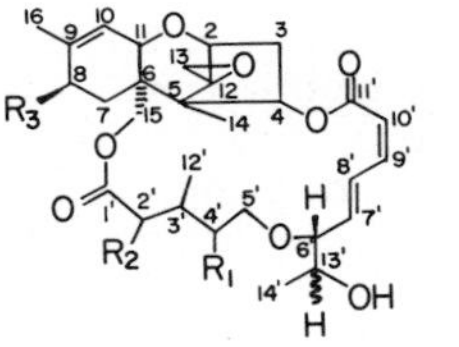

Roridin A: $R_1 = R_3 = H$; $R_2 = OH$
B3, B7: $R_2 = R_3 = OH$; $R_1 = H$
B1, B2: $R_1 = R_3 = OH$; $R_2 = H$

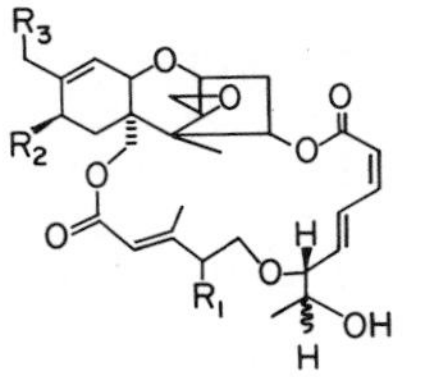

Roridin E: $R_1 = R_2 = R_3 = H$
B13, B14: $R_1 = R_2 = OH$; $R_3 = H$
B15, B16: $R_1 = R_3 = OH$; $R_2 = H$

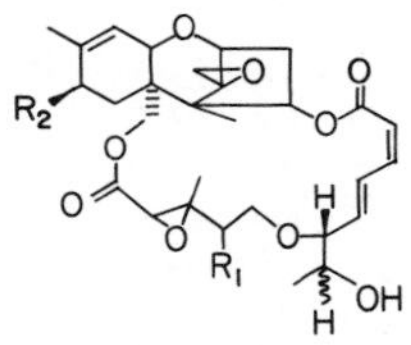

Roridin D: $R_1 = R_2 = H$
B4, B6: $R_1 = R_2 = OH$

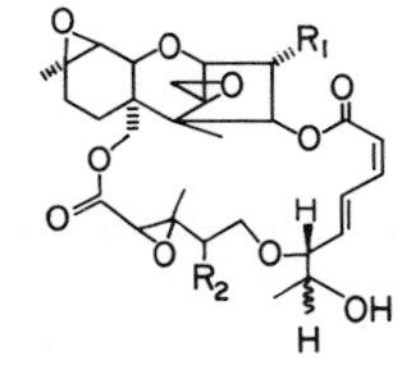

B5, B8: $R_1 = H$; $R_2 = OH$
B11, B12: $R_1 = OH$; $R_2 = H$
B17, B18: $R_1 = R_2 = H$

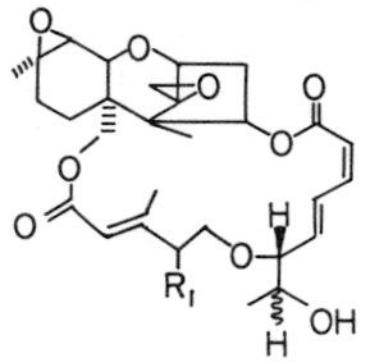

B9, B10: $R_1 = OH$

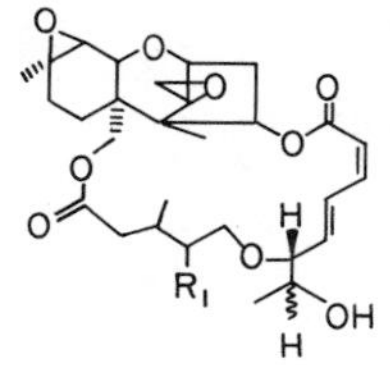

B19, B20: $R_1 = OH$

Figure 1. Structures of roridins and baccharinoids (B1–B20) isolated from B. megapotamica.

near Curitaba, we isolated 10 mg of roridin E and 4 mg of
roridin A. A preliminary study (50) has found
Myrothecium verrucaria and M. roridum associated with the
roots of some of these plants. However, a great many
other common soil fungi (e.g. Fusarium, Trichoderma,
Penecillium, etc.) also were found (50), and, as yet, we
do not have a clear picture of the relationship of
Myrothecium species with Baccharis species.

Our findings suggest that some Baccharis species
form an association with Myrothecium species which produce roridins. These roridins are taken up by the plant
and, in the case of B. coridifolia stored, but in the
case of B. megapotamica, the roridins are oxidized
(metabolized) (45) to the baccharinoids.

These findings have generated a great many questions
about the chemical and biochemical relationship between
Baccharis and Myrothecium. One point is clear: the
roridins and baccharinoids are both exceedingly phytotoxic compounds toward most plants (41). When tested in
the wheat coleoptile bioassay, which measures growth
inhibition (37), roridins A and iso E and baccharinoid B4
all showed significant growth inhibition at 10^{-6} M concentration (46) which makes these compounds as active as
chaetoglobosin K, the previously most active compound
tested in this bioassay (51). Verrucarins A and J (38)
and trichoverrin B (43) are an order of magnitude more
active than roridin A in this bioassay (46). What makes
certain Baccharis species immune to the toxic effects of
these trichothecenes whereas other species of Baccharis
as well as the vast majority of other plants (41) quite
sensitive to the toxic effects of trichothecenes is
entirely unclear.

Our most recent work with B. megapotamica has been
with the isolation of large quantities of baccharinoids
from a 1800 Kg collection. The workup of the crude
extract (ca. 30 Kg of black tarry material) of this plant
material was conducted by Dr. Fred Boettner of
Polyscience, Inc. This near Herculean task required tremendous quantities of solvents, column packings and time
to complete. An outline of the fractionation scheme is
presented in Figure 2. Fractions F6-F11 contain a large
number of baccharinoids. To date, we have characterized
over twenty macrocyclic trichothecenes found in B. megapotamica, including roridins A, D, and E, 3α-hydroxy-9β,
10β-epoxyroridin D (B11,B12 -- see Figure 1), and a large
number of congeners which possess either an 8β-hydroxyl
group or a 9β,10β-epoxy group. These latter baccharinoids have a number of variations in their structures
including the following: a C2' or C4' hydroxyl, a
2',3'-double bond with or without a hydroxyl at C4', a
2',3'-epoxide with or without a hydroxyl at C4'. In
addition, careful examination has invariably shown that

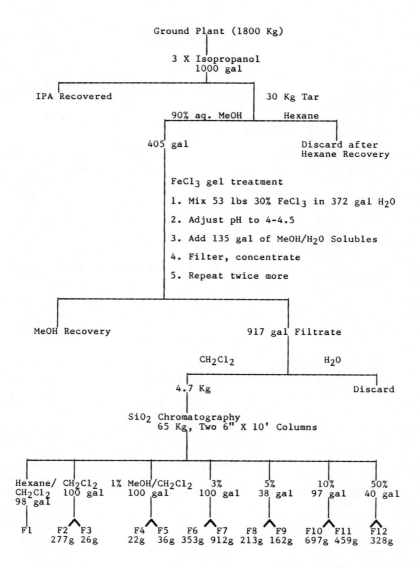

Figure 2. Large scale extract of B. megapotamica.

these compounds exist as diastereomers, epimeric at C13'. The principal isomers have the R configuration at C13', but with sufficient labor, we can always isolate at least small quantities of the C13'S epimer. Needless to say, the presence of so many closely related congeners including diastereomers has presented formidable problems for the isolation and purification of these compounds. We have relied very heavily on chromatographic techniques the description of which will be presented elsewhere.

Baccharis has proven to be a fascinating plant. In the future, we plan to investigate the fungus associated with this plant, as it is found Brazil, and unravel the interaction between the plant and the fungus.

Acknowlegments. This work was supported in part by the United States Department of Agriculture under cooperative aggreement 8-301-1109-20290-BBJ. BBJ wishes to acknowledge support from the National Institutes of Health-NCI (Grant No. CA 25967) and the U.S. Army (Contract No. DMAD 17-82-C-2240).

Literature Cited

1. Khosla, S. N.; Sobti, S. N. Pesticides, 1979, 13, no. 7, 21.
2. Towers, G. H. N.; Rodriguez, E.; Bennett, F. D.; Subba Rao, P. V. J. Scient. Ind. Res. 1977, 36, 672.
3. Herz, W. In "The Biology and Chemistry of the Compositae", Heywood, V. H., Harborne, J. B., Turner, B. L., Eds.; Academic Press: New York, 1977; pp 567.
4. Rice, E. L. In "Allelopathy"; Kozlowski, T. T.; Ed.; Academic Press: New York, 1974; pp 1.
5. Sarma, K. K. V.; Giri, G. S.; Subrahmanyam, K. Trop. Ecol. 1976, 17, 76.
6. Rajan, L. Current Science 1973, 42, 729.
7. Kanchan, S. D. Current Science 1975, 44, 358.
8. Kanchan, S. D.; Jayachandra Plant and Soil 1980, 55, 67.
9. Kanchan, S. D.; Jayachandra Plant and Soil 1980, 55, 61.
10. Rodriguez, E.; Dillon, M. O.; Mabry, T. J.; Mitchell, J. C.; Towers, G. H. N. Experientia 1976, 32, 236.
11. Ranade, S. Maharastra Med. J. 1971, 17, 495.
12. Lonkar, A.; Jog, M. K. Contact Dermatitis Newslett. 1972, 11, 291.
13. Mitchell, J. W.; Livingston, G. A. Agricultural Handbook 1968, 336, Agricultural Research Service, U. S. Department of Agriculture, U. S. Government Printing Office, Washington, D. C.
14. Herz, W.; Watanabe, H.; Miyazaki, M.; Kishida, Y. J. Am. Chem. Soc. 1962, 84, 2601.

15. Herz, W.; Miyasaki, M.; Kishida, Y. Tetrahedron Letters 1961, 2, 82.
16. Herz, W.; Watanabe, H. J. Am. Chem. Soc. 1959, 81, 6088.
17. Picman, A. K.; Towers, G. H. N.; Subba Rao, P. V. Phytochemistry 1980, 19, 2206.
18. Rodriguez, E.; Towers, G. H. N.; Mitchell, J. C. Phytochemistry 1976, 15, 1573.
19. Khosla, S. N.; Sobti, S. N. Indian J. For. 1981, 4, 56.
20. Khosla, S. N.; Singh, K.; Sobti, S. N. Indian J. For. 1980, 3, 261.
21. Char, M. B. S.; Shankarabhat, S. Experientia 1975, 31, 1164. 22. Heathcock, C. H; Tice, C. M.; Germroth, T. L. J. Am.Chem. Soc. 1982, 104, 6081.
23. Kupchan, S. M.; Streelman, D. R.; Jarvis, B. B.; Dailey, R. G., Jr.,; Sneden, A. T. J. Org. Chem. 1977, 42, 4221.
24 Douros, J.; Suffness, M. Cancer Chemother. Pharmacol. 1978, 1, 91.
25. Suffness, M.; Douros, J. Methods Cancer Res. 1979, 14, 73.
26. Suffness, M., personal communication.
27. Ueno, Y. (ED) "Trichothecenes - Chemical, Biological, and Toxicological Aspects"; Kodensha Scientific, Tokyo, 1983.
28. Mirocha, C. J.; Pawlosky, R. A.; Chatterjee, K.; Watson, S.; Hayes, W. J. Assoc. Off. Anal. Chem. 1983, 66, 1485.
29. Ember, L. R. Chem. & Eng. News 1984, 62 (2), 8.
30. Kaneko, T.; Schmitz, H.; Essery, J. M.; Rose, W.; Howell, H. G.; O'Herron, F. A.; Nachfolger, S.; Huftalen, J.; Bradner, W. T.; Partyka, R. A.; Doyle, T. W.; Davies, J.; Cundliffe, E. J. Med. Chem. 1982, 25, 579.
31. Bamburg, J. R.; Strong, F. M. In "Microbial Toxins"; Kadis, S., Ciegler, A., Ajl, S. J., Eds.; Academic Press: New York 1971; Vol. 7, p 207.
32. Bamburg, J. R. In "Mycotoxins and Other Fungal Related Food Problems"; Rodricks, J. V., Ed.; American Chemical Society: Washington, DC, 1976; Adv. Chem. Ser., No. 149, p 144.
33. Doyle, T. W.; Bradner, W. T. In "Anticancer Agents Based on Natural Product Models"; Cassidy, J. M., Douros, J. D., Eds.; Academic Press: New York, 1980; p 43.
34. Ueno, Y. Adv. Nutr. Sci. 1980, 3, 301.
35. Ong, C. W. Heterocycles 1982, 19, 1685.
36. Bamburg, J. R. Prog. Mol. Subcell. Bio. 1983, 8, 41.
37. Cole, R. J.; Dorner, J. W.; Cox R. H.; Cunfer, B. M.; Cutler, H. G.; Stuart, B. P. J. Natur. Prod. 1981, 44, 324.
38. Tamm, Ch. Fortschr. Chem. Org. Naturst. 1974, 31, 63.
39. Tamm, Ch. In "Mycotoxins in Human and Animal Health"; Rodricks, J. V., Hesseltine,C. W., Mehlamn, M. A.,

Eds; Pathotox Publishers: Park Forest South, IL, 1977, p 209.

40. Tamm, Ch.; Breitenstein, W. In "The Biosynthesis of Mycotoxins, A. Study in Secondary Metabolism"; Steyn, P. S., Ed.; Academic Press: New York, 1980, p. 69.

41. Jarvis, B. B.; Mazzola, E. P. Acc. Chem. Res. 1982, 15, 388.

42. Jarvis, B. B.; Eppley, R. M.; Mazzola, E. P. In "Trichothecenes-Chemical, Biological, and Toxicological Aspects", Ueno, Y., Ed.; Kodansha Scientific: Tokyo, 1983, p. 20.

43. Jarvis, B. B.; Stahly, G. P.; Pavanasasivam, G.; Midiwo, J. O.; DeSilva, T.; Holmlund, C. E.; Mazzola, E. P.; Geoghegan, R. F., Jr. J. Org. Chem. 1982, 47, 1117.

44. Jarvis, B. B.; Midiwo, J. O.; Mazzola, E. P. J. Med. Chem. 1984, 27, 239.

45. Jarvis, B. B.; Midiwo, J. O., Tuthill, D.; Bean, G. A. Science (Washington, DC) 1981, 214, 460.

46. Jarvis, B. B.; Cutler, H. G.; unpublished results.

47. Busam, L.; Habermehl, G. G. Naturwissenschaften 1982, 69, 392.

48. Bean, G. A. (U. Md.); Kommedahl, T.; Kurtz H. (U. Mn).

49. Tokarnia, C. H.; Döbereiner, J. Pesq. agropec. bras., Ser. Vet. 1976, 11, 19 & 27.

50. Kommedahl, T., private communication.

51. Cutler H. G.; Crumley, F. G.; Cox, R. H.; Cole, R. J.; Vormer, J. W.; Springer, J. P.; Latterell, F. M.; Thean J. E.; Rossi, A. E.; J. Agric. Food Chem. 1980, 28, 139.

RECEIVED June 27, 1984

Effects of Allelochemicals on Mineral Uptake and Associated Physiological Processes

NELSON E. BALKE

Department of Agronomy, University of Wisconsin, Madison, WI 53706

Allelopathic inhibition of mineral uptake results from alteration of cellular membrane functions in plant roots. Evidence that allelochemicals alter mineral absorption comes from studies showing changes in mineral concentration in plants that were grown in association with other plants, with debris from other plants, with leachates from other plants, or with specific allelochemicals. More conclusive experiments have shown that specific allelochemicals (phenolic acids and flavonoids) inhibit mineral absorption by excised plant roots. The physiological mechanism of action of these allelochemicals involves the disruption of normal membrane functions in plant cells. These allelochemicals can depolarize the electrical potential difference across membranes, a primary driving force for active absorption of mineral ions. Allelochemicals can also decrease the ATP content of cells by inhibiting electron transport and oxidative phosphorylation, which are two functions of mitochondrial membranes. In addition, allelochemicals can alter the permeability of membranes to mineral ions. Thus, lipophilic allelochemicals can alter mineral absorption by several mechanisms as the chemicals partition into or move through cellular membranes. Which mechanism predominates may depend upon the particular allelochemical, its concentration, and environmental conditions (especially pH).

Interference in the growth of one plant by another plant can result from either competition for nutrients, water, and light or from chemicals released from one plant (donor) that affect the second plant (receiver). The latter phenomenon is known as allelopathy ([1]) and may be as important as competition for influencing plant growth in both natural and agricultural ecosystems. Many naturally-occurring compounds (primarily secondary metabolites) produced by plants have been found to alter the growth of plants ([1]). Although micro-organisms also produce compounds that can

0097–6156/85/0268–0161$06.00/0

affect plant growth, this review will be concerned with only
allelochemicals known to be synthesized by higher plants.

Although many physiological and biochemical processes in
plants are affected by various allelochemicals, in most instances
the details of the mechanism of action of a particular
allelochemical have not been elucidated. Because soil mediates the
transfer of most allelochemicals (except perhaps volatile compounds)
from a donor to a receiver, plant roots are often the first tissues
to contact an allelochemical. Thus, it is not surprising that root
growth and development are inhibited in many instances of
allelopathy (1-3) One of the primary physiological functions of
plant roots is the absorption of mineral nutrients. Therefore, it
is logical that the influence of allelopathic interactions on
mineral absorption by plant roots has been investigated.

Although the definition of allelopathy includes stimulation as
well as inhibition of growth by allelochemicals (1, 4), allelochemicals that definitively affect mineral absorption by plant roots
have been found to primarily inhibit, rather than stimulate, the
process. The first part of this review presents evidence that
alteration of mineral absorption is a physiological mechanism of
allelopathy. Possible physiological and biochemical bases for the
inhibition of mineral absorption by allelochemicals are then
discussed.

Influence of Fertility Level on Allelopathic Interactions

The first finding to suggest that mineral absorption might be
inhibited by allelochemicals was a report in 1912 by Schreiner and
Skinner (5) showing that the inhibition of growth by phenolic
compounds could be modified by adding nutrients to the water
culture. Glass (6) also found that addition of mineral nutrients
decreased the inhibitory activity of a mixture of phenolics found in
soil under bracken fern (Pteridium equilinum L. Kuhn). More
specifically, nitrogen fertilizer helped alleviate the inhibition of
growth in birdsfoot trefoil (Lotus corniculatus L.) caused by
extracts of tall fescue (Festuca arundinacea Schreb.) (7), and
soluble phosphorus fertilizer largely overcame the deleterious
effects of goldenrod (Solidago canadensis L.) extracts on nutrition
and growth of maple (Acer saccharum Marsh.) seedlings (8).
Increased levels of nitrogen and phosphorous reduced the inhibition
of barley (Hordeum vulgare L.) growth caused by p-coumaric and
vanillic acids (8a). Although other explanations are possible,
these effects of fertilizers suggest that inhibition of mineral
absorption was responsible for the observed inhibition of growth.

Alteration of Mineral Concentration in Intact Plants

One of the primary bases for the hypothesis that allelochemicals
affect mineral uptake rests with the fact that the concentration of
minerals in receiver plants can be altered by donor plants by a
mechanism other than competition for minerals. In general, studies
in support of this hypothesis have involved measuring the amount of
mineral present per weight of a particular plant part following
absorption of minerals by the roots of the intact receiver.
Although for many of these studies one can question whether

competition was truly eliminated and only allelopathic effects were being observed, some studies have clearly eliminated competition by physically separating the donor from the receiver. These latter, more convincing, experiments have used four basic approaches: (1) leachates from roots of donors growing separately from receivers, (2) donor residue incorporated into the medium in which receivers are grown, (3) extracts of donor residue applied to receivers, and (4) specific allelochemicals added to the growth medium of receivers.

Donor and Receiver Grown Together. Chambers and Holm (9) reported that one common bean (Phaseolus vulgaris L.) plant absorbed less PO_4^{3-} (^{32}P) when grown in association with other bean, redroot pigweed (Amaranthus retroflexus L.) or green foxtail (Setaria viridis L.) plants than when grown alone. Although this result could be interpreted as resulting from competition for ^{32}P, one donor bean plant reduced ^{32}P uptake as much as two, three, or four bean plants. Also, pigweed absorbed seven times as much ^{32}P as bean but had less effect on ^{32}P uptake by the receiver than did donor bean plants, which absorbed less ^{32}P than pigweed. Thus, although competition was not eliminated in this study, the results suggested that allelopathy was responsible for the decreased ^{32}P uptake.

Two reindeer-moss species (Cladonia rangiferina L. and Cladonia alpestris L.) decreased PO_4^{3-} concentration in jack pine (Pinus banksiana Lamb.) and white spruce (Picea glauca Moench) (10). In addition, C. rangiferina decreased N concentration of the jack pine. However, K^+, Ca^{2+}, and Mg^{2+} concentrations in both receivers were not altered by either donor. Although mycorrhizae were associated with receiver plants in all the treatments, the authors concluded that impairment of either mycorrhizal function in absorbing PO_4^{3-} or PO_4^{3-} translocation in the receiver was responsible for decreased PO_4^{3-} uptake. They suggested that allelopathy might be responsible.

Two studies have suggested that quackgrass (Agropyron repens L.) can reduce K^+ uptake by maize (Zea mays L.) plants. Bandeen and Buchholtz (11) found that K^+, but not N or PO_4^{3-}, content of corn was decreased by quackgrass growing with maize. Because high levels of fertilization did not overcome the effect of quackgrass, the authors concluded that competition was not responsible (11, 12). Furthermore, when part of the receiver's root system was placed in nutrient solution (split-root technique), the receiver grew much better. The authors concluded that quackgrass caused an allelopathic inhibition of K^+ and N uptake by maize (12).

All the above studies suffer from the possibility that competition influenced the results. Any time the donor and receiver are grown together, the possibility of competition cannot be dismissed totally. Thus, the donor and receiver must be grown separately to convincingly eliminate competition.

Donor and Receiver Grown Separately. When donor and receiver plants are grown separately, a method must be devised to transfer the allelochemical(s) from the donor to the receiver. This can be accomplished by applying leachates from the donor to the receiver. One method of doing this is to arrange pots of the donor and pots of the receiver in a "stairstep" arrangement (13) so that nutrient

solution flows out the bottom of a donor pot into the top of a
receiver pot. The pots can be separated sufficiently to eliminate
competition for light between the donor and the receiver. A simpler
method is to flush donor pots with nutrient solution and apply the
leachate to receiver pots. Both of these techniques have been used
to study allelopathic alteration of mineral absorption.

Leachates from sand in which hoop pine (Araucaria cunninghamii
Ait.) slash pine (Pinus elliottii Engl.), or crows ash (Flindersia
australis R. Br.) was growing decreased the PO_4^{3-} concentration of
hoop pine seedlings growing in sand or soil (14). N and K^+ status
was not affected significantly. Similarly, when leachates from
roots of Anthoxanthum odoratum, Lolium perenne, Plantago lanceolata,
or Trifolium repens were applied to pots containing each of the same
four species individually, some of the exudates decreased PO_4^{3-}
uptake some receivers whereas other exudates stimulated PO_4^{3-}
uptake (15).

Using the stairstep design, Walters and Gilmore (16) found that
leachates from the rhizosphere of fescue altered the mineral content
of sweetgum (Liquidambar styraciflua L.) seedlings. K^+ and Mg^{2+}
concentrations in the leaves increased and PO_4^{3-} concentration in the
roots decreased whereas N and Ca^{2+} were not affected.

Donor Residue Incorporated into Growth Medium of Receiver. Because
plant litter, in addition to root exudates, is a potential source of
allelochemicals (17), several studies have investigated the
influence of residues of plants on mineral absorption by receivers.
Two extensive studies have been done by Bhowmik and Doll (18, 19).
In a field study, ten weed and crop residues were incorporated
individually into plots into which maize or soybean (Glycine max L.
Merr.) were planted subsequently. Most of the residues had no
effect on N, PO_4^{3-}, or K^+ concentration in the two crops. However,
sunflower (Helianthus annus L.) residue increased PO_4^{3-} content in
soybean; barnyardgrass (Echinochloa crus-galli L. Beauv.), yellow
foxtail (Setaria lutescens Wiegel), and soybean residue decreased N
content in maize; common lambsquarter (Chenopodium album L.) and
redroot pigweed increased K^+ content in both maize and soybean;
sunflower residue increased PO_4^{3-} content in soybean; and
barnyardgrass increased K^+ content in soybean (18). A similar
study conducted in a controlled-environment facility with five weed
residues against maize and soybean again showed little effect on N
and K^+ uptake (19). Ragweed (Ambrosia artemisifolia L.) decreased
N content in soybean; ragweed and velvetleaf (Abutilon theophrasti
Medic.) increased PO_4^{3-} content in soybean. However, all five
residues increased K^+ content in both crops.

Young and Bartholomew (20) incorporated Namarthria altissima
(Poir.) Stepf. and Hubb. root residue into soil. The tops of
Desmodium intorturm (Mill.) Urb. that grew in that soil in the
greenhouse contained less PO_4^{3-}, although N, K^+, and Mg^{2+}
levels were not affected. The authors concluded that allelopathic
substances were responsible for the decrease in PO_4^{3-} content.

In another greenhouse study, sunflower debris increased the
PO_4^{3-} concentration of redroot pigweed but did not alter the
concentrations of N or K^+ (21). When nutrient solutions were
added, the PO_4^{3-} and N responses were the same, but K^+ was
slightly increased by the debris.

Although these studies utilizing incorporated debris are valuable because they show the potential for allelochemicals to be released from plant litter, they suffer from a disadvantage. The amount of debris added and its carbon to nitrogen ratio might lead to alterations in nutrient contents in the soil as the result of proliferation or shifts in populations of micro-organisms. Thus, a control in which a material of similar C/N ratio but lacking allelochemicals needs to be included for such studies to be conclusive. The above studies did not include such controls and thus are not definitive.

<u>Leachates of Donor Residue</u>. Use of leachates of donor plant residue results in much less total material being put into the growth medium of the receiver. Thus, this is a more refined manner in which to test for allelopathic inhibition of mineral absorption.

Walters and Gilmore (16) incorporated either roots or leaves from fescue into the sand in the donor pots of a stairstep design. The subsequent circulation of nutrient solution leached chemicals from the donor debris and carried them into the receiver pots. K^+, Mg^{2+}, and Ca^{2+} contents in the leaves of sweetgum were increased, PO_4^{3-} content was decreased, and N content did not change when leachates from fescue roots or leaves were applied. Fescue roots and leaves produced the same results on N and PO_4^{3-} content in sweetgum roots as leaves, but K^+, Mg^{2+}, and Ca^{2+} content in sweetgum roots was variable.

Mulches of aster (<u>Aster novae-angliae</u> L.) or goldenrod placed on the top of pots in which sugar maple were grown resulted in lower PO_4^{3-}, Ca^{2+}, and Mg^{2+} content and higher K^+ content in the maple (<u>8</u>). The authors concluded that allelochemicals were leached from the mulches and caused the altered mineral contents in the receiver. PO_4^{3-} content in needles of red pine (<u>Pinus resinosa</u> Ait.) was reduced when red pine trees were watered with aqueous extracts of <u>Lonicera tatarica</u> or <u>Solidago gigentea</u> foliage (<u>22</u>).

<u>Specific Chemicals</u>. Very few studies have tested directly the influence of specific allelochemicals on mineral uptake by plants. In all the previous papers cited in this review, no attempt was made to isolate and identify the chemicals responsible for the alteration of mineral content in the receiver. Obviously, if we are to prove that some allelochemicals act by inhibiting mineral absorption, individual chemicals must be tested (<u>23</u>).

Olmsted and Rice (<u>24</u>) reported that chlorogenic and gallotannic acids decreased the amount of K^+ and Ca^{2+} removed from nutrient solutions by seedlings of redroot pigweed during 12 days of growth. Because chlorogenic acid is a phenolic compound found in sunflower, Hall et al. (<u>21</u>) investigated the influence of chlorogenic acid on mineral uptake by redroot pigweed. The compound increased N, decreased PO_4^{3-}, and did not affect K^+ concentration during 7 weeks of pigweed growth.

Another phenolic compound, caffeic acid, altered the mineral content of <u>Argyrodendron trifoliolatum</u> after 6 months (<u>25</u>). Zn^{2+}, Mn^{2+}, and PO_4^{3-} contents increased at 10 ppm caffeic acid, but 50 ppm decreased Mn^{2+} and PO_4^{3-} compared to their concentrations at 10 ppm.

Collectively, these experiments show that mineral concentrations

in plants can be altered when plants grow in situations where
allelopathy may exist. The concentration of many different minerals
can be altered, but for any particular mineral, a particular
donor/receiver combination may produce an increase, a decrease, or
no change in the mineral concentration in the receiver. This is not
surprising because both competition and concentrations of natural
products will differ from donor to donor and even within a single
donor depending upon growth conditions. Thus, it is difficult to
reach any general conclusions from experiments with living tissue,
debris, or leachates from donors because the identity and quantity
of the potential allelochemicals present are not known.
Furthermore, the long time periods of these experiments (12 days to
10 months) make it impossible to conclude that allelochemicals
directly alter mineral absorption. An equally feasible explanation
is that the chemicals inhibit growth and this indirectly alters the
concentration of minerals in the tissues.
        One way to address the direct nature of the allelopathic
alteration of mineral absorption is to measure absorption over short
periods of time. Anthoxanthum adoratum, Lolium perenne, Plantago
lanceolata, or Trifolium repens plants grown in leachates from
living Plantago lanceolata roots for 3 days subsequently absorbed
less PO_4^{3-} during a 4-hour exposure to ^{32}P than did untreated
roots (15). Intact roots of soybean absorbed less ^{32}P during a
90-minute exposure to ferulic acid than did unexposed roots (26).
Ferulic acid did not affect the percentage of absorbed ^{32}P that
was translocated to the shoot. Thus, both an allelopathic leachate
(15) and a pure allelochemical (26) affected ^{32}P absorption during
short enough time periods that growth would have been minimal.
These two experiments provide the strongest data showing that
allelochemicals can inhibit mineral absorption by intact plants.

Inhibition of Mineral Absorption in Excised Roots. More conclusive
evidence that allelochemicals can inhibit mineral absorption has
been obtained using purified allelochemicals and excised plant roots
as the experimental system (Table 1). Use of excised roots
eliminates the possibility that exists with intact plants that
inhibition of translocation rather than absorption is responsible
for decreased mineral content. Use of purified allelochemicals
rather than plant debris or leachates allows more definitive
conclusions to be reached regarding the capacity of allelochemicals
to inhibit mineral absorption.
        Several general characteristics of the results compiled in Table
I are worthy of mention. Compared to the variety of chemicals
postulated to be involved in allelopathy (1), few specific compounds
have been tested for inhibition of mineral absorption. The most
extensively studied compounds are the phenolic acids, probably
because of their being ubiquitously found in nature (1). Also,
several flavonoids are inhibitory to mineral absorption (Table I).
Both of these groups of compounds are often cited as being
responsible for allelopathic interactions between plants.
        Absorption of both cations and anions can be inhibited by
various allelochemicals. Most studies have been conducted with
$^{86}Rb^+$ (a tracer for K^+) or PO_4^{3-} but Cl^- and NO_3^- absorption
(34) and even phenolic glycoside absorption (27) was shown to be
inhibited by several flavonoids . The lack of specificity of these

Table I.   Effects of naturally-occurring phenolic compounds on mineral absorption in excised roots.

CHEMICAL CLASS Chemical	Plant Species[a]	Mineral ion	Chemical Treatment Conc.	Time	Ref.
SIMPLE PHENOLS					
Hydroquinone	Hv	K^+	5 mM	3 hr	(27)
PHENOLIC ACIDS [b]					
BENZOIC					
Salicylic acid	Hv	PO_4^{3-}	500 μM	3 hr	(28-30)
	Hv	K^+	250 μM	3 hr	(31)
	As	K^+	500 μM	1 hr	(32)
CINNAMIC					
Ferulic acid	Hv	PO_4^{3-}	500 μM	3 hr	(28, 30)
	Hv	K^+	250 μM	3 hr	(31)
	Gm	PO_4^{3-}	500 μM	1 hr	(26)
	As	K^+	500 μM	1 hr	(32)
NAPHTHOQUINONES					
Juglone	As	K^+	100 μM	10 min	(33)
FLAVANONES					
Naringenin	Ta	Cl^-, NO_3^-			
		PO_4^{3-}	10 μM	4 hr	(34)
ISOFLAVONES					
Genistein	Ta	Cl^-, NO_3^-	10 μM	4 hr	(34)
FLAVONOLS					
Kaempferol	As	K^+	100 μM	10 min	(33)
DIHYDROCHALCONES					
Phloretin	As	K^+	100 μM	10 min	(33)

[a] Abbreviations of plant species: As, Avena sativa; Gm, Glycine max; Hv, Hordeum vulgare; Ta, Triticum aestivum.

[b] Several additional benzoic and cinnamic acid derivatives were tested (28-31).

chemicals towards individual minerals needs to be studied further.

Another limitation to the studies in Table I is the small number of plant species tested. Primarily monocotyledonous plants have been studied, although McClure et al. (26) found ferulic acid inhibitory in soybean. The restriction of studies to monocots is probably because the mechanism of mineral absorption has been more fully elucidated with monocots. Harper and Balke (32) reported some minor differences in the inhibition of K^+ absorption by salicylic acid among oats (Avena sativa L.), wheat (Triticum aestivum L.), barley, and maize roots.

Chemical concentrations of 0.1 to 0.5 mM were used routinely for these studies (Table I), but concentrations as low as 10 μM (naringenin and genistein) (34) and as high as 5 mM (hydroquinone) (27) were inhibitory. Phenolic compounds in the 10 to 100 μM concentration range have been extracted from soil (35, 36). Thus, it would appear that most of these studies have been performed at the upper limit of phenolic concentrations that can be expected in soils. However, because several phenolics are usually found together in soils and many phenolics inhibit mineral absorption (28, 31), additive or even synergistic inhibition from mixtures of phenolics can be expected. Also, pH has a dramatic effect on the degree of inhibition produced by a given concentration of a phenolic acid. As pH was decreased, K^+ absorption was inhibited more by salicylic acid (Figure 1). Thus, at acidic pH's much lower concentrations of phenolic acids are required to inhibit mineral absorption.

The short time periods (10 min to 4 hr) over which absorption was measured (Table 1) helps support the hypothesis that certain allelochemicals inhibit mineral absorption directly. Under acidic conditions (pH 4.0) salicylic acid inhibited K^+ absorption within 1 min (32). The degree of inhibition remained constant over time when salicylic acid inhibited K^+ absorption (32) and when vanillic acid inhibited PO_4^{3-} absorption (28). Thus, at least phenolic acids appear to inhibit absorption rapidly and consistently.

Two additional characteristics of the inhibition of mineral absorption by phenolic acids were observed. The inhibition of both PO_4^{3-} absorption (27) and K^+ absorption (31, 32) was reversed when the phenolic acid was removed from the absorption solution. Harper & Balke (32) found this reversibility to be dependent upon pH; the lower the pH, the less the reversal. Also, kinetic plots of the inhibition of mineral absorption showed that the phenolic acids did not competitively inhibit either PO_4^{3-} (26, 28) or K^+ (31) absorption. Rather, ferulic acid inhibited PO_4^{3-} absorption in a noncompetitive (26) or uncompetitive (28) manner and p-hydroxybenzoic acid inhibited K^+ absorption in an uncompetitive manner (31).

Two studies have used single cells to study the effect of phenolic acids on mineral absorption. In sterile cell cultures of Paul's Scarlet rose, 100 μM ferulic acid inhibited Rb^+ absorption in about 10 min when the cells were 4-5 days old (37). Uptake from 0.2 mM RbCl was inhibited about 25% and absorption from 5.0 mM RbCl was inhibited 45%. Absorption by 10-day-old cells was affected little. Salicylic acid at 10 μM inhibited PO_4^{3-} absorption by Scenedesmus, a unicellular green alga (38). These studies show that allelochemicals inhibit mineral absorption in cellular systems as well as tissue systems (Table I).

Mechanism(s) of Action

It would be valuable to determine how allelochemicals inhibit
mineral absorption because such knowledge might indicate how to
control the inhibition. Several recent reviews outline the
fundamental biochemical and biophysical aspects of mineral
absorption by plant roots (39-41), and the reader is referred to
them for detailed discussion of the process of mineral absorption.
Mineral ions move into cells in response to an electrical potential
difference (PD) that is maintained across the plasma membrane and
tonoplast. Part (diffusion potential) of this PD arises from
different permeabilities of cations and anions through the
membranes, but the major portion (electrogenic potential) of the
potential is produced by an electrogenic pump. The electrogenic
pump moves electrical charge (ions) across the membrane. These
pumps require metabolic energy and are believed to be ATPase (ATP
phosphohydrolase) enzymes located in the plasma membrane and
tonoplast. In root cells these ATPases utilize ATP produced by
mitochondria and are believed to pump H^+ out of the cytoplasm.
Based on this model of active mineral absorption, one can
hypothesize several ways that allelochemicals could inhibit mineral
absorption: (1) alter the PD, (2) inhibit ATPases, (3) decrease
cellular ATP content, and (4) alter membrane permeability to ions.

Alteration of Electrical Potential (PD).

Study of the influence of
allelochemicals on the electrical potentials across plant cell
membranes has been restricted to phenolic acids. Glass and Dunlop
(42) reported that at pH 7.2, 500 µM salicylic acid depolarized the
electrical potential in epidermal cells of barley roots. The
electrical potential changed from −150 mV to −10 mV within 12 min.
Recovery of the PD was very slow over about 100 min when the
salicylic acid was removed. As the concentration of the
allelochemical was increased, the extent of depolarization
increased, but the time required for depolarization and recovery
were constant.

Salicylic acid depolarized PD in epidermal cells of oat roots
also. At pH 4.5, 500 µM salicylic acid caused a transient
hyperpolarization followed by a dramatic depolarization to about −45
mV (Figure 2). Removal of salicylic acid produced a transient,
partial repolarization. At pH 6.5, salicylic acid did not affect
PD. These results with different pH's are consistent with the
influence of salicylic acid on K^+ absorption in oat roots (32).

Other lipophilic weak acids have been shown to alter PD in plant
cells. Benzoic and butyric acids (1 µM) rapidly depolarized the PD
in oat coleoptile cells at pH 6.0 to about −100 mV (43). Higher
concentrations (10 mM) of butyrate produced hyperpolarization.
Butyrate also hyperpolarized apical cortical cells of maize roots
(44). The hyperpolarization was accentuated at pH 5.5 compared to
pH 6.4.

In cells of the fungus Neurospora crassa, 5 mM butyrate (pH 5.8)
produced a transient hyperpolarization followed by depolarization
(45). The chemical also increased the pH of the cytoplasm of the
cells. Whether phenolic acids also increase cytoplasmic pH is not

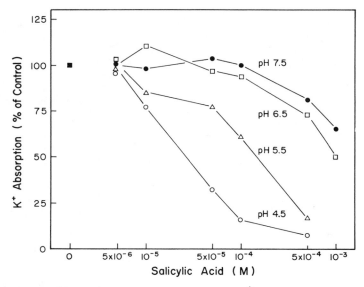

Figure 1. Influence of salicylic acid on K^+ absorption by excised oat roots at four pH values. Reproduced with permission from Ref. 32. Copyright 1981, American Society of Plant Physiologists.

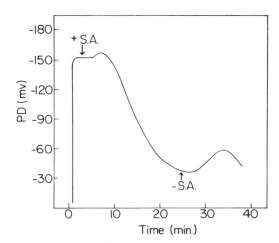

Figure 2. Influence of 500 μM salicylic acid on the electrical potential (PD) of epidermal cells of excised oat roots at pH 4.5. Microelectrode was inserted in a cell at 1 min; salicylic acid was added at 2.5 min and removed by flushing at 25 min.

known. In general, the concentrations required to change the pH
were greater than those required to depolarize root cells (42).

Benzoic acid derivatives also altered the electrical potential
across the cell membrane in neurons of the marine mollusk Navanax
inermis (46). Salicylic acid (1–30 mM) caused a depolarization very
rapidly (1–2 min) and decreased the ionic resistance across the
membrane. As pH was decreased, more salicylic acid was required to
reverse the effect of pH on the membrane potential (47). This
result is contradictory to the influence of pH on the amount of
salicylic acid required to affect mineral absorption in roots (32).
The ability of a series of salicylic and benzoic acid derivatives to
increase PD correlated with their octanol/water partition
coefficients and pK_a values (48). The authors proposed that the
organic acid anions bound directly to membranes to produce the
observed results.

The influence of these phenolic acids on electrical potentials
may reflect effects on either the diffusion potential or the
electrogenic potential of plant root cells. Influence on the
electrogenic component could result from inhibition of ATPases which
generate the electrogenic component or from reductions in the
substrate (ATP) for the ATPases.

Effects of Allelochemicals on ATPases. Several flavonoid compounds
inhibit ATPase activity that is associated with mineral absorption.
Phloretin and quercetin (100 μM) inhibited the plasma membrane
ATPase isolated from oat roots (33). The naphthoquinone juglone was
inhibitory also. However, neither ferulic acid nor salicylic acid
inhibited the ATPase. Additional research has shown that even at 10
mM salicylic acid inhibits ATPase activity only 10–15% (49). This
lack of activity by salicylic acid was substantiated with the plasma
membrane ATPase isolated from Neurospora crassa (50); however, the
flavonols fisetin, morin, myricetin, quercetin, and rutin were
inhibitory to the Neurospora ATPase. Flavonoids inhibited the
transport ATPases of several animal systems also (51–53). Thus, it
appears that flavonoids but not phenolic acids might affect mineral
transport by inhibiting ATPase enzymes.

Effects of Allelochemicals on ATP Supply. Allelochemicals might
decrease the ATP content of tissue by either increasing ATP
utilization or decreasing ATP production. Some allelochemicals that
inhibit mineral absorption decrease ATP content of plant tissues.
Salicylic acid decreased the ATP content of oat roots in a pH
dependent manner (Figure 3). This result suggested that
mitochondrial production of ATP was decreased in the tissue. On the
other hand, Tillberg (38) found that salicylic acid and cinnamic
acid increased the ATP content of Scenedesmus. Various flavonoids
inhibited ATP production by mitochondria isolated from cucumber
(Cucumis sativus L.) hypocotyls (54). Flavones such as kaempferol
were more inhibitory than the corresponding flavanones. Substituted
cinnamic acids such as caffeic acid were not inhibitory.
Mitochondria from cucumber roots, pea (Pisum sativum L.) roots, and
maize coleoptiles reacted in a manner similar to mitochondria from
cucumber hypocotyls. Hence, it appears that various allelochemicals
can produce different effects on ATP production.

Because mitochondria are the primary source for ATP production

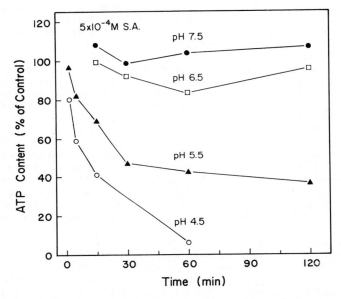

Figure 3.   Influence at 500 μM salicylic acid on ATP content of
excised oat roots at four pH values.   ATP determined by luciferin/
luciferase assay.

in nonchlorophyllous tissues, alteration of mitochondrial activity by allelochemicals could affect ATP production. Various allelochemicals have been found to inhibit respiration (O_2 consumption) by mitochondria (55-57) and uncouple oxidative phosphorylation from respiration (56, 58). In addition, O_2 consumption by tissue was affected by allelochemicals (6, 55, 56, 60-62). Thus, allelochemicals can decrease ATP production by inhibiting mitochondrial activities. Because mitochondrial reactions involved in ATP production take place on intramitochondrial membranes, it is quite possible that the manner by which allelochemicals inhibit mineral transport across the plasma membrane and the tonoplast and the manner by which they inhibit mitochondrial activities are similar. A strong possibility is that allelochemicals alter the permeability of membranes in both instances.

Alteration of Membrane Permeability. The ability of allelochemicals to alter membrane permeability and thus inhibit mineral absorption has been investigated in detail with only phenolic acids. Salicylic acid induced the efflux of PO_4^{3-} (28) and K^+ (42) from barley roots, but p-hydroxybenzoic acid did not cause the efflux of K^+ (31). At pH 6.5 salicylic acid produced only a slight, short-lived increase in K^+ efflux from oat roots, but at pH 4.5, the compound caused dramatic and extensive loss of K^+ from the tissue (Figure 4). Salicylic acid caused K^+ to leak from yeast cells also (63, 64). In addition to minerals, organic metabolites that absorb light at 260 nm leaked from oat roots in the presence of salicylic acid at pH 4.5 but not pH 6.5 (32).

Two hypotheses have been proposed to explain how phenolic acids directly increase membrane permeability. The first is that the compounds solubilize into cellular membranes, and thus cause a "loosening" of the membrane structure so that minerals can leak across the membrane (28-30, 42). Support for this hypothesis comes from the fact that the extent of inhibition of electrical potentials correlates with the log P (partition coefficient of a compound between octanol and water) for various benzoic and cinnamic acid derivatives (Figure 5).

In both Navanax neurons (65) and an artificial phospholipid bilayer membrane (66), salicylic acid (1-30 mM) increased K^+ permeability but decreased Cl^- permeability resulting in a net increase in membrane conductance. To account for the selective effect of salicylic acid (and other benzoic acids) on the two permeabilities, it was proposed that the anions of the organic acids adsorb to membranes to produce either a negative surface potential (66) or an increase in the anionic field strength of the membrane (47, 48).

It remains to be determined if one or both of these hypotheses are correct for plant roots. One feature of the inhibition of absorption by salicylic acid (and probably other phenolic acids) that may be relevant to this point is whether the neutral acid or the anion is responsible for the inhibition. In oat roots, the amount of neutral acid present when salicylic acid caused 50% inhibition of K^+ absorption was constant regardless of pH (Figure 6). However, the concentration of anion present changed several orders of magnitude. This result suggests the neutral acid is the species

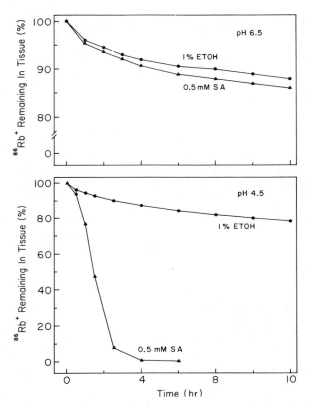

Figure 4.   Retention of K^+ ($^{86}Rb^+$) in excised oat roots in the
presence and absence of 0.5 mM salicylic acid at pH 6.5 and
pH 4.5.   The control contained 1% ethanol (ETOH).

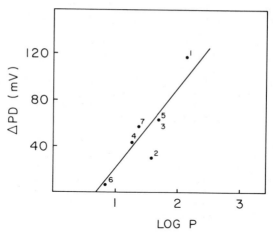

Figure 5. Relationship of depolarization of membrane electrical potential (E) in excised roots of barley 250 μM benzoic acids against their partition coefficients (log P). The numbers correspond to specific benzoic acid derivatives: (1) salicylic, (2) p-hydroxybenzoic, (3) protocatechuic, (4) gentisic, (5) vanillic, (6) gallic, (7) syringic acids. Adapted with permission from Ref. 42. Copyright 1974, American Society of Plant Physiologists.

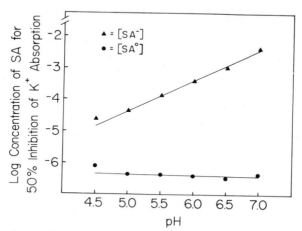

Figure 6. Relationship between the concentrations of neutral salicylic acid (SA^o) and salicylate anion (SA$^-$) present when K$^+$ (^{86}Rb$^+$) absorption by excised oat roots was inhibited 50% at six pH values.

that is important for the inhibition and supports, but does not
prove, the hypothesis that "leaks" are produced in plant root
membranes by salicylic acid.

A third possibility is that mineral ions leak out of tissue in
the presence of phenolic acids, not because membrane permeability is
altered, but rather because the driving force that maintains high
ion concentrations in cells (i.e. PD) is dissipated by the
chemicals. Without an electrical potential, ions would distribute
solely according to their chemical concentrations. Thus, most ions
would leak out of cells to reach chemical equilibrium with the
external environment.

It is quite possible that phenolic acids may produce more than
one effect on the cellular processes responsible for mineral
absorption. The potential sites of action discussed above all
involve cellular membranes in some way. Which mechanism of action
is predominant in a given situation may depend upon the
concentration of allelochemicals present and the conditions (e.g.
pH) of the plant/chemical interaction.

Summary and Conclusions

Although several allelochemicals (primarily phenolic acids and
flavonoids) have been shown to inhibit mineral absorption, only the
phenolic acids have been studied at the physiological and
biochemical levels to attempt to determine if mineral transport
across cellular membranes can be affected directly rather than
indirectly. Similar and even more definitive experiments need to be
conducted with other allelochemicals that are suspected of
inhibiting mineral absorption. Membrane vesicles isolated from
plant cells are now being used to elucidate the mechanism of mineral
transport across the plasma membrane and tonoplast (67, 68). Such
vesicle systems actively transport mineral ions and thus can serve
as simplified systems to directly test the ability of
allelochemicals to inhibit mineral absorption by plant cells.

By no means have all the potential allelochemicals produced by
plants been isolated and identified. There are certainly unknown
compounds that are active at very low concentrations. When these
compounds are purified, physiological bases for their action will
need to be determined. To say that inhibition of mineral absorption
is the sole, primary mechanism of action of allelochemicals is
unjustified. However, undoubtedly some allelochemicals can inhibit
mineral transport and alter other membrane phenomena.

Acknowledgments

Supported by the College of Agricultural and Life Sciences,
University of Wisconsin, Madison and SEA/USDA Grant
78-59-2551-0-1-032-1 from the Competitive Research Grants Office.

Literature Cited

1. Rice, E.L. "Allelopathy". 2nd Ed.; Academic Press: Orlando,
    FL, 1984.
2. Cou, C-H.; Young, C-H. J. Chem. Ecol. 1975, 1, 183-93.

3. Nielsen, K.F.; Cuddy, T.; Woods, W. Can. J. Plant Sci. 1960, 40, 188-97.
4. Molisch, H. "Der Einfluss einer Pflanze auf die andere-Allelopathie": Fischer, Jena, 1937.
5. Schreiner, O.; Skinner, J.J. Bot Gaz. 1912, 54, 31-48.
6. Glass, A.D.M. Can. J. Bot. 1976, 54, 2440-44.
7. Luu, K.T.; Matches, A.G.; Peters, E.J. Agron. J. 1982, 74, 805-8.
8. Fisher, R.F.; Wood, R.A.; Glavicic, M.R. Can J. Forest Res. 1978, 8, 1-9.
8a. Stowe, L.G.; Osborn, A. Can. J. Bot. 1980, 58, 1149-53.
9. Chambers, E.E.; Holm, L.G. Weeds 1965, 13, 312-14.
10. Fisher, R.F. Forest Sci. 1979, 25, 256-60.
11. Bandeen, J.D.; Buchholtz, K.P. Weeds 1967, 15, 220-24.
12. Buchholtz, K.P. In "Biochemical Interactions Among Plants"; Natl. Acad. Sci.: Washington, D.C., 1971; pp. 86-89.
13. Wilson, R.E.; Rice, E.L. Bull. Torrey Bot. Club 1968, 95, 432-48.
14. Bevenge, D.I. Plant and Soil 1968, 29, 263-73.
15. Newman, E.I.; Miller, M.H. J. Ecol. 1977, 65, 399-411.
16. Walters, D.T.; Gilmore, A.R. J. Chem. Ecol. 1976, 2, 469-79.
17. Borner, H. Bot. Rev. 1960, 26, 393-424.
18. Bhowmik, D.C.; Doll, J.D. Agron. J. 1982, 74, 601-6.
19. Bhowmik, P.C.; Doll, J.D. Agron. J. 1984, 76, 283-88.
20. Young, C.C.; Bartholomew, D.P. Crop Sci. 1981, 21, 770-74.
21. Hall, A.B.; Blum, U.; Fites, R.C. J. of Chem. Ecol. 1983, 9, 1213-22.
22. Norby, R.J.; Kozlowski, T.T. Plant and Soil 1980, 57, 363-74.
23. Fuerst, E.P.; Putnam, A.R. J. Chem. Ecol. 1983, 9, 937-44.
24. Olmsted, C.E., III; Rice, E.L. Southwestern Naturalist 1970, 15, 165-73.
25. Chandler, G.; Goosem, S. New Phytol. 1982, 92, 369-80.
26. McClure, P.R.; Gross, H.D.; Jackson, W.A. Can. J. Bot. 1978, 56, 764-67.
27. Glass, A.D.M.; Bohm, B.A. Planta 1971, 100, 93-105.
28. Glass, A.D.M. Plant Physiol. 1973, 51, 1037-41.
29. Glass, A.D.M. In "Mechanisms of Regulation of Plant Growth", Bieleski, R.L.; Ferguson, A.R.; Cresswell, M.M., Eds.; Royal Soc. New Zealand: Wellington, 1974; pp. 159-64.
30. Glass, A.D.M. Phytochem. 1975, 14, 2127-30.
31. Glass, A.D.M. J. Exp. Bot. 1974, 25, 1104-13.
32. Harper, J.R. and Balke, N.E. Plant Physiol. 1981, 68, 1349-53.
33. Balke, N.E. Ph.D. Thesis, Purdue University, W. Lafayette, IN, 1977.
34. Stenlid, G. Physiol. Plant. 1961, 14, 659-70.
35. Whitehead, D.C. Nature 1964, 202, 417-18.
36. McPherson, J.K.; Chou, C-H.; Muller, C.H. Phytochem. 1971, 10, 2925-33.
37. Danks, M.L.; Fletcher, J.S.; Rice, E.L. Amer. J. Bot. 1975, 62, 749-55.
38. Tillberg, J-E. Physiol. Plant. 1970, 23, 647-53.
39. Lüttge, U.; Higinbotham, N. "Transport in Plants"; Springer-Verlag: New York, 1979.

40. Leonard, R.T. In "Advances in Plant Nutrition"; Tinker, P.B.; Läuchli, A., Eds.; Praeger Scientific: New York, 1984; in press.
41. Balke, N.E. In "Weed Physiology", Vol. I, Duke S.O., Ed.; CRC Press; Boca Raton, FL, 1984; in press.
42. Glass, A.D.M.; Dunlop, J. Plant Physiol. 1974, 54, 855-58.
43. Bates, G.W.; Goldsmith, M.H.M. Planta 1983, 159, 231-37.
44. Marre, M.T.; Romani, G.; Marre, E. Plant, Cell and Environment 1983, 6, 617-23.
45. Sanders, D.; Hansen, U-P.; Slayman, C.L. Proc. Natl. Acad. Sci. 1981, 78, 5903-7.
46. Barker, J.L.; Levitan, H. Science 1971, 172, 1245-47.
47. Barker, J.L.; Levitan, H. Adv. in Neurology 1974, 4, 503-11.
48. Levitan, H.; Barker, J.L. Science 1972, 178, 63-64.
49. Balke, N.E.; Harper, J.R.; Groose, R.W., unpublished data.
50. Bowman, B.J.; Mainzer, S.E.; Allen, K.E.; Slayman, C.W. Biochem. Biophys. Acta 1978, 512, 13-28.
51. Fewtrell, C.M.S.; Gomperts, B.D. Nature 1977, 265, 635-36.
52. Robinson, J.D. Molec. Pharmac. 1969, 5, 584-92.
53. Carpenedo, F.; Bortignon, C.; Bruni, A.; Santi, R. Biochem. Pharmac. 1969, 18, 1495-1500.
54. Stenlid, G. Phytochem. 1970, 9, 2251-56.
55. Muller, W.H.; Lorber, P.; Haley, B.; Johnson, K. Bull. Torrey Bot. Club 1969, 96, 89-95.
56. Koeppe, D.E. Physiol. Plant. 1972, 27, 89-94.
57. Koeppe, D.E.; Miller, R.J. Plant Physiol. 1974, 54, 374-78.
58. Demos, E.K.; Woolwine, M.; Wilson, R.H.; McMillan, C. Am. J. Bot. 1975, 62, 97-102.
59. Muller, W.H.; Lorber, P.; Haley, B. Bull. Torrey. Bot. Club 1968, 95, 415-22.
60. Patrick, Z.A.; Koch, L.W. Can. J. Bot. 1958, 36, 621-46.
61. Marinos, N.G.; Hemberg, T. Physiol. Plant. 1960, 13, 571-81.
62. Dedonder, A.; Van Sumere, C.F. Z. Pflanzenphysiol. 1971, 65, 70-80.
63. Scharff, T.G.; Perry, A.C. Proc. Soc. Exp. Biol. and Medicine 1976, 151, 72-77.
64. Hoeberichts, J.A.; Hulsebos, Th.J.M.; Van Wezenbeek, P.M.G.F.; Borst-Pauwels, G.W.F.H. Biochem. Biophys. Acta 1980, 595, 126-32.
65. Levitan, H.; Barker, J.L. Science 1972, 176, 1423-25.
66. McLaughlin, S. Nature 1973, 243, 234-36.
67. DuPont, F.M.; Bennett, A.B.; Spanswick, R.M. Plant Physiol. 1982, 70, 1115-19.
68. Churchill, K.A.; Sze, H. Plant Physiol. 1983, 71, 610-17.

RECEIVED August 20, 1984

Effects of Allelochemicals on Plant–Water Relationships

F. A. EINHELLIG, M. STILLE MUTH, and M. K. SCHON[1]

Department of Biology, University of South Dakota, Vermillion, SD 57069

Bioassays with grain sorghum [Sorghum bicolor (L.) Moench.] seedlings grown under summer glasshouse conditions demonstrated that leaf diffusive resistance increased and water potential decreased following treatments with ferulic acid, p-coumaric acid, and extracts from several allelopathic weeds. During the week following treatment with 0.5 mM ferulic or p-coumaric acid, sorghum leaf resistances indicated that stomates were almost closed. Effects on water potential were found at a lower treatment level, with 0.25 mM ferulic or p-coumaric acid-treated sorghum having midday leaf water potentials of approximately -10 bars, compared to -5 bars for controls. Water potential changes resulted from reductions in both osmotic potential and turgor pressure. Aqueous extracts from Kochia scoparia, Helianthus tuberosus, and Xanthium pensylvanicum caused growth reductions in sorghum that correlated with high diffusive resistances and low water potentials, with these effects found using extracts from 1 g fresh-leaf material in 60 ml of nutrient medium. Growth of sorghum in soil containing dried residue from these weeds was also reduced. The data suggest one mechanism of allelopathic action is a disruption of plant water balance.

The process of allelopathy implies that compounds (allelochemicals) of plant origin are released into the environment, subsequently modifying the growth and development of other plants. Definitive explanations of this process require an understanding of how allelochemical function. Some inhibitory allelochemicals indirectly influence the growth of vascular plants by their effects on nitrification and nitrogen fixation (1, 2), disease resistance or susceptibility (3), or mycorrhizal fungi (4). Alternatively, more direct allelochemical effects may occur.

[1] Current address: Mount Marty College, Yankton, SD 57078

Plants susceptible to inhibitory allelochemicals typically have reduced or delayed germination and stunted seedling growth. Some evidence indicates seedling growth is more sensitive to allelochemicals than seed germination (5), and long-term seedling growth may be the best indicator of sensitivity to a chemical (6). However, identifying the growth regulatory mechanisms at the system or cellular level has been elusive. Allelochemicals appear to alter a variety of physiological processes (7, 8), and it is difficult to separate primary from secondary effects. Determining the mechanisms of allelopathic action in a particular field situation is further confused by the fact that several different allelochemicals may be involved. While these substances may act in concert to reduce growth (5, 9, 10), in most cases the interaction of combinations of allelochemicals on growth or physiological processes has not been documented.

Investigation of allelopathy has often implicated phenolic derivatives of coumarin, cinnamic acid, and benzoic acid as growth inhibitors (7, 11). Early evidence indicated that scopoletin and several phenolic acids regulate growth by controlling the level of indole-3-acetic acid (12, 13). While there has been no comprehensive assessment of the physiological effect of a single substance, compounds from this group of chemicals have been reported to alter seedling metabolism in a variety of other ways. These include changes in photosynthetic rate (14, 15), stomatal function (16), chlorophyll content (17), respiratory rate (18, 19), flow of carbon (20), mineral uptake (21-24), and membrane permeability (25). Glass and Dunlop (25) reported that several phenolic acids cause membrane depolarization, an action that would logically have a direct effect on ion uptake and plant water balance. These actions could then trigger other metabolic effects, as suggested in Figure 1.

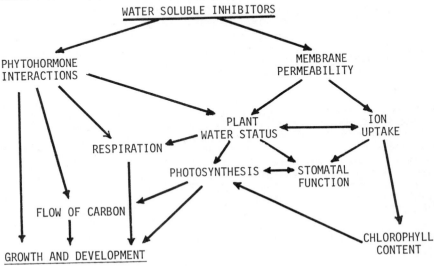

Figure 1.   Hypothetical action sequence suggesting allelochemical involvement in plant processes. Each arrow suggests a negative impact.

Experiments were designed to test the hypothesis that phenolic acids alter the water balance of seedlings, and that changes in plant water status would be correlated with inhibition of seedling growth. The allelopathic potential of several weeds common to agricultural fields was also tested using seedling bioassays, and water relationships of the bioassay plants were monitored.

Materials and Methods

Plant materials and growth conditions. Grain sorghum [Sorghum bicolor (L.) Moench., hybrid 701, Gurney's Nursery, Yankton, SD.] was used for the tests in these experiments. Grain sorghum was chosen because it is sensitive to phenolic acids (9, 17), and its growth may be influenced by allelochemicals from weeds in agricultural fields (26). Plants were grown under summer glasshouse conditions with normal variations in temperature and light, using supplemental lighting to obtain a 16 hr photoperiod.

Experiments with phenolic acids. Seedlings were grown in nutrient solution augmented with either ferulic acid (FA) or p-coumaric acid (pCA) for testing effects on water balance. Seeds were germinated in vermiculite for a 6 to 8 day period, then individual seedlings were transplanted into opaque-plastic vials (80 ml) containing nutrient medium. The nutrient medium was one-third strength Hoagland's (27) modified to contain 2.5 times the normal iron supplied as sodium ferric diethylenetriamine pentaacetate (Sequestrene 330). Seedlings were selected for uniformity and phenolic acid treatments were initiated 2 or 3 days following transplant. In each experiment there was a control and three phenolic acid regimes with 25 seedlings per group. Treatments were prepared by dissolving pCA or FA (Sigma Chemical Co.) in fresh nutrient medium to obtain concentrations of 0.25, 0.5, and 0.75 mM. The solutions were replaced with freshly prepared medium 4 days after treatment. Growth and water status of seedlings were monitored daily for 6 days following treatment.

Water status of the seedlings was determined each afternoon by obtaining leaf diffusive resistance, water potential, and osmotic potential. Diffusive resistance was measured on both the adaxial and abaxial surfaces of the youngest fully expanded leaf for six randomly selected plants in each treatment using a Lambda Model LI-60 meter and a narrow aperture sensor. Total leaf resistance (R) was calculated from the component resistances (r) as follows:

$$\frac{1}{R} = \frac{1}{r_{abaxial}} + \frac{1}{r_{adaxial}}$$

Light conditions each day were determined as an average of the intensity before and after taking the resistance readings.

Leaf water potential and osmotic potential were measured using a Wescor Dewpoint Microvoltmeter (Model HR-33) coupled with C-51 and C-52 sample chambers. Two plants from each group were sampled each day by taking two 7-mm diameter leaf disks from each plant, one for water potential and one for osmotic potential. Plants from which leaf disks were obtained were discarded. The water potential of a leaf disk was read following a 2-hr equilibration period in a sample

chamber. Osmotic potential of the leaf solution was determined after
freeze rupturing the tissue and saturating a filter paper disk which
was equilibrated in a sample chamber for 30 min. Turgor pressure was
calculated as the difference between water potential and osmotic potential. The average for each water parameter was computed for the
6-day period to facilitate analysis of the effects of the treatments.
    The dry weights (104 C, 48 hr) of ten plants from each treatment
group were taken at the termination of each experiment in order to
compare growth effects with plant water status. Dry weight data were
analyzed using analysis of variance (ANOVA) and Duncan's multiple-
range test. Diffusive resistance and water potential were evaluated
using the t-test. Each of these and subsequent experiments was replicated.

Aqueous-weed extract experiments. Fresh shoot material from the
leaves and small stems of Kochia [Kochia scoparia (L.) Schrad.],
Jerusalem artichoke (Helianthus tuberosus L.), and cocklebur
(Xanthium pensylvanicum Wallr.) were collected from plants in the
field. Aqueous extracts were made by boiling 10 g fresh weight of
plant material in 100 ml of water for 10 min., grinding in a Waring
blender, filtering through cheesecloth and Whatman No. 4 paper, and
adjusting the final volume to 100 ml (28). Thus 1 ml of the extract
contained the water-soluble substances from 0.1 g of leaf material.
The extract pH was adjusted to 5.5. Dilutions of an aqueous extract
were incorporated into nutrient medium for growing seedlings so that
the final ratios of fresh shoot or leaf weight to nutrient medium
(w/v) in the experiments ranged from 1:20 to 1:120. The nutrient
medium was a modified full-strength Hoagland's as described by
Einhellig and Schon (29), and all treatments had the same nutrient
concentration. The medium having the lowest osmotic potential among
the treatments was the Kochia 1:20 treatment, which was -1.4 bars.
The osmotic potentials for other treatments were in a range from -1
bar to -0.6 bar for the control.
    Preparation of seedlings for treatments with extract-amended
nutrient solution was similar to that described for testing the effects of phenolic acids, except 40 plants were used per treatment and
no replacement of the nutrient solution was made during the treatment
period. Data collection procedures were modified in that only abaxial leaf resistance was obtained and water potential was determined
from four plants each day. Prior work established that abaxial resistance provided an adequate indicator of stomatal effects. The
data were analyzed as described in experiments with pCA and FA.

Experiments with weed residue in soil. The effects of dried shoot
and leaf material from Kochia, Jerusalem artichoke, and cocklebur
were determined by growing sorghum in soil amended with these residues. Actively growing plants were collected from the field in
August, air dried in the glasshouse for one week, and the leaves and
smaller stems ground to pass through a 1-mm screen. This residue was
mixed into a fertile silty-clay soil having characteristics previously reported (29). The plant residue was added to air-dried soil so
that treatment soils contained either 2.5, 1.25, 0.63, or 0.31% weed
residue. These rates were estimated on the basis of the amount of
plant material found from sampling several quarter-meter quadrats
in areas where a particular weed was abundant. In the first of the

two trials conducted with each residue, the control soil was amended with 2.5% sphagnum moss as a nontoxic organic matter control (28). The growth containers used in the soil studies were the same plastic vials (3.2-cm diameter x 10-cm deep) utilized in the other experiments, and 80 g of soil was added to each container.

Thirty to 40 soil pots were used for each treatment group. Three sorghum seeds were planted 1-cm deep in the soil of each pot, and seedlings were thinned to one per pot after emergence. The soil pots were watered to field capacity each day by a weight procedure. Surface evaporation was reduced by covering the soil surface with a plastic disk. After emergence of the seedlings, a hole was punched in the plastic for the shoot. The growth period from germination to harvest was two weeks. Abaxial leaf resistance, water potential, and osmotic potential were obtained from sorghum on two or three days during the second week of growth. The water used by each plant was recorded and after harvest the transpiration ratio was computed. All data were analyzed as previously described.

Results

Experiments with phenolic acids.   Treatments with pCA and FA caused similar effects on the growth of grain sorghum seedlings. Seedlings grown for several days with either 0.75 or 0.5 mM pCA or FA in the medium had a wilted appearance, some chlorosis of the lower leaves, more visible anthocyanin accumulation than controls, and stunted growth. These conditions were most extensive in the 0.75 mM treatments, and these plants had folded leaves and proliferation of adventitious roots as the experiment progressed. The harvest dry weights show that both 0.75 and 0.5 mM levels of the two phenolic acids significantly inhibited growth and this inhibition was concentration dependent (Table I). The weights of plants treated with 0.25 mM pCA or FA were not significantly below the controls at the end of the 6 days of treatment, but a comparison to controls indicates that the 0.25 mM pCA and FA treatments may have had a slight effect.

The effects of pCA on plant water stress were similar in the replicate experiments, thus only the data from the second trial has been presented (Table II). Treatments with 0.75 and 0.5 mM pCA elevated leaf resistance of the sorghum seedlings after one day of treatment, and high resistances persisted through the duration of the experiment. Only a minor increase was induced by 0.25 mM pCA. Some of the day to day variation in resistance reflects differences in light intensity and temperature. All pCA treatments depressed sorghum-leaf water potential for the duration of the experiment. The summary data of the 6 days shows that even 0.25 mM pCA-treated seedlings had water potentials twice as negative as the controls, and the other treatments further lowered water potential. This depression of leaf water potential was due to both a lowering of osmotic potential and turgor pressure. The occasional negative turgor pressure calculated apparently occurred from the combined error arising in the measurement procedures, but the overall data confirm the loss of turgor that the wilted appearance of seedlings treated with 0.5 and 0.75 mM pCA suggested.

The effects of FA on the water status of grain sorghum seedlings were similar to that observed from pCA treatments, as illustrated in

TABLE I

Effects of Phenolic Acids on the Dry Weights of Grain Sorghum

Cpd.	Trial	Phenolic Acid Treatment				
		Control	0.25 mM	0.5 mM	0.75 mM
		-- mg ± SE --				
pCA	1	291.8 ± 22.4a	270.2 ± 15.1a	114.4 ± 8.3b	69.3 ± 4.8c
	2	151.3 ± 28.9a	135.8 ± 7.1a	86.3 ± 4.9b	65.1 ± 3.0c
FA	1	173.1 ± 10.6a	159.7 ± 11.2a	106.3 ± 6.9b	92.7 ± 6.1b
	2	217.6 ± 13.7a	213.6 ± 9.2a	125.9 ± 13.6b	91.3 ± 5.9c

Note: Values in a row not followed by the same letter are significantly
different, $P < 0.05$, ANOVA and Duncan's multiple-range test.

TABLE II

Effects of p-Coumaric Acid on the Water Status of Grain Sorghum

Days Treatment	pCA Treatments			
	Control	0.25 mM	0.5 mM	0.75 mM
	Total Leaf Resistance (Sec/cm ± SE)			
1	4.2 ± 0.4	5.7 ± 0.2[b]	32.5 ± 6.5[b]	47.7 ± 2.3[b]
2	5.3 ± 1.0	6.1 ± 0.6	15.1 ± 4.3[a]	48.0 ± 1.7[b]
3	4.6 ± 0.3	19.3 ± 7.4	28.2 ± 6.1[b]	31.1 ± 7.0[b]
4	4.6 ± 0.3	5.2 ± 0.4	10.3 ± 1.6[b]	36.0 ± 5.4[b]
5	5.6 ± 0.7	7.3 ± 0.7	20.2 ± 8.0	45.2 ± 3.5[b]
6	3.3 ± 0.3	6.5 ± 3.4	3.3 ± 0.2	7.7 ± 1.2[b]
6-day $\overline{X}$	4.6 ± 0.3	8.4 ± 2.2	18.3 ± 4.5[a]	36.0 ± 6.3[b]
	Water Potential (bars)			
1	- 5.5	-11.0	-14.2	-15.7
2	- 5.9	-11.0	-13.6	-16.0
3	- 3.7	-14.2	-12.7	-10.2
4	- 4.1	- 9.0	-10.4	- 8.1
5	- 2.8	- 9.7	-10.4	-10.5
6	- 5.4	- 8.3	- 9.7	- 9.5
6-day $\overline{X}$	- 4.6 ± 0.5	-10.6 ± 0.9[b]	-11.8 ± 0.8[b]	-11.7 ± 1.4[b]
	Osmotic Potential (bars)			
1	-10.9	-14.2	-12.4	-14.0
2	-11.3	-12.7	-12.4	-15.4
3	-10.4	-16.1	-11.3	-14.1
4	-10.7	-13.3	-12.4	-12.8
5	- 9.2	-15.1	-15.1	-14.1
6	- 8.7	-12.8	-12.5	-14.1
6-day $\overline{X}$	-10.2 ± 0.4	-14.0 ± 0.6[b]	-12.7 ± 0.4[b]	-14.1 ± 0.3[b]
	Turgor Pressure (bars)			
1	5.4	3.2	- 1.8	- 1.7
2	5.4	1.6	- 1.2	- 0.6
3	6.7	1.9	- 1.4	3.9
4	6.6	4.3	2.0	4.7
5	6.4	5.4	4.7	3.6
6	3.3	4.5	2.8	4.6
6-day $\overline{X}$	5.6 ± 0.5	3.5 ± 0.6[a]	0.9 ± 1.1[b]	2.4 ± 1.2[a]

Note -- Values each day are the mean of six seedlings for resistance
and two for water potential. Resistance data and the 6-day
mean for water potential were analyzed with the t-test.

[a]Differs significantly from the control, $P < 0.05$, [b]$P < 0.01$.

data from trial 2 (Table III). Ferulic acid caused an increase in
total leaf resistance and the extent of this action was concentration dependent. However, a treatment level effect was not measured
for the water potential parameters. Each day plants treated with all
levels of FA had a depressed leaf water potential, with the average
over the period near -12 bars, compared to approximately -5 bars for
controls. Similar to the findings with pCA, the low water potentials of groups of FA-treated seedlings resulted from both a lower
osmotic potential and reduced turgor pressure.

Aqueous-weed extract experiments. Observations in numerous agricultural fields indicated that Kochia, Jerusalem artichoke and cocklebur
might cause part of their interference with crop growth because of
their allelopathic nature. When the nutrient medium for growing sorghum was amended with aqueous-extracts from these weeds, sorghum
growth was inhibited and the extent of growth inhibition correlated
with the quantity of extract in the medium (Table IV). Kochia
amendments of 1 g fresh weight in 20 ml of medium (1:20) reduced
growth in both trials, and the 1:60 treatment significantly inhibited
sorghum growth in the second trial. Preliminary tests indicated the
strong toxicity of extracts from Jerusalem artichoke and cocklebur,
and lower dilutions of these extracts were tested. Even the lowest
dilution tested from these two weeds, 1:120, caused significant inhibition of sorghum growth. Plants grown with the several extracts
exhibited chlorosis, necrotic spots on leaves, leaf folding, and
reduced root development during the treatment period. The most obvious symptoms occurred with the higher extract amendments, and several plants in these groups were near death by the termination of the
experiment.
    Data summarizing the effects of the inhibitory weed extracts on
sorghum water status was taken from the second trial in each case
(Table V). All of these extract treatments had reduced sorghum
growth, and all but one had some effect on plant water balance.
Kochia amendments of 1:20 resulted in diffusive resistances that indicated stomates were closed, and the water potentials of the sorghum
seedlings approached -20 bars. The water potential of plants in the
1:60 Kochia treatment was also significantly below that of control
plants as the result of the low osmotic potential. Sorghum plants
grown in the 1:60 extracts from Jerusalem artichoke had significantly
higher diffusive resistances than controls and a depressed water potential. However, no significant differences in water status were
found in seedlings grown with Jerusalem artichoke treatments of 1:120.
Sorghum treated with the lowest cocklebur extract had higher resistances and lower water potentials than control plants, and these effects were more extreme in the higher treatments.

Experiments with weed residue in soil. Weed residue from each of the
three weeds had a negative effect on the growth of grain sorghum at
some level in the spectrum of soil treatments (Table VI). With one
exception, the degree of growth reduction in sorghum was correlated
with quantity of dried-weed residue mixed into the soil. Growth of
plants in soil amended with 0.63% Kochia was significantly less than
controls in the second trial, and the higher amendments reduced
growth in both trials. Residue from Jerusalem artichoke caused similar effects to that found from Kochia, with the 0.63% treatment level

TABLE III

Effects of Ferulic Acid on the Water Status of Grain Sorghum

Days Treatment	Control	0.25 mM	0.5 mM	0.75 mM
		FA Treatments		

Total Leaf Resistance (Sec /cm $\pm$ SE)

Days Treatment	Control	0.25 mM	0.5 mM	0.75 mM
1	3.7 + 0.5	9.4 + 2.8	13.2 + 2.6[b]	19.4 + 8.2[b]
2	3.8 $\mp$ 0.4	4.7 $\mp$ 0.2[b]	14.8 $\mp$ 4.0[b]	45.8 $\mp$ 1.4[b]
3	3.3 $\mp$ 0.3	6.8 $\mp$ 1.0[b]	20.2 $\mp$ 4.5[b]	31.4 $\mp$ 5.5[b]
4	4.1 $\mp$ 0.3	6.0 $\mp$ 0.6[a]	10.7 $\mp$ 3.2[b]	34.1 $\mp$ 5.6[b]
5	2.8 $\mp$ 0.2	3.9 $\mp$ 0.2[b]	5.8 $\mp$ 0.7[b]	31.0 $\mp$ 6.1[b]
6	5.1 $\mp$ 0.5	6.7 $\mp$ 0.6[a]	36.2 $\mp$ 8.1[b]	38.4 $\mp$ 6.2[b]
6-day $\overline{X}$	3.8 $\mp$ 0.3	6.3 $\mp$ 0.8[a]	16.8 $\mp$ 4.3[a]	33.4 $\mp$ 3.6[b]

Water Potential (bars)

Days Treatment	Control	0.25 mM	0.5 mM	0.75 mM
1	- 5.0	-10.2	-12.7	-10.9
2	- 6.6	-12.5	-15.8	-15.6
3	- 4.4	-12.0	-10.2	- 9.3
4	- 5.2	-12.8	-11.6	-12.5
5	- 4.7	-11.3	-11.3	-13.9
6	- 4.8	-11.4	- 9.4	- 9.7
6-day $\overline{X}$	- 5.1 $\pm$ 0.3	-11.7 $\pm$ 0.4[b]	-11.8 $\pm$ 0.9[b]	-12.0 $\pm$ 1.0[b]

Osmotic Potential (bars)

Days Treatment	Control	0.25 mM	0.5 mM	0.75 mM
1	-14.1	-16.0	-12.2	-16.7
2	-13.2	-14.5	-13.3	-16.2
3	-11.4	-14.1	-11.6	-13.0
4	-14.5	-15.0	-12.2	-15.6
5	-12.1	-16.6	-14.9	-16.7
6	-10.4	-13.1	-14.2	-14.5
6-day $\overline{X}$	-12.6 $\pm$ 0.7	-14.9 $\pm$ 0.5[a]	-13.1 $\pm$ 0.5	-15.5 $\pm$ 0.6[b]

Turgor Pressure (bars)

Days Treatment	Control	0.25 mM	0.5 mM	0.75 mM
1	9.1	5.8	- 0.5	5.8
2	6.6	2.0	- 2.5	0.6
3	7.0	2.1	1.4	3.7
4	9.3	2.2	0.6	3.1
5	7.4	5.3	3.6	2.8
6	5.6	1.7	4.8	4.8
6-day $\overline{X}$	7.5 $\pm$ 0.6	3.2 $\pm$ 0.8[b]	1.2 $\pm$ 1.1[b]	3.5 $\pm$ 0.7[b]

Note -- Values each day are the mean of six seedlings for resistance
and two for water potential. Resistance data and the 6-day
mean for water potential were analyzed with the t-test.

[a]Differs significantly from the control, $P < 0.05$,   [b]$P < 0.01$.

TABLE IV

Effects of Aqueous-Weed Extracts on the Dry Weight of Grain Sorghum

Weed Extract	Trial	Fresh Weight of Tissue in Nutrient Medium (g:ml)			
		Control	1:120	1:60	1:20
		------------- mg ± SE-------------			
Kochia	1	231.0 ± 10.2a	NT[1]	202.6 ± 16.6a	101.9 ± 4.1b
	2	264.4 ± 8.1a	NT	95.5 ± 5.5b	71.2 ± 4.1c
Jerusalem artichoke	1	138.7 ± 4.9a	90.0 ± 4.0b	75.7 ± 2.7c	NT
	2	166.1 ± 6.1a	104.5 ± 2.9b	59.0 ± 3.9c	NT
Cocklebur	1	238.3 ± 9.2a	NT	65.1 ± 3.7b	37.2 ± 1.8c
	2	235.3 ± 9.2a	117.0 ± 7.1b	72.1 ± 5.0c	NT

Note: Values in a row not followed by the same letter are significantly different, $p < 0.05$, ANOVA and Duncan's multiple-range test. Kochia data adapted with permission from Einhellig and Schon (29), Copyright 1982, the National Research Council of Canada.

[1]NT = No Test

TABLE V

Effects of Aqueous-Weed Extracts on the Water Status of Grain Sorghum

Weed Extract in Nutrient Medium (g:ml)	Abaxial Leaf Resistance (sec/cm)	Water Potential (bars)	Osmotic Potential (bars)	Turgor Pressure (bars)
Kochia				
Control	15.2 ± 2.1	$- 5.5 \pm 0.5$	-11.2 ± 0.3	5.8 ± 0.4
1:60	21.0 ± 2.5	$- 8.3 \pm 0.8^b$	-14.9 ± 1.1^b	6.6 ± 0.7
1:20	$> 100^c$	-19.7 ± 6.6^c	-23.6 ± 3.1^c	3.9 ± 4.0
Jerusalem artichoke				
Control	19.9 ± 1.5	$- 6.1 \pm 0.4$	-10.5 ± 0.2	4.3 ± 0.5
1:120	27.0 ± 3.7	$- 6.6 \pm 0.5$	-10.8 ± 0.5	4.2 ± 0.4
1:60	57.0 ± 7.4^c	$- 8.8 \pm 0.7^c$	-11.4 ± 0.4^b	2.8 ± 0.8
Cocklebur				
Control	16.6 ± 0.9	$- 5.5 \pm 0.3$	$- 9.4 \pm 0.3$	3.9 ± 0.5
1:120	26.2 ± 1.9^c	$- 9.5 \pm 1.4^b$	-11.8 ± 0.5^b	2.0 ± 1.7
1:60	57.6 ± 8.2^c	-16.1 ± 1.5^c	-13.5 ± 0.9^c	$-1.7 \pm 1.6b$

Note -- Values are the 6-day mean $\pm$ SE with diffusive resistance taken from six seedlings each day and water potential from four seedlings each day. Kochia data adapted with permission from Einhellig and Schon (29), Copyright 1982, the National Research Council of Canada.

[a]Differ significantly from the control of the group, $P < 0.05$;
[b]$P < 0.01$; [c]$P < 0.001$.

TABLE VI

Effects of Dried-Weed Residue in Soil on Dry Weight of Grain Sorghum

Trial		Control	Percentage Weed Residue in Soil		
			0.63%	1.25%	2.5%
			------------ mg ± SE ------------		
Kochia	1	171.7 ± 3.5a	162.1 ± 5.8a	129.9 ± 6.7b	107.9 ± 7.1c
	2	141.4 ± 4.9a	124.3 ± 4.8b	122.6 ± 4.6b	98.5 ± 3.1c
Jerusalem artichoke	1	69.4 ± 1.5a	70.0 ± 2.0a	66.7 ± 1.3ab	63.8 ± 1.7b
	2	78.4 ± 1.8a	67.0 ± 2.4b	69.5 ± 1.5b	60.9 ± 1.8c

Trial		Control	Percentage of Weed Residue in Soil		
			0.31%	0.63%	1.25%
Cocklebur	1	98.2 ± 1.6a	95.5 ± 1.8a	92.7 ± 2.3a	85.4 ± 1.6b
	2	135.2 ± 2.9a	104.3 ± 2.8c	116.9 ± 2.6b	105.1 ± 2.2c

Note: Values in a row not followed by the same letter are significantly different, $p < 0.05$, ANOVA and Duncan's multiple-range test. Kochia data adapted with permission from Einhellig and Schon (29), Copyright 1982, the National Research Council of Canada.

near the threshold for inhibition. Experiments with cocklebur encompassed a lower sequence of residue treatments. Soil amended with 1.25% cocklebur inhibited the growth of grain sorghum seedlings, and in one of the two replicates significant inhibition was found as low as 0.31%. Thus the residue of cocklebur appeared to have higher toxicity than <u>Kochia</u> or Jerusalem artichoke.

Leaf resistance and water potential parameters ascertained during the second week of seedling growth in residue-amended soil provided evidence of the effects of the weed residue on water balance. Data in Table VII was taken on the 10th day for the second trial with each plant. In the <u>Kochia</u> experiments, the leaf resistance, water potential, and osmotic potential of treated plants departed from the control values according to the quantity of extract in the soil. However, only plants grown in soil amended with 2.5% <u>Kochia</u> had resistances significantly higher and water potentials significantly below the controls. These findings are consistent with the trends found on other days of measurement. In contrast, residue from Jerusalem artichoke did not appear to alter the diffusive resistance or water potential of the sorghum seedlings. The water parameters of plants grown with cocklebur in the soil also were not statistically different from control plants. However, in the several days monitored in the two replicate experiments, the water potential of the 1.25% cocklebur-treated plants averaged 2 bars below the controls. Also, the transpiration ratio (data not shown) was higher in sorghum plants grown with the higher residues of both Jerusalem artichoke and cocklebur, indicating the residue had some impact on water relationships.

Discussion

The results of these experiments support the hypothesis that inhibitory allelochemicals may interfere with the water balance of seedlings. Alterations in the water status of grain sorghum seedlings were caused both by known phenolic acids and by allelopathic weeds.

Plants grown with growth-inhibitory dilutions of either pCA or FA had total leaf resistances that indicated partial stomatal closure, and the leaf water potential was severely depressed. These data do not allow precise cause and effect relationships to be determined, but the correlations are evident. Water potential was a sensitive and early indicator of stress from pCA or FA, as shown by the fact that 0.25 mM-treated seedlings had significantly reduced water potentials even though seedling growth was not stunted over the treatment period. Although some change in water potential appears to be possible without an effect on short-term growth, it is likely that a depression of water potential over an extended period would be detrimental.

Other work in our laboratory has shown that soybean seedlings subjected to growth-inhibitory treatments of pCA or FA have a depressed water potential and an increased diffusive resistance (<u>30</u>). Patterson (<u>31</u>) tested ten phenolic acids on three week old soybeans and found that the six compounds inhibiting growth also reduced stomatal conductance. Only half of these affected the water potential, indicating stomatal inhibition can be independent of an interference with water potential. Action of phenolic acids may also vary according to the age of the seedlings. We have found that the youngest

TABLE VII

Effects of Dried-Weed Residue on the Water Status of Grain Sorghum

Treatment of Weed Residue in Soil	Abaxial Leaf Resistance (sec/cm)	Water Potential (bars)	Osmotic Potential (bars)	Turgor Pressure (bars)
Kochia				
Control	20.6 ± 1.3	- 5.9 ± 0.7	- 8.6 ± 0.5	2.6 ± 0.7
0.63%	26.2 ± 4.2	- 6.1 ± 1.0	- 9.8 ± 0.3[a]	3.7 ± 1.2
1.25%	30.7 ± 4.5	- 8.5 ± 1.2	-11.6 ± 0.7[a]	3.0 ± 0.8
2.50%	50.0 ± 0.4[a]	-10.3 ± 1.6[a]	-12.7 ± 0.7[b]	2.4 ± 1.0
Jerusalem artichoke				
Control	25.5 ± 0.9	- 6.9 ± 1.3	- 9.9 ± 0.9	2.9 ± 1.1
0.63%	30.4 ± 3.7	- 8.5 ± 1.5	- 9.2 ± 1.0	0.7 ± 2.4
1.25%	29.1 ± 2.2	- 6.0 ± 0.7	- 9.2 ± 1.0	3.3 ± 0.8
2.5%	26.9 ± 2.0	- 7.5 ± 0.9	-10.1 ± 0.6	2.6 ± 0.4
Cocklebur				
Control	24.9 ± 1.7	- 7.3 ± 1.1	-10.2 ± 1.0	2.9 ± 1.4
0.31%	26.5 ± 2.8	- 7.6 ± 0.7	-10.5 ± 0.3	2.9 ± 0.7
0.63%	22.3 ± 1.4	- 6.7 ± 1.3	- 8.7 ± 0.6	2.0 ± 1.6
1.25%	26.4 ± 4.1	- 9.0 ± 1.0	-10.3 ± 0.4	1.2 ± 1.3

Note: Values were obtained on approximately the 10th day after
        germination. Resistance values are the mean ± SE of six
        seedlings and water potential values are from four seedlings.
        Kochia data adapted with permission from Einhellig and Schon
        (29), Copyright 1982, the National Research Council of Canada.

[a]Differ significantly from the control of the group, $P < 0.05$;

[b]$P < 0.01$.

seedlings are the most sensitive. Earlier investigations demonstrated that scopoletin, chlorogenic acid, and tannic acid caused stomatal closure in tobacco plants (16, 32). Thus, different test species and several different compounds show that interference with some aspect of water relationships often occurs from allelochemical stress. The extent of this interference may depend on other environmental factors (33, 34). Einhellig and Eckrich (34) found a significant interaction between FA stress and environmental temperature, with greater inhibition from FA at higher growth temperatures. It is likely that a temperature stress would cause water deficits in the plants, and the effects of FA would be magnified under these conditions.

Relating the effects caused by specific allelochemicals to those caused by an allelopathic plant is complicated because the inhibitory substances released from a plant are often unknown, and generally several different compounds are involved. However, the actions of the weeds studied in our investigations have certain parallels to those found from pCA and FA. The allelopathic nature of Kochia, Jerusalem artichoke, and cocklebur was established, since both aqueous extracts and weed residues reduced sorghum growth. The data show a concentration dependency characteristic of allelopathy, and some difference in toxicity among the three weeds was observed with cocklebur the most toxic.

Simultaneous with growth reductions, Kochia and cocklebur extracts interfered with stomatal function and water potential, and this correlation was also found with 1:60 Jerusalem artichoke treatments. The osmotic potential of the nutrient medium which contained the weed extracts was less than -1 bar in all cases except for the 1:20-Kochia treatments, indicating growth and water balance changes were not due to an osmotic effect (35, 36). Einhellig and Schon (29) found Kochia extracts had similar effects on soybean seedlings. Apparently, allelochemicals in the extract are the source of inhibition. Lodhi (37) reported that Kochia contains FA, chlorogenic acid, caffeic acid, myricetin, and quercetin. It is also likely that allelochemicals other than phenolics are involved. Identification of inhibitors from cocklebur has not been completed, so it is not possible to predict what combination of cocklebur allelochemicals are active. Other plants which interfere with water content concurrent with suppression in growth of the receiver species are Celtis laevigata Willd. (38), Abutilon theophrasti Medic. (36), and Helianthus annuus L. (39). Thus there is evidence from several studies that interference from allelopathic plants may impact on plant–water relationships.

Soil amended with 1.25% or more Kochia, Jerusalem artichoke, or cocklebur did not support normal growth of grain sorghum, and lower residue amendments were inhibitory in the second trial with each weed. The fact that a slightly lower residue level inhibited growth in the replicates without sphagnum moss in the control soil suggests some interference may have occurred from sphagnum. Minor differences in the threshold for toxicity between replicates may also result from variations in the glasshouse growth environment (34), or from the chance encounters of roots with areas of different toxicity in the rhizosphere (40).

Kochia residue levels of 2.5% significantly altered sorghum water balance, and plants grown in soil with cocklebur residues above the growth-inhibition threshold showed a trend toward elevated leaf resistances and lower water potentials than controls. A lower growth-inhibition threshold than found for measurable effects on water balance may be the result of the combined action of several inhibitory substances, with certain ones disrupting other aspects of metabolism. Numerous allelochemicals probably result from the decomposition of these weeds, and it is not likely they all have the same mode of action. Turner and Rice (41) noted that FA in soil may be transformed to vanillic acid, an inhibitor that has less impact on stomatal function and water potential (9, 31). In any case, the alteration in water balance that does occur is one mechanism of growth inhibition. Even the higher quantities of weed residue that were tested in these experiments are found in certain field situations, and their allelopathic action must be considered as one factor limiting crop production (42).

In summary, our results indicate that allelopathic stress may result in a change in plant water status. The action of specific allelochemicals, such as pCA and FA, lowers water potential by reducing turgor and osmotic pressure, and causes partial stomatal closure in young seedlings. Different allelochemicals in a complex matrix may have alternative modes of action, but any interference with water relationships would be damaging. Effects on water balance are likely to impede other physiological processes, with the combined action causing growth reduction.

Acknowledgment

This research was supported in part by a grant from the Office of Water Research and Technology, U. S. Department of Interior, grant No. B-061-SDAK-3556.

Literature Cited

1. Rice, E. L. Bot. Rev. 1979. 45, 15-109.
2. Rice, E. L.; Lin, C. Y.; Huang, C. M. J. Chem. Ecol. 1981. 6, 333-4
3. Patrick, A. A.; Toussoun, T. A. In "Ecology of Soil-Borne Plant Pathogens"; Baker, K. F.; Snyder, W. C.; Eds., Univ. of Calif. Press, Berkeley, 1965. p. 440-59.
4. Brown, R. T.; Mikola, P. Acta Forestalia Fennica. 1974. 141, p. 23.
5. Einhellig, F. A.; Schon, M. K.; Rasmussen, J. A. J. Plant Growth Reg. 1982. 1, 251-8.
6. Rietveld, W. S. J. Chem. Ecol. 1983. 9, 295-308.
7. Rice, E. L. "Allelopathy"; Academic Press, Inc.: Orlando, Florida, 1984, 2nd ed.; p. 422.
8. Horsley, S. B. Proc. Fourth N. Am. Forest Biol. Wksp. 1977. pp. 93-136.
9. Einhellig, F. A.; Rasmussen, J. A. J. Chem. Ecol. 1978. 4, 425-36.
10. Williams, R. D.; Hoagland, R. E. Weed Sci. 1982. 20, 206-12.
11. Putnam, A. R. Chem. Eng. News. 1983. 61, 34-45.
12. Andreae, W. A. Nature. 1952. 170, 83-4.
13. Zenk, M. H.; Muller, G. Nature. 1963. 200, 761-3.
14. Einhellig, F. A.; Rice, E. L.; Risser, P. G.; Wender, S. H. Bull. Torrey Bot. Club. 1970. 97, 22-33.

15. Kadlec. K. D. M. A. Thesis, Univ. of South Dakota, Vermillion. 1973. p. 42.
16. Einhellig, F. A.; Kaun, L. Y. Bull. Torrey Bot. Club. 1971. 98, 155-62.
17. Einhellig, F. A.; Rasmussen, J. A. J. Chem. Ecol. 1979. 5, 815-24.
18. Van Sumere, C. F.; Cottenie, J.; DeGreef, J.; Kint, J. Recent Add. Phytochem. 1971. 4, 165-221.
19. Demos, E. K.; Woolwine, M.; Wilson, R. H.; McMillan, C. Am. J. Bot. 1975. 62, 97-102.
20. Danks, M. L.; Fletcher, J. S.; Rice, E. L. Am. J. Bot. 1975. 62, 311-17.
21. Glass, A.D.M. Plant Physiol. 1973. 51, 1037-41.
22. Glass, A.D.M. J. Exp. Bot. 1974. 25, 1104-13.
23. Danks, M. L.; Fletcher, J. S.; Rice, E. L. Am. J. Bot. 1975. 62, 749-55.
24. Harper, J. R.; Balke, N. E. Plant Physiol. 1981. 68, 1349-53.
25. Glass, A.D.M.; Dunlop, J. Plant Physiol. 1974. 54, 855-8.
26. Rasmussen, J. A.; Einhellig, F. A. Am. Midl. Nat. 1975. 94, 478-83.
27. Hoagland, D. R.; Arnon, D. I. Calif. Agric. Ext. Serv. Circ. 347. 1950.
28. Wilson, R. E.; Rice, E. L. Bull. Torrey Bot. Club. 1968. 95, 432-48.
29. Einhellig, F. A.; Schon, M. K. Can. J. Bot. 1982. 60, 2923-30.
30. Einhellig, F. A.; Stille, M. L. Bot. Soc. Am. Misc. Pub.157. 1979. pp. 40-1.
31. Patterson, D. T. Weed Sci. 1981. 29, 53-8.
32. Einhellig, F. A. Proc. S. D. Acad. Sci. 1971. 50, 205-9.
33. Bhowmik, P. C.: Doll, J. D. J. Chem. Ecol. 1983. 9, 1263-80.
34. Einhellig, F. A.; Eckrich, P. C. J. Chem. Ecol. 1984. 10, 161-70.
35. Stowe, L. G. J. Ecol. 1979. 67, 1065-85.
36. Colton, C. E.; Einhellig, F. A. Am. J. Bot. 1980. 67, 1407-13.
37. Lodhi, M.A.K. Can. J. Bot. 1979. 57, 1083-8.
38. Lodhi, M.A.K.; Nickell, G. L. Bull. Torrey Bot. Club. 1973. 100, 159-65.
39. Schon M. K.; Einhellig, F. A. Bot. Gaz. 1982. 143, 505-10.
40. Patrick, Z. A. Soil Sci. 1971. 111, 13-18.
41. Turner, J. A.; Rice, E. L. J. Chem. Ecol. 1975. 1, 41-58.
42. Einhellig, F. A. Proc. Plant Growth Reg. Soc. Am. 1981. 8, 40-51.

RECEIVED June 12, 1984

Mechanisms of Allelopathic Action in Bioassay

G. R. LEATHER[1] and F. A. EINHELLIG[2]

[1] Agricultural Research Service, U.S. Department of Agriculture, Frederick, MD 21701
[2] Department of Biology, University of South Dakota, Vermillion, SD 57069

Bioassays to detect a wide range of concentrations
of allelochemicals were developed to follow allelopathic activity during compound identification and
to determine the biochemical mechanism(s) of plant
growth inhibition by the allelochemicals. Comparison of several bioassays for sensitivity to
phenolic acids, flavanoids, and coumarins showed
that the growth and reproduction of cultured Lemna
species was inhibited at concentrations as low as
50 μM. Other assays in order of decreasing
sensitivity were: sorghum seedling growth in
nutrient culture, seed germination, and radicle
elongation. The Lemna assay was developed using
24-well culture dishes that provided six
replications of each treatment. Beginning with
three to five fronds per well, the rate of
vegetative reproduction and growth rate were
determined over a seven-day culture period. Lemna
growing in treatments containing high concentrations
of phenolic acids (1000 μM) failed to produce new
fronds and lacked chlorophyll. Lower concentrations
(to 250 μM) reduced the growth rate 50% over a
seven-day period. Determination of chlorophyll in
Lemna minor and anthocyanin in Lemna obscura,
increased the sensitivity of this bioassay to 0.5 nM
concentrations of allelochemicals. The Lemna assays
were also useful for determining bioactive fractions
extracted from allelopathic plants.

Bioassays are useful tools for detecting physiological activity of
substances (allelochemicals) in plant and soil extracts and for
following activity as extracts are purified and the components
separated into various fractions. Frequently, a bioassay detects
physiological activity at concentrations much lower than the
sensitivity of chemical tests. The choice of a bioassay depends
upon the amount of chemical available for testing, the suspected
physiological activity of the allelochemical, and the sensitivity

0097–6156/85/0268–0197$06.00/0

needed for detection. A commonly used assay has entailed the use
of leachates from sand-nutrient cultures of suspected allelopaths
to irrigate target plants, using a stair-step array in which pots
containing donor plants are placed at a higher level than those
holding the target plants (1). Other variations of sand culture
include incorporation of plant residues into the rooting medium,
irrigation with nutrient solutions containing plant extracts, and
irrigation through surface-applied plant residues (2, 3). A great
number of investigators of allelopathy have used effects on seed
germination and seedling morphology in bioassays (3, 4, 5, 6).
Although germination bioassays are simple and expedient, they
often have been used without proper replication and statistical
analysis (7).

Other kinds of bioassays have been used to detect the
presence of specific allelochemical effects (8), effects on
N_2-fixation (9), the presence of volatile compounds (10) and of
inhibitory substances produced by marine microalgae (11). Putnam
and Duke (12) have summarized the extraction techniques and
bioassay methods used in allelopathy research. Recent
developments in high performance liquid chromatography (HPLC)
separation of allelochemicals from plant extracts dictates the
need for bioassays with sensitivity to low concentrations of
compounds contained in small volumes of eluent. Einhellig et al.
(13) described a bioassay using Lemna minor L. growing in tissue
culture cluster dish wells that maximizes sensitivity and
minimizes sample requirements.

The reported (14) mechanisms of action of allelochemicals
include effects on root ultrastructure and subsequent inhibition
of ion absorption and water uptake, effects on hormone-induced
growth, alteration of membrane permeability, changes in lipid and
organic acid metabolism, inhibition of protein synthesis and
alteration of enzyme activity, and effects on stomatal opening and
on photosynthesis. Reduced leaf water potential is one result of
treatment with ferulic and p-coumaric acids (15). Colton and
Einhellig (16) found that aqueous extracts of velvetleaf (Abutilon
theophrasti Medic.) increased diffusive resistance in soybean
[Glycine max. (L.) Merr.] leaves, probably as a result of stomatal
closure. In addition, there was evidence of water stress and
reduced quantities of chlorophyll in inhibited plants.
Einhellig and Rasmussen (17) reported that in addition to ferulic
and p-coumaric acids, vanillic acid reduced chlorophyll content of
soybean leaves but did not affect chlorophyll in grain sorghum
[Sorghum bicolor (L.) Moench.]. It is not known whether these
reported mechanisms are primary or secondary events in the
inhibition of plant growth by allelochemicals.

We investigated the value of different bioassays for
elucidating the mechanisms of action of allelochemicals in
inhibiting plant growth.

Materials and Methods

Seed germination. Tests for allelochemical inhibition of wild
mustard [Brassica kaber (DC.) L.C. Wheeler var. pinnatifida
(Stokes) L.C. Wheeler] seed germination were made according to the
methods of Leather (3). Germination of white clover (Trifolium

repens L.) and radish (Raphanus sativus L.) seeds was evaluated in 5-cm petri dishes with one ml of test solution which saturated a 4.9 cm disc of Whatman No. 3 filter paper. Each experiment contained 6 replications of 50 clover or 25 radish seeds. The germinator was maintained at 25° C with 8 hr of fluorescent light.

Radicle elongation. Pre-germinated sorghum seeds were placed ten each at the top of a 4.5 by 5.0 cm piece of Whatman No. 3 filter paper saturated with 1 ml of test solution. The seeds were positioned with the emerging radicle distal to the paper and aligned with the long axis of the paper. The paper with seeds was placed between the glass plates of a sandwich chromatography chamber and the plates clamped. The plates were placed at a 45° angle in a germinator for 4-6 hr. After the radicles made contact with the paper, the tip position was indicated by an ink mark on the glass. Measurements of radicle elongation were made 24 hr later. This procedure was a modification of the bioassay described by Parker (18).

Sorghum seedling growth. Seeds of sorghum were germinated in vermiculite under greenhouse conditions (35/25° C day/night temperature, 14 hr supplemental light from metal halide lamps). After 6 days, the seedlings were transplanted to 80-ml opaque vials filled with full strength Hoagland and Arnon's (19) solution containing 1.6 times the normal concentration of iron. After 3 days acclimatization, seedlings of uniform size were selected and treated with an allelochemical. Each treatment was replicated 15 times in a completely randomized design. The plants were harvested 7 days after treatment and dry weights of the roots and shoots determined.

Lemna bioassay. A bioassay for allelochemicals using Lemna minor L. (duckweed) growing in 24-well tissue culture cluster dishes was previously described by Einhellig et al. (13). In addition, tests were made with allelochemicals to determine their effects on chlorophyll production by L. minor. Chlorophyll content of duckweed was determined after 7 days in culture using the method described by Einhellig and Rasmussen (17). In other experiments, total anthocyanin production by L. obscura cultured as above was assayed. Harvested L. obscura were soaked in 3 ml of 0.1 M HCl for 4 hr in the dark. The solution was filtered and the absorbance of the supernatant at 550 nm was determined. The plant residue was dried at 70° C for 24 hr and weighed. Anthocyanin content was expressed as µg of anthocyanin per mg dry weight of L. obscura. In preliminary studies, we observed that ethanol reduced anthocyanin production in this bioassay. Since many crude fractions need to be solubilized in alcohol or other solvent (20), we examined the effects by adding 5 µl of ethanol solution to each well. All Lemna experiments contained 6 replications of each treatment and were repeated at least one time.

Results and Discussion

Seed germination. Germination of wild mustard seeds with a dormancy level of about 50 percent was stimulated after 10 days to

near 70 percent by 500 and 100 µM ferulic acid (Table I). High
concentrations (750 and 1500 µM) of benzoic acid inhibited wild
mustard germination throughout the 10 day test, but at 150 µM
germination was delayed at 3 days and by 10 days stimulation was
apparent.

TABLE I.   Effect of Ferulic (FA) and Benzoic (BA) Acids on
the Germination of Wild Mustard Seeds

Treatment	Germination[1]	
	3 day	10 day
	---------------- % ----------------	
H$_2$O Control	50.6 a	54.6 a
100 µM FA	60.6 a	69.7 c
500 µM FA	58.4 a	67.0 bc
1000 µM FA	58.4 a	60.0 ab
H$_2$O Control	50.6 c	54.6 c
150 µM BA	41.0 b	68.6 de
750 µM BA	12.0 a	12.0 b
1500 µM BA	3.0 a	3.0 a

[1] Means in each column followed by the same letter are not
significantly different at the 5% level of probability according
to Duncan's Multiple Range test.

The rate of germination of white clover seed was stimulated
by 100 µM ferulic acid through the initial 24 hr of the test
(Table II). Germination of clover was delayed by 500 µM ferulic
acid, but after 36 hr no differences were observed at 100 to 500 µM
concentrations. High concentrations (1000–2000 µM) were consistently inhibitory. Radish seed germination was not affected by
ferulic acid levels tested except at 24 hr when those in 500 µM
were less than controls.

TABLE II.   Effect of Ferulic Acid (FA) on the Germination
of White Clover and Radish Seed

Treatment	White clover			Radish		
	18 hr	24 hr	36 hr	18 hr	24 hr	36 hr
	-------------------- % germination[1] --------------------					
H$_2$O Control	30.0 bc	51.4 b	62.0 b	58.0 a	78.0 bc	86.7 a
100 µM FA	43.0 d	62.4 c	72.6 b	56.0 a	76.7 bc	88.7 a
250 µM FA	37.0 cd	71.0 c	74.6 b	44.7 a	67.3 ab	82.7 a
500 µM FA	19.6 a	47.6 ab	62.0 b	51.3 a	58.0 a	86.7 a
1000 µM FA	19.4 a	41.2 a	45.6 a	49.3 a	74.7 bc	85.3 a
2000 µM FA	21.4 ab	39.4 a	45.4 a	49.3 a	83.3 c	92.0 a

[1] Means in each column followed by the same letter are not
significantly different at the 5% level of probability according to
Duncan's Multiple Range test.

Radicle elongation. The radicle elongation test was the least
sensitive bioassay. Only very high (2000 µM) concentrations of
ferulic acid inhibited the radicle elongation of pre-germinated
grain sorghum (Table III). In earlier investigations we observed
inhibition of radicle elongation in this test using extracts of

sunflower (<u>Helianthus</u> <u>annuus</u> L.) at a level of 4 g/100 ml H_2O whereas log dilutions of the extract were not active (unpublished data).

TABLE III.   Effect of Ferulic Acid (FA) on Radicle
Elongation of Pre-germinated Grain Sorghum

Control	250 µM FA	500 µM FA	1000 µM FA	2000 µM FA
		mm + SEM after 24 hr		
25.8 + 1.0	28.7 + 1.7	24.8 + 1.7	23.1 + 1.0	16.6 + 0.9*

* Differs significantly from the control, P = 0.05

<u>Grain sorghum seedlings</u>. The growth of grain sorghum in nutrient culture was inhibited by 250 µM concentrations of salicylic acid (Table IV). Threshold concentrations were variable under greenhouse conditions but were consistently lower than in seed germination or radicle elongation experiments. Einhellig and Rasmussen (<u>17</u>) reported the threshold concentration for inhibition of grain sorghum by vanillic, p-coumaric, and ferulic acids, to be 500 µM. A recent study shows that 200 µM ferulic acid inhibits grain sorghum growth when day temperature is 37° C (21).

TABLE IV.   Effect of Salicylic Acid (SA) on Grain
Sorghum Seedlings Growing in Nutrient Solution

Concentration SA (µM)	Experiment 1	Experiment 2
	mg dry wt.[1]	
0	209.8 cd	227.9 cd
50	213.7 d	220.1 cd
100	184.6 bc	260.9 d
250	165.8 b	206.8 bc
500	76.0 a	172.4 b
1000	57.9 a	66.5 a

[1] Values are the means of 15 plants harvested 7 days after treatment. Means in each column followed by the same letter are not significantly different at the 5% level of probability using Duncan's Multiple Range test.

<u>Lemna bioassay</u>. The growth of <u>L</u>. <u>minor</u> in tissue culture dishes has been reported to have been inhibited by 50 µM concentrations of allelochemical (<u>13</u>). Results of experiments designed to compare the activity of different allelochemicals on the growth of <u>L</u>. <u>minor</u> are shown in Tables V and VI. All the compounds except chlorogenic acid decreased the final frond number and dry weight when supplied at 1000 µM concentration. However, 250 µM chlorogenic acid inhibited both of the measured growth parameters. The 250 µM concentration of esculetin was also inhibitory. These results compare with those reported previously for ferulic acid (<u>13</u>). Another important feature of this bioassay is the ability to detect stimulatory levels of compounds. This was shown in tests with catechin where stimulation of growth and frond production was measured at levels as low as 50 µM (Tables V and VI).

TABLE V. Comparison of Allelochemicals on the Final
Frond Number of L. minor Growing for 7 Days in
24-well Tissue Culture Cluster Dishes

Concentration µM	Catechin	Esculetin	Chlorogenic acid	Scopoletin	Vanillic acid
			Final frond no.[1]		
0	35.7 c	39.5 a	52.5 a	54.5 ab	43.5 a
50	59.3 a	39.3 a	45.7 ab	43.5 b	NS[2]
100	49.0 b	37.0 a	41.8 abc	55.0 ab	42.8 a
250	42.3 bc	23.0 b	36.0 bcd	58.0 a	36.2 ab
500	35.8 c	11.2 c	39.3 bcd	28.3 c	24.5 b
1000	10.3 d	4.0 c	40.8 abc	4.2 d	4.0 c

[1] Values are the means of 6 replications harvested 7 days after
treatment. Means in a column followed by the same letter are not
significantly different at the 5% level using Duncan's Multiple
Range test.
[2] No sample.

TABLE VI. Comparisons of Allelochemicals on the Dry Weight
of L. minor Growing for 7 Days in 24-well
Tissue Culture Cluster Dishes

Concentration µM	Catechin	Esculetin	Chlorogenic acid	Scopoletin	Vanillic acid
			Dry wt. (mg)[1]		
0	2.9 c	3.3 a	4.5 ab	3.9 ab	3.7 ab
50	5.4 a	3.6 a	4.8 a	3.4 b	NS[2]
100	4.5 b	3.1 a	3.6 ab	4.5 a	3.8 a
250	4.3 b	1.4 b	2.8 c	4.5 a	3.7 ab
500	3.2 c	0.6 b	4.2 ab	2.9 b	2.4 b
1000	1.9 d	0.3 c	4.6 a	0.6 c	0.2 c

[1] Values are the means of 6 replications harvested 7 days after
treatment. Means in each column followed by the same letter are not
significantly different at the 5% level using Duncan's Multiple
Range test.
[2] No sample.

Additional research with Lemna species has indicated that
parameters other than direct growth measurements may be more
sensitive in this bioassay. Table VII compares growth and
chlorophyll content of L. minor as affected by catechin. Final
frond number was inhibited by 1000 µM catechin and stimulated at
lower concentrations of 50 and 100 µM. Chlorophyll content on a
per-frond basis, however, was consistently inhibited by catechin and
was concentration dependent to 100 µM.

Lemna obscura fronds contain high quantities of anthocyanin
that are extractable by soaking the fronds in 0.1 M HCl. L. obscura
growth in the culture dish bioassay was similar to that of L. minor
and appeared to be more sensitive to low levels of allelochemicals.
Anthocyanin concentration was affected by low concentrations of
salicylic acid (Table VIII). Final frond number and dry weight were
consistently reduced by 100 µM and 500 µM concentrations of
salicylic acid whereas anthocyanin formation was inhibited by 50 µM
concentrations of salicylic acid and stimulated by 0.5 µM.

TABLE VII.   Effect of Catechin on the Growth and Chlorophyll
Content of <u>Lemna minor</u> L. Growing in 24-well
Tissue Culture Cluster Dishes

Concentration catechin, µM	Final frond no.	Chlorophyll (µg/frond)
	--------- % of control ---------	
50	125.1*	100.6
100	127.9*	80.8
250	110.0	61.8*
500	98.7	49.8*
1000	26.5*	21.3*

* Differs significantly from the control, P = 0.05.

TABLE VIII.   Effect of Salicylic Acid (SA) on Growth and
Anthocyanin Content in <u>Lemna obscura</u> L. Cultured
in 24-well Tissue Culture Cluster Dishes

Concentration SA, (µM)	Final frond no.	Final dry wt. (mg)	Anthocyanin (µg/mg dry wt.)
	----------------- % of control -----------------		
0.5	96.3	93.1	134.4*
1.0	93.1	104.5	88.4*
5.0	84.4*	93.7	94.0
10.0	100.5	127.2*	102.7
50.0	88.7*	123.5*	48.5*
100.0	62.4*	84.6*	30.6*
500.0	13.3*	56.2*	0.0*

* Differs significantly from the controls, P = 0.05 by t test.

The use of ethanol or other organic solvents for dissolving
plant allelochemicals to permit bioassay may influence the results
of sensitive bioassays. We examined the effects of several
different concentrations of ethanol on frond production and
anthocyanin content in <u>L. obscura</u>. The results show that both
final frond number and anthocyanin content were decreased when the
culture medium contained 0.133 or greater percent ethanol (v/v)
(Table IX).

TABLE IX.   Influence of Ethanol Amendments on Final Frond
Number and Anthocyanin Content of <u>L. obscura</u> Growing
in 24-well Tissue Culture Cluster Dishes

	Percent ethanol (v/v)[1]							
	0.0	0.017	0.033	0.067	0.133	0.200	0.267	0.333
Final frond no.	39.5	35.6	34.5	33.3	31.4	27.8	27.6	25.6
Anthocyanin (µg/mg dry wt.)	3.2	3.4	3.4	3.2	2.0	1.8	1.8	1.5

[1]Results are the means of two experiments, each value replicated
6 times.

The results of our experiments with several bioassays for the determination of allelochemical activity show differences in threshold levels for inhibition or stimulation. We have demonstrated that anthocyanin production by L. obscura growing in 24-well tissue culture cluster dishes is the most sensitive to allelochemicals. Other assays in order of decreasing sensitivity were: 1) chlorophyll production and growth of L. minor in tissue culture dishes, 2) sorghum seedling growth in nutrient culture, 3) seed germination, and 4) radicle elongation.

Selection of a bioassay depends on the growth parameters to be measured and the quantity of allelochemical available. For example, the sorghum seedling bioassay used in our studies may be the best for determining the uptake, distribution, and metabolism of known allelochemicals where quantity of compound is not a factor and the use of radiolabeled materials would be expedient. Additional investigation is required to determine if effects on these pathways are primary or secondary mechanisms of action and if the same mechanisms are applicable to more anatomically complex plants and under a range of environmental conditions.

The Lemna bioassay, as previously described (13), is particularly useful for the determination of bioactive fractions collected during HPLC of extracts from allelopathic plants. Activity has been detected in the Lemna bioassay with HPLC fractions of unknown allelochemicals from sunflower using quantities as low as 5 µl of a 28 ppm w/v solution (see Saggese et al. this publication).

Literature Cited

1. Bell, D. T., Koeppe, D. F. Agron. J. 1972, 64, 321.
2. Rice, E. L. "Allelopathy"; Academic: New York, 1974; pp. 353.
3. Leather, G. R. Weed Sci. 1983, 31, 37.
4. Lawrence, T., Kilcher, M. R. Can. J. Plant Sci. 1962, 42, 308.
5. Guenzi, W. T., McCalla, T. M. Proc. Soil Sci. Am. 1962, 26, 456.
6. Macfarlane, M. J., Scott, D., Jarvis, P. N. Zealand J. Agric. Res., 1982, 25, 503.
7. Lehle, F. R., Putnam, A. R. Plant Physiol. 1982, 69, 1212.
8. Fay, P. K., Duke, W. B. Weed Sci. 1977, 25, 224.
9. Kapustka, L. A., Rice, E. L. Soil Biol. Biochem. 1976, 8, 497.
10. Muller, C. H., Chou, C. H. In "Phytochemical Ecology"; Harborne, J. B., Ed.; Academic: London, 1972; pp. 201-16.
11. Chan, A. T., Andersen, R. J., LeBlanc, M. J., Harrison, P. J. Marine Biology. 1980, 59, 7.
12. Putnam, A. R., Duke, W. B. "Allelopathy in Agroecosystems"; Ann. Rev. Phytopathol 16; Annual Reviews Inc. 1978; pp. 431-51.
13. Einhellig, F. A., Leather, G. R., Hobbs, L. L. J. Chem. Ecol. 1984, 10, (in press).
14. Rice, E. L. Bot. Rev. 1979, 45, 15.
15. Einhellig, F. A., Stille, M. L. Bot. Soc. Amer. Misc. Publ. 1979, 157, 40.
16. Colton, C. E., Einhellig, F. A. Amer. J. Bot. 1980, 67, 1407.

17.   Einhellig, F. A., Rasmussen, J. A.   J. Chem. Ecol. 1979, 5, 815.

18.   Parker, C.   Weeds. 1966, 14, 117.

19.   Hoagland, D. R., Arnon, D. I.   "The water-culture method for growing plants without soil," Calif. Agric. Exp. Stn. Manual 347, 1950.

20.   Davis, D. G., Wergin, W. P., Dusbabek, K. E.   Pestic. Biochem. and Physiol. 1978, 8, 84.

21.   Einhellig, F. A., Eckrich, P. C.   J. Chem. Ecol. 1984, 10, 161.

RECEIVED June 12, 1984

Phytotoxic Compounds Isolated and Identified from Weeds

G. F. NICOLLIER[1,3], D. F. POPE[1], and A. C. THOMPSON[2]

[1]Center for Alluvial Plains Studies, Delta State University, Cleveland, MS 38733
[2]Boll Weevil Research Laboratory, Midsouth Area, U.S. Department of Agriculture, Agricultural Research Service, Mississippi State, MS 39762–5367

Water and methanol extracts of ninety weed and crop species were bioassayed at 3,30 and 300 ppm by weight for their effect on turnip (<u>Brassica</u> <u>rapa</u> L. 'Purple Top') root growth. Eighteen species were significantly inhibitory and six species were significantly stimulatory to turnip root growth. Several species showing activity were chosen for further chemical and biological investigation. Coumarin, <u>o</u>-coumaric acid and melilotic acid were found to be the major active compounds in white sweet clover (<u>Melilotus alba</u> Desr.). Investigation of related compounds showed that activity was inhanced by increasing the side chain length and by inserting ortho-substituted hydroxyl groups, while blockage of the ortho-hydroxyls decreased activity. Rhizome exudates and pure compounds identified in johnsongrass (<u>Sorghum</u> <u>halapense</u> (L.) Pers.) were active against radish (<u>Raphanus</u> <u>soliocus</u> L.) and tomato (<u>Lycopersicon</u> <u>asculentum</u> Miller) root growth as well as growth of nine bacterial species. Leaf extracts and leachates of Illinois bundleflower (<u>Desmanthus</u> <u>illinoensis</u> (Michaux) MacM.) were active against radish and tomato root growth, bacterial growth and tomato root growth, bacterial growth and tobacco budworm (<u>Heliothis</u> <u>virescens</u> F.) larval growth. Leaf extract activity varied widely due to the physiological age of the plant.

Ninety weed and crop species common to the southeastern U. S. were tested for phytotoxin content to discover candidates for an allelopathy testing program. Plants were collected in the spring, summer and fall in a three county area of northeastern Mississippi. Plants were maintained in sealed plastic

[3]Current address: Ciba-Geigy AG, CH-04002, Basel, Switzerland

bags at approximately 10°C until arrival at the laboratory
where they were immediately extracted. Plant materials were
ground in distilled water using a blender and allowed to stand
for 15 minutes. The extract was centrifuged, filtered and
dried under vacuum at 40°C. The water extracted plant residue
was further extracted with methanol under reflux for 3 hours.
The methanol extract was filtered off under vacuum and dried
under vacuum at 40°C. The residues from the water and methanol
extracts were dissolved separately in distilled water and
diluted to concentrations of 3, 30, and 300 ppm for bioassay.
These solutions were evaluated for phytotoxicity using a single
bioassay system. Germinated trunip (Brassica rapa L. 'Purple
Top') seed with roots 2-5 mm in length were placed on germination blotter paper in 9 cm square petri dishes (15
seedlings per dish) wetted with 9 ml of distilled water or test
solution. Root and shoot lengths were measured after incubation at 25°C for 48 hours. Bioassay results were interpreted
using Duncan's Multiple Range Test at the 5% level.

From the ninety weeds assayed, extracts from 18 gave
significant inhibition of turnip root growth, while 6 gave
significant stimulation (Table I.).

Table I.
Plants Yielding Inhibitory Extracts

Sida spinosa L.	Melilotus alba Desr.
Penstemon australis Small	Rudbeckia amplexicaulis Vahl.
Polygonum pensylvanicum L.	Sonchus asper (L.) Hill
Sorghum halapense (L.) Pers.	Lactuca canadensis L.
Carduus spinosissimus Walter	Euphorbia maculata L.
Desmodium canescens	Oenothera laciniata Hill
Desmanthus illinoesis (Michaux) MacM.	Rorippa sessiliflora (Nuttal Hitchcock)
Eupatorium rotundifolium L.	Vicia dasycarpa Tenore
Plantago lanceolata L.	Smilax bona-nox L.
Scutellaria parvula Michaux	

Plants Yielding Stimulatory Extracts

Geranium carolinianum L.	Conyza canadensis (L.) Cronq.
Helianthus hirsutus Raf.	Erigeron strigosus Muhl.
Erigeron philadelphicus L.	Boltonia diffusa Ell.

White Sweet Clover (Melilotus alba Desr.)

White sweet clover extracts were also active against
radish and tomato root growth (unpublished data). We found
that coumarin, o-coumaric acid and melilotic acid accounted for
the major portion of the crude extract activity (1). Thirteen
structurally similar compounds were investigated to determine
structure-activity relationships. Increasing the number of
carbons between the phenyl group and the carbonyl group on the
side chain phenyl substituted carboxylic acids increased the
inhibition of root growth (Figure 1). Ortho-substituted
hydroxyphenyl aliphatic acids were always more active than the

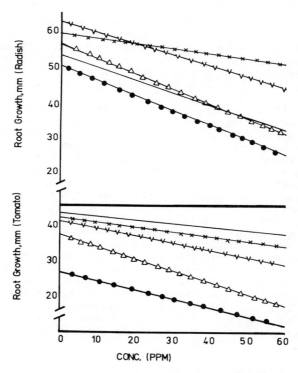

Figure 1. Linear regression plot of tomato and radish seedling root growth inhibition with varying concentrations of phenyl aliphatic acids. Benzoic acid (V); phenylacetic acid (X); 3-phenylpropanoic acid (Δ); 4-phenylbutanoic acid (●); trans-cinnamic acid (---) (1).

corresponding para-substituted acids (Figure 2). Blockage of
the ortho-hydroxyl group essentially eliminated the phytotoxic
properties. Two new flavonoids (melitin and clovin, Figure 3)
were characterized and studied, but were found to account for
only minor portions of the activity of crude white sweet clover
extracts.

Johnsongrass [Sorghum halapense (L.) Pers.]

Johnsongrass root and rhizome exudates and extracts have
been shown to be toxic to many plants species (2, 3, 4). Also,
Parks and Rice (5) found that some soil algae are inhibited by
rhizome extracts. This suggested that johnsongrass's competitive ability may be increased by it's ability to influence
the soil microflora and thereby indirectly affect other higher
plants.
    To clarify the nature and activity of these toxins, rhi-
zome exudates (washings), compounds isolated from rhizomes
(dhurrin, taxiphyllin, p-hydroxybenzaldehyde), and structurally
similar compounds were tested against radish and tomato root
growth and growth of nine species of bacteria. Figures 4 and 5
show the results of the radish and tomato bioassay of pure compounds. Dhurrin and toxiphyllin showed little activity in
these assays, whereas p-hydroxybenzaldehyde was active in the
radish bioassay.
    Dhurrin and taxiphyllin (Figure 6) were active against
growth of most of the bacteria tested, whereas p-hydroxybenzaldehyde was active against only 2 species (Table II). These
compounds might act to inhibit potentially destructive
microorganisms in the rhizosphere. While it is known that root
exudates affect the kind, frequency and proportions of the rhizosphere microflora; the rhizosphere microflora may also
markedly affect plant growth (6). Johnsongrass may be able to
influence the rhizosphere microflora in order to protect itself
from pathogens and stimulate growth of species favorable to its
own growth.

Desmanthus illinoensis (Michaux) MacM.

Desmanthus leaf extracts were strongly active in the survey bioassay. We have reported the isolation and identification of several flavonoids from the leaves along with
their growth inhibition and antibacterial properties (7).
Although all the flavonoids identified inhibited tomato root
growth, their concentration in the crude leaf extract does not
appear to be sufficient to account for the activity of the
crude extract. The 2"-gallic acid ester of myrcitrin was the
most active of the flavonoids tested in the tomato root growth
bioassay, bacterial bioassay and tobacco budworm larval growth
bioassay.
    Toxicity During Growing Season. It was observed that the
toxicity level of leaf leachates and extracts varied according
to the times sampled in the field and greenhouse. To determine

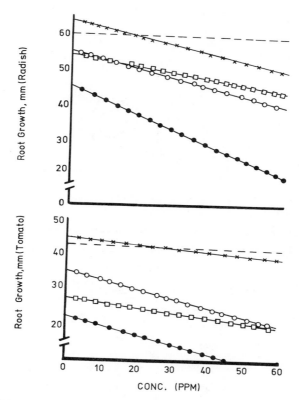

Figure 2. Linear regression plot of root growth inhibition of tomato and radish seedlings by o and p-hydroxyphenyl) acids. o-Hydroxycinnamic acid, (o-hydroxyphenyl) acetic acid (□); melilotic acid, 2-(o-hydroxyphenyl) butanoic acid (●); o-hydroxy-benzoic acid, (p-hydroxyphenyl)acetic acid (0); p-hydroxycinnamic acid, p=hydroxybenzoic acid, 3-(p-hydroxyphenyl) propanoic acid (X); o-coumaryl glucoside, water control (---) (1).

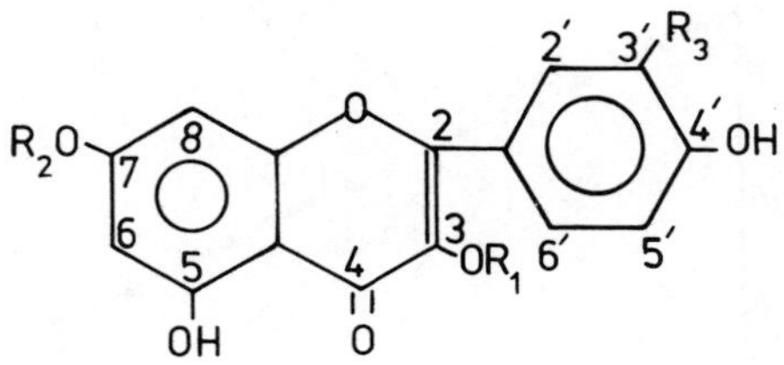

R_1 = galactoglucosyl (1→6) R_1 = rhamnogalactosyl (1→6)

R_2 = rhamnorhamnosyl (1→3) R_2 = rhamnose

R_3 = H R_3 = H

              Melinin   Robinin

               R_1 = rhamnogalactosyl (1→6)
               R_2 = rhamnose
               R_3 = OH

                    Clovin

Figure 3. Chemical structures of compounds isolated from M. alba
flowers (1).

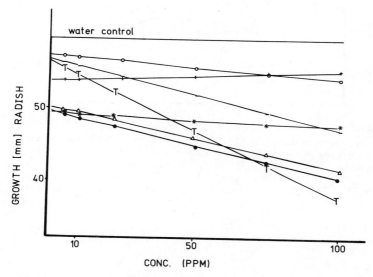

Figure 4. Regression analysis of radish root growth inhibition with concentration (ppm). (T) p-Hydroxybenzaldehyde, $y=56.34e^{0.10x}$; (-) prunasin, $y=56.61e^{-0.08x}$; (Δ) p-hydroxymandelonitrile, $y=50.41e^{-0.08x}$; (o) dhurrin, $y=54.91e^{-0.04x}$; (•) mandelonitrile, $y=49.62e^{0.02x}$; (*) NaCN (pH 5.5), $y=49.65e^{0.002x}$; (+) taxiphyllin, $y=53.81e^{-0.02x}$ (8). Copyright 1983, American Chemical Society.

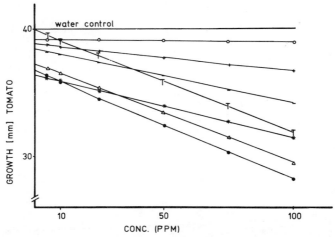

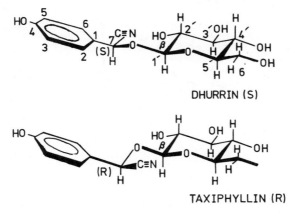

Figure 5. Regression analysis of tomato root growth inhibition with concentration (ppm). (T) p-Hydroxybenzaldehyde, $y=41.39e^{-0.01x}$; (-) prunasin, $y=38-51e^{-0.04x}$; (Δ) p-Phydroxymandelonitrile, $y=37.77e^{-0.08x}$; (O) dhurrin, $y+39.21e^{-0.002x}$; ($\bullet$) madelonitrile, $y=36.77e^{-0.08x}$; (*) NaCN (ph 5.5), $y=36.38e^{-0.05x}$; (+) taxiphyllin, $y=38.87e^{-0.02x}$. (8) Copyright 1983, American Chemical Society.

DHURRIN (S)

TAXIPHYLLIN (R)

Figure 6. Structures of dhurrin (S) and taxiphyllin (R). (8) Copyright 1983, American Chemical Society.

Table II.

Inhibitio of bacterial growth by durrin, taxiphyllin, prunasin, p-hydroxybenzaldehyde and johnsongrass root washings

Bacteria	Inhibition Zone (mm) + SD				
	Dhurrin*	Taxi-phyllin*	p-hydroxy-benzaldehyde	Prunasin	Root washings**
Pseudomonas mendocina	0	0	0	0	—
Pseudomonas maltophilia	8.0±0.4	8.0±0.1	0	0	—
Pesudomonas delaphiddi	10.0±0.5	10.0±2.0	12.0±1.0	0	8.0±0.5
Bacillus sphaericus	8.0±1.6	8.0±0.5	0	0	—
Bacillus subtilis	12.0±1.0	14.0±1.0	8.0±0.5	0	—
Bacillus thuringiensis	12.0±0.5	10.0±1.0	0	8.0±0.5	8.0±0.5
Enterobacter aerogenes	8.0±0.1	0	0	0	—
Enterobacter cloacae	10.0±0.5	9.0±0.5	0	0	—
Staphylococcus aureus	9.0±0.4	9.0±0.5	0	0	—

* 100 μg applied to a bacteriological test paper disk.

** 1000 μg were applied to a bacteriological test paper disk in this case.

if phytotoxicity showed a regular change, samples from field
grown plants were collected on a regular basis throughout the
growing season. Leachates and extracts that showed 20% inhibition (compared to the water control) in the tomato assay, and
25% inhibition in the radish assay were considered active.

Methods. Each week for 18 weeks during the growing
season, starting at two weeks after emergence of new leaves,
five D. illenoensis plants were collected, placed in plastic
bags and covered with ice for transportation to the laboratory.
The plants were separated into leaves, flowers and fruit.
Leaves from the five plants were combined and divided into two
equal weight samples. One sample was leached in distilled
water for 2 hours at a ratio of 15 gms leaf fresh weight to 100
ml water. The leaf leachate was decanted and used directly in
the bioassay. The second sample was ground in a blendor with
distilled water at a ratio of 1 gm leaf fresh weight to 100 ml
water. After standing for 15 min, the mixture was filtered and
the filtrate was used directly in the bioassay.

Five ml aliquots of leachate or extract were pipetted onto
3 sheets of germination paper in a petri dish. Twenty five
tomato or radish seeds were placed in each dish. Each
treatment/seed combination was replicated five times. The
assays were incubated at 20°C; radish roots length was measured
at 96 hrs and tomato at 168 hrs.

Results and Discussion. Figure 7 shows the results of the
tomato root growth bioassay of water leachates and extracts of
leaves. The extract gave strong inhibitory activity at weeks
3, 4 and 5; less activity at weeks 7 through 11, and 2 peaks of
activity at 14 and 16 weeks. Also shown are the results of an
identical bioassay of leaves from greenhouse-grown plants 2
years earlier. The curves are aligned at the same physiological plant age; that is, week 1 represents two weeks after the
first leaves were observed growing from the base of the plant.
The similarity of the field and greenhouse extract bioassays
indicates that the phytotoxicity is due to a function of the
physiological age of the plant, since it is likely that the
environmental parameters affecting the plants were markedly
different. The increased activity of the leachate at weeks 11
through 14 compared to the extract, may indicate the biological
state of the plant. Biochemical conditions in the plant
extract may destroy the activity, whereas by leaching the activity could be obtained. This can be important for isolation
and identification of the active component(s), since the
leachate will usually have fewer interfering chemical components compared to the extract.

The radish bioassay, shown in Figure 8, does not show the
strong activity at weeks 3 through 5, and the leachate activity
is significantly different from the extract only at week-1.
The general trend of activity is similar to the tomato bioassay
at other times.

Flowers were produced from weeks 8 through 11 and fruits
from weeks 9-18. The greatest inhibition from the flower

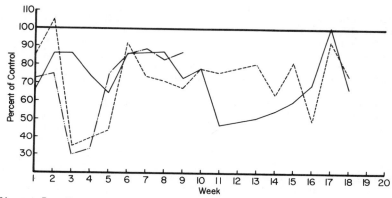

Figure 7. Tomato root growth bioassay of water leachates and extracts of <u>D. illinoensis</u>. (-) leaf leachate; (---) leaf extract; (-..-) leaf extract from greenhouse plants.

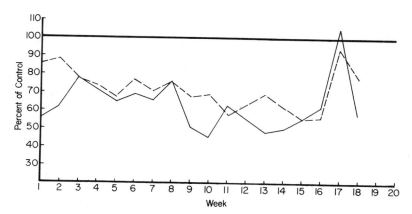

Figure 8. Radish root growth bioassay of water leachate and extract of <u>D. illinoensis</u>. (-) leachate; (---) leaf extract.

leachates and extracts was observed at weeks 8 and 11 in the
tomato assay and weeks 9 and 11 in the radish bioassay. The
Desmanthus fruit leachates and extracts showed significant
inhibition in the tomato and radish bioassay at weeks 14 and 15.

Literature Cited

1. Nicolier, G. F.; Thompson, A. C. J. Agric. Food Chem.
    1982, 30, 760-64.
2. Abdul-Wahab, A. S. "The toxicity of Johnsongrass excre-
    tions: A mechanism of root competition" Master's Thesis,
    Louisiana State University, Baton Rouge, LA. 1964.
3. Abdul-Wahab, A. S.; Rice, E. L. Bull. Torrey Bot. Club
    1964, 94, 489-97.
4. Kovacs, M. F. "Dhurrin (p-hydroxy mandelonitrile-β-O-
    glucoside): An allelopath identified in johnsongrass
    (Sorghum halapense (L.) Pers.) rhizome exudates", Ph.D.
    Thesis, University of Maryland, College Park, MD.   1972.
5. Parks, J. M.; Rice, E. L.   Bull Torrey Bot. Club 1969, 96,
    345-60.
6. Russell, R. S. "Plant Root Systems: Their Function and
    Interaction with the Soil"; McGraw-Hill: - New York, NY
    1977; pp 131-35.
7. Nicollier, G. F.; Thompson, A. C. J. Natural Prod.   1983,
    46, 112-17.
8. Nicollier, G. F.; Pope, D. F.; Thompson, A. C.
    J. Agric. Food Chem. 1983, 31, 744-48.

RECEIVED August 10, 1984

Phytotoxicity of Root Exudates and Leaf Extracts of Nine Plant Species

DANIEL F. POPE[1], A. C. THOMPSON[2], and A. W. COLE[3]

[1] Center for Alluvial Plains Studies, Delta State University, Cleveland, MS 38733
[2] Boll Weevil Research Laboratory, Midsouth Area, Agricultural Research Service, U.S. Department of Agriculture, Mississippi State, MS 39762–5367
[3] Mississippi State University, Mississippi State, MS 39762

Root exudates of common purslane [Portulaca oleracea L.], prickly sida [Sida spinosa L.], johnsongrass [Sorghum halapense (L.) Pers.], bahiagrass [Paspalum notatum Flugge 'Pensacola'], cotton [Gossypium hirsutum L. 'Stoneville 213'], soybean [Glycine max (L.) Merr. 'Bragg'], cogongrass [Imperata cylindrica (L.) Beauv.], lantana [Lantana camara L.] and Illinois bundleflower [Desmanthus illinoensis (Michx.) MacMill. Ex B. L. Robins] were tested for phytotoxicity in germination, root growth and height growth bioassays. Root exudates of common purslane, prickly sida, johnsongrass, bahiagrass and lantana significantly slowed soybean height increase. Prickly sida root exudate reduced radish (Raphanus sativus L. 'Champion') and tomato (Lycopersicon esculentum Miller 'Homestead 24') root growth. Root exudates of soybean, johnsongrass, cogongrass and prickly sida reduced 24 hour beet (Beta vulgaris L. 'Asgrow Wonder') germination. Bahiagrass, soybean, cotton, prickly sida and johnsongrass root exudates reduced 48, 72 and 96 hour okra (Abelmoschus esculentus L. 'Clemson Spineless') germination. Leaf extracts of cogongrass, soybean, johnsongrass and Illinois bundleflower reduced tomato root growth.

Allelopathy has been defined as "any direct or indirect beneficial or detrimental effect by one plant on another through production of chemical compounds that escape into the environment" (1). These compounds may move into the environment by volatilization, leaching, exudation or decomposition of plant parts. Decomposition of plant parts is perhaps the most widely studied mode of release of active compounds (1). However, many species release substantial quantities of potentially allelopathic compounds in root exudates and leaf leachates (2,3). Collection and bioassay of these compounds is an important part of the assessment of the allelopathic potential of a given species. Several methods have been used to grow plants in containers to facilitate the collection of potentially allelopathic compounds. Most of these techniques involve growing the plants in single pots or in a series of pots connected by tubing, as in the stairstep method (1). This method is useful for studying the effect of root exudates on bioassay plants grown in pots alternating with pots of donor species. Tang and Young (4) used glass pots with a hydrophobic resin to trap organics from a recirculating nutrient solution. Many times it is necessary to collect relatively large quantities of root exudates or leaf leachates for simultaneous biological and chemical work. Growing plants in individual containers can be very time consuming and laborious when much material is needed. For this reason, we designed a plant growth system that could be easily constructed from inexpensive materials and would allow collection of root exudates and leaf leachates with a minumum of time and labor.

Technique for Collection of Root Exudates and Leaf Leachates

The first systems were built from 10 cm ID PVC drain pipe and fittings. A complete unit consisted of 8 T-joints, 9 16.5 cm long pipe sections, and 1 elbow joint (Figure 1). The T-joints, with perpendicular arms erect, were joined together in series by the pipe sections. The elbow joint was placed on one end to serve as a receptacle for water and nutrient solutions. The components were fitted together without glue or caulk to facilitate cleaning and storage. The pipe section at the end opposite the elbow joint was covered with a glued-on plastic screen mesh to retain the soil mix and allow drainage and collection of the root exudates. An aluminum hood fastened on the drainage end excluded light, dust and prevented algal growth.

Another advantage of this system is that sections of transparent pipe may easily be fixed in the upright arms of the T-joints, enclosing the plants in a miniature "greenhouse" and allowing collection of volatiles from the plants.

Systems built later were composed of 305 cm lengths of 10

Figure 1. System for collection of root exudates and leaf extracts.

cm pipe without T-joints. A line of holes was drilled along
the length of the pipe and the species under study were planted
in these holes. This system worked well and avoided the use of
T-joints, a major expense in the earlier systems.

Each system was supported on a wooden framework so that
the slope from elbow end to the screened end could be adjusted
to provide adequate drainage. The pipes could be rotated
in the frame so that the plants lay parallel to the ground; the
plant tops were sprayed with water which ran into a pan under
the plants. In this way "leaf leachates" were quickly and conveniently collected.

The soils used in these systems were lightweight mixes
consisting of perlite, coarse sand and vermiculite.
These lightweight mixes were found necessary to provide adequate drainage. Chemical analysis of the root exudates was
simplified by use of those relatively inert compounds.
However, the low buffer and ion exchange capacity of these
mixes necessitated careful monitoring of pH and nutrient conditions.

Biological Activity of Root Exudates and Leaf Extracts of Nine Weed and Crop Species

Nine plant species (prickly sida, johnsongrass, Illinois
bundleflower, lantana, common purslane, cogongrass, bahiagrass,
cotton 'Stoneville 213', and soybean 'Bragg' were chosen for a
study of the biological activity of their root exudates and
leaf extracts. The first three species were chosen from those
showing activity in a screening bioassay; the next four were
chosen from literature reports of their interference with
crops, and the last two (cotton and soybean) were chosen
because of their widespread usage as crops. (5)

Materials and methods. Each species studied was grown in
the greenhouse in a growth unit constructed from 10 cm PVC
drain pipe and T-fittings, as described previously. The growth
units were filled with an artificial soil mix of perlite/coarse
sand/coarse vermiculite 3/2/1 by volume. On the second and
fifth day of each week four liters of a full strength Peter's
Hydro-sol^R solution plus calcium nitrate (36.7 g/37.8 liters
plus 17.0 g Ca(NO$_3$)$_2$ /37.8 liters) were added to each growth
unit. Tap water was used as needed at all other times. On
the first day of each week, eight liters of tap water were
applied to each growth unit to wash the root systems. This
effluent is termed the "root exudate," and the first liter was
collected for use in the germination and root elongation
bioassays. The next three liters collected from each growth
unit were used in the soybean bioassay. The remaining
effluent was discarded. An identically treated growth unit
with no plants was used as a source of effluent for the
control treatment for root exudate bioassays. The pH and

osmality of all effluents was recorded. Leaves of plants growing in the units were collected weekly for leaf extract bioassays.

Soybean bioassays of root exudates. Four soybean seeds ('Bragg') were planted in each of 100 12.5 cm plastic pots filled with an artificial soil mix consisting of perlite/coarse sand/coarse vermiculite: 3/2/1 by volume. After one week the plants were thinned to two per pot and the treatments were begun. The experimental design was a completely randomized design with 10 replications (pots) per treatment. On the first day of each week each pot was watered with 300 ml effluent from the appropriate growth units. On the fifth day of each week all pots were watered with Peter's Hydro-sol solution with $Ca(NO_3)_2$. At other times the pots were watered as needed with tap water. On the second and fifth day of each week the height of the soybeans (base to apical bud) was measured.

Root elongation bioassay of root exudates. Five ml aliquots of the root exudates were pipetted onto three layers of Anchor[R] germination paper in a 10 by 10 by 1.5 cm plastic petri dish. Twenty five radish or tomato seeds were placed in a 5x5 array in each petri dish. Radish seeds were incubated at 20C for 96 hours; tomato seeds were incubated at 20C for 168 hours, before the root length was measured. Experimental design was a completely randomized design with three replications (dishes) per treatment per bioassay seed species. The bioassay was repeated each week for 23 weeks.

Seed germination bioassays of root exudates. Five ml aliquots of the effluent from each growing unit were pipetted onto three layers of Anchor[R] germination paper in a 10 by 10 1.5 cm plastic petri dish. Twenty five radish, tomato, okra, carrot, or beet seeds were placed in a 5x5 array in each dish. Each test was incubated at 20C and germination counts were taken at 24, 48, 72, 96 and 168 hours. Seed were counted as germinated when two mm of root was visible. The experimental design was a completely randomized design with two replications (dishes) per treatment per bioassay seed species. The experiment was repeated each week for 23 weeks.

Tomato root growth bioassay of leaf extracts. Three hundred mg samples of fully expanded leaves were taken from each plant studied. Each sample was ground with a Polytron[R] in 30 ml of distilled water and the extract was filtered. Five ml aliquots of each extract were pipetted onto three layers of germination paper in a 10 by 10 by 1.5 cm plastic petri dish. Distilled water was used as a control treatment. Twelve tomato seeds were placed in a 3x4 array in each dish, and incubated at 20C for 168 hours, prior to root measurement. The experimental design was a completely randomized design with five replications (dishes) per treatment except the control which had 10 replications. The experiment was repeated each week for 9 weeks.

RESULTS AND DISCUSSION

Soybean bioassay of root exudates. The data shown in
Table I are the slopes of the linear regression of soybean
height over time for each treatment. Comparison of the treatment slopes to the control slope indicates whether a particular
treatment causes soybean plants to increase in height faster
(higher value) or slower (lower value) than the soybeans in the
control treatment. Observations of height and dry weight
generally produce similar results, but height may be measured
repeatedly on the same plants, allowing growth rate comparisons
to be made (6). The rate of height increase is important in
plant competitive ability, especially in crop plants where
quick development of a complete canopy shades the soil,
slowing weed germination and growth, and limiting the competitiveness of the weed with the crop (7,8).

Common purslane, prickly sida, johnsongrass, bahiagrass,
and lantana root exudates significantly slowed the growth of
soybean. Prickly sida and johnsongrass are considered weed
pests of soybean in some areas. Prickly sida is evidently not a
strong competitor but nevertheless reduces yields in soybeans
at sufficient densities (8). Toxic root exudates may play a
role in this yield reduction. Johnsongrass is a strong competitor and its root exudations have been reported to inhibit
soybeans and other species (9,10). Bahiagrass is a popular
pasture grass in the southern United States. It grows well in
variable soil and moisture conditions and is not easily invaded
by other species. Bahiagrass will invade coastal bermudagrass
fields, because of the fast growth rate, prostrate growth and
thick sod formed by bahiagrass (11). However, the results of
the soybean bioassay and other bioassays reported here indicate
that toxic root exudates may be, in part, responsible for the
ability of bahiagrass to interfere with other species.

Both lantana and common purslane are weed pests in some
crops. Common purslane root exudates inhibited soybean growth
but showed stimulation in other bioassays reported in this
paper. This may be a concentration and bioassay species
dependent effect, since the effect of any toxin varies
according to concentration and bioassay species (12). Lantana
root exudates showed the strongest inhibitory effect on
soybean growth of any of the species tested. Lantana is not
found in soybean fields but is a weed pest in some areas in
citrus orchards. Holm (13) included lantana, johnsongrass,
and cogongrass in his list of the worst weed pests in the
world.

Root elongation bioassay of root exudates. Only prickly
sida root exudate significantly affected radish or tomato root
elongation (Table II). It significantly inhibited radish root
elongation and significantly stimulated tomato root elongation.

Table I. Soybean growth bioassay of root exudates

Root Exudates	Slope of Linear Regression of Soybean Height Over Time (mm/day) [1]
Illinois bundle flower	11.2 a
Cogongrass	11.0 ab
Control	10.9 abc
Soybean	10.5 bcd
Cotton	10.4 cde
Common purslane	10.2 de
Prickly sida	10.1 e
Johnsongrass	10.0 ef
Bahiagrass	10.0 ef
Lantana	9.6 f

[1] Figures followed by the same letters do not differ at the 1% level of significance by LSD.

Table II. Radish and tomato root growth bioassay of root exudates

Root Exudates	23 Week Mean Root Length (mm) [1]	
	Radish	Tomato
Common purslane	50 a	64 ab
Cogongrass	49 a	65 ab
Control	49 a	63 a
Illinois bundle flower	49 a	64 ab
Cotton	49 a	64 ab
Lantana	49 a	64 ab
Soybean	49 a	65 ab
Johnsongrass	48 ab	65 ab
Bahiagrass	47 ab	65 ab
Prickly sida	45 b	66 b

[1] Figures followed by the same letter do not differ at the 1% level of significance by LSD.

Comparison of the results of these two bioassays reported in table II illustrates the importance of the choice of the bioassay species, as well as the variability of results to be expected when "phytotoxins" are released into the environment. Treatments applied to the radish and tomato seeds were identical, but most treatments tended to inhibit radish growth and stimulate tomato growth.

Seed germination bioassay of root exudates. Bioassay results are presented as a 23 week mean for each germination count time (Table III, IV, V, VI). Means were separated by LSD after data normalization by the inverse sine transformation.

All observed activity in these bioassays was a delay of germination rather than a reduction of total germination within the time frame of the experiment. In no case did any root exudate significantly decrease germination of any species at the final (168 hr) count. This is in agreement with the study of Williams and Hoagland (12) on the effect of phenolic compounds on seed germination. However, the effects of even a small retardation of germination can be greatly amplified by competition for light, nutrients, etc. (10).

Prickly sida root exudate retarded okra and radish seed germination and cotton root exudate also slowed okra germination. Elmore et al. (8) found that pot grown pairwise combinations of several members of the Malvaceae family [cotton, prickly sida and velvetleaf (Abutilon theophrasti Medic.)] gave lower dry matter yields per pot than would have been predicted from pure stand yields. This is in contrast to higher total dry matter yields per pot found in combinations of these three species and other highly competitive species including purple nutsedge (Cyperus rotundus L.). Other reports indicate that velvetleaf may release phytotoxins; the bioassays reported here indicate that prickly sida and cotton may also release phytotoxins that affect members of the Malvaceae as well as other families (14,15).

Johnsongrass root and rhizome exudate retarded okra and radish germination, but showed no influence on carrot, tomato, or beet germination. Although radish germination at 24 hours was significantly inhibited, radish root length at 96 hours was not significantly different from the control. Kovacs (16) found that johnsongrass rhizome exudates inhibited germination of cotton and wheat but had no effect on tomato, barley, millet or soybean, which agrees with the results reported here. Rice (17) found that activity of johnsongrass rhizome exudates varied from strongly toxic to no effect depending on the species used for bioassay.

Soybean root exudates were inhibiting to okra, radish and tomato germination but not to carrot or beet germination. Although radish and tomato germination was significantly slowed, soybean root exudate activity in the tomato and radish

Table III. Radish and tomato germination bioassay of root exudates

Root Exudates	Mean Percent Germination [1]	
	Radish—24 hr.	Tomato—72 hr.
Common purslane	87.8 a	81.6 a
Cotton	87.4 a	79.1 ab
Control	87.0 a	82.1 ab
Bahiagrass	87.0 a	80.7 ab
Illinois bundle flower	86.2 ab	81.3 a
Lantana	85.8 ab	78.4 ab
Soybean	84.6 bc	76.2 b
Johnsongrass	84.4 bc	78.7 ab
Cogongrass	84.0 bc	81.4 a
Prickly sida	83.4 c	79.4 ab

[1] Figures followed by the same letter are not significantly different at the 1% level by LSD.

Table IV. Carrot germination bioassay of root exudates

Root Exudates	Mean Percent Germination [1]	
	72 hours	96 hours
Bahiagrass	38.5 ab	69.6 ab
Common purslane	39.3 ab	68.8 ab
Lantana	41.7 a	70.3 a
Soybean	38.3 ab	70.4 a
Cogongrass	35.9 b	67.0 ab
Illinois bundle flower	35.8 b	65.1 b
Control	38.2 ab	68.3 ab
Cotton	37.5 ab	66.8 ab
Johnsongrass	36.4 b	66.3 ab
Prickly sida	36.5 b	68.1 ab

[1] Figures followed by the same letter are not significantly different at the 1% level by LSD.

Table V. Beet germination bioassay of root exudate

Root Exudates	Mean Percent Germination [1]		
	48 hours	72 hours	96 hours
Common purslane	17.0 a	58.5 ab	79.3 ab
Prickly sida	15.7 ab	57.9 ab	78.4 ab
Control	15.1 abc	61.2 a	79.6 ab
Cogongrass	15.0 abc	60.2 a	80.4 a
Johnsongrass	14.9 abc	58.1 ab	78.4 ab
Soybean	15.5 bcd	58.9 ab	77.5 ab
Illinois bundle flower	14.2 bcd	55.5 b	75.2 b
Bahiagrass	13.7 bcd	58.8 ab	79.7 ab
Cotton	13.6 bcd	58.3 ab	76.9 ab
Lantana	12.2 d	55.4 b	77.0 ab

[1] Figures followed by the same letter are not significantly different
   at the 1% level by LSD.

Table VI. Okra germination bioassay of root exudates

Root Exudates	Mean Percent Germination [1]			
	48 hours	72 hours	96 hours	168 hours
Control	6.2 a	19.0 a	42.4 a	82.7 abc
Illinois bundle flower	5.7 ab	18.6 a	38.7 abc	83.8 ab
Cogongrass	5.6 ab	15.9 ab	38.2 abcd	84.6 a
Lantana	5.0 abc	14.7 abc	35.8 bcde	81.2 bc
Common purslane	4.2 abc	15.3 abc	39.8 ab	83.5 ab
Bahiagrass	4.1 abc	12.1 bcd	35.8 bcde	81.7 abc
Soybean	3.5 bc	12.2 bcd	33.7 de	80.2 c
Cotton	3.4 bc	13.7 bcd	35.4 bcde	84.3 ab
Prickly sida	3.0 c	9.4 d	31.7 e	83.1 abc
Johnsongrass	2.9 c	11.2 cd	34.0 cde	83.6 ab

[1] Figures followed by the same letter are not significantly different at the 1% level by LSD.

root growth bioassay was not significant. The time elapsing from germination inhibition to the measurement of root growth may be sufficient to allow early germination differences to be obscured by inherent variability of seedling growth. Although many crop plants have been shown to release phytotoxins, little work has been done toward breeding crops with the aim of biological weed control. Many legumes have been shown to release compounds that strongly affect the growth of other species; soybeans may have considerable potential for development of a "self weeding" crop.

Bahiagrass reduced okra germination at 72 and 96 hours. Lantana reduced beet germination at 48 and 72 hours. Cogongrass reduced radish germination at 24 hours. Evidently only cogongrass has been investigated previously for its allelopathic potential (18,19).

Tomato root growth bioassay of leaf extracts. Leaf extracts of Illinois bundle flower, johnsongrass, soybean and cogongrass were found to significantly inhibit tomato root growth (Table VII). Johnsongrass and cogongrass have been previously reported to contain phytotoxins in their leaves. Abdul-Wahab and Al-Naib (18) reported that cogongrass leaves contained scopolin, scopoletin, chlorogenic acid and isochlorogenic acid, all of which have been reported to be phytotoxic (1,17).

Abdul-Wahab and Rice (9) found that boiling water extracts of johnsongrass leaves inhibited germination and seedling growth of several bioassay species at concentrations 80 times that used in this present study.

Soybean leaf extracts have not previously been reported to be phytotoxic; however many other leguminous species have been shown to be active (1,17). The possible allelopathic activity of soybean indicated by these bioassays is potentially economically important, since soybean is a major crop world wide.

The extracts of Illinois bundleflower leaves showed significant activity in the bioassay; however, the activity varied considerably over the nine week test period (data reported elsewhere in this volume).

The results of the bioassays reported here indicate that some of the species studied may release phytotoxins into the environment. Further work is needed to identify any active compounds and determine their influence in the field.

Table VII. Tomato root growth bioassay of leaf extracts

Leaf Extracts	Nine Week Mean Root Length (mm) [1]
Control	56 a
Cotton	53 ab
Lantana	52 ab
Bahiagrass	50 ab
Prickly sida	50 ab
Common purslane	49 ab
Cogongrass	48 b
Soybean	47 b
Johnsongrass	46 b
Illinois bundle flower	38 c

[1] Figures followed by the same letter are not significantly different at the 1% level by LSD.

Literature Cited

1. Rice, E. L. Bot. Rev. 1979, 45, 15–109.
2. Rovira, A. D. Bot. Rev. 1969, 35, 35–57.
3. Tukey, H. B., Jr. Biochemical Interactions Among Plants (U. S. Nat. Comm. for IBP, eds.) Nat. Acad. Sci., Washington, D. C. 1971, pp 25–32.
4. Tang, Chung–Shih; Young, Chiu–Chung. Plant Physiol. 1982, 69, 155–160.
5. Pope, Daniel F.; Thompson, A. C.; Cole, A. W. Proc. Southern Weed Sci. Soc. 1983, p 357.
6. Horowitz, M. Weed Res. 1976, 16, 209–215.
7. Murphy, T. R.; Gosset, B. J. Weed Sci. 1981, 29, 610–615.
8. Elmore, C. D.; Brown, M. A.; Flint, E. P. Weed Sci. 1983, 31, 200–207.
9. Abdul–Wahab, A. S.; Rice, E. L. Bull. Torrey Bot. Club 1967, 94, 486–497.
10. Lolas, P. C.; Coble, H.C. Weed Sci. 1982, 30, 589–593.
11. Smith, A. E. Weed Sci. 1983, 31, 88–92.
12. Williams, R. D.; Hoagland, R. E. Weed Sci. 1982, 30, 206–212.
13. Holm, L. Weed Sci. 1969, 17, 113–118.
14. Colton, C. E.; Einhellig, F. H. Am. J. Bot. 1980, 67, 1407–1413.
15. Elmore, C. D. Weed Sci. 1980, 28, 658–660.
16. Kovacs, M. F. "Dhurrin (p–hydroxy–mandelonitrile–B–D–glucoside): An allelopath identified in johnsongrass (Sorghum halapense (L.) Pers.) rhizome exudate. Ph. D. Thesis, Univ. of Maryland. 1972.
17. Rice, E. L. "Allelopathy"; Academic Press. 1974.
18. Abdul–Wahab, A. S.; Al–Naib, F. A. G. Bull. Iraq. Nat. Hist. Mus. 1972, V, 17–24.
19. Sajise, P. E.; Lales, J. S. Kalikasan Phillip. J. Biol. 4, 155–164.

RECEIVED August 6, 1984

The Effect of Root Exudates on Soybeans
Germination, Root Growth, Nodulation, and Dry-Matter Production

DANIEL F. POPE[1], A. C. THOMPSON[2], and A. W. COLE[3]

[1]Center for Alluvial Plains Studies, Delta State University, Cleveland, MS 38733
[2]Boll Weevil Research Laboratory, Midsouth Area, Agricultural Research Service,
  U.S. Department of Agriculture, Mississippi State, MS 39762–5367
[3]Mississippi State University, Mississippi State, MS 39762

Root exudates of common cocklebur [Xanthium pensylvanicum Wallr.], ivyleaf morningglory [Ipomoea
hederacea (L.) Pers.], hemp sesbania [Sesbania
exaltata (Raf.) Cory], redroot pigweed [Amaranthus
retroflexus L.], sicklepod [Cassia obtusifolia L.],
smooth crabgrass [Digitaria ischaemum (Schreb.) Muhl.],
soybean [Glycine max Merr. 'Centennial'], johnsongrass
[Sorghum halapense (L.) Pers.], and bermudagrass
[Cynodon dactylon (L.) Pers.], were tested for their
effect on soybean germination and growth. Cynodon
root exudates stimulated soybean germination;
Cassia inhibited soybean germination. Xanthium
reduced dry weight of soybean nodules and stems, Glycine
reduced dry weight of soybean nodules, roots, leaves,
and stems, and Sesbania root exudates reduced dry
weight of soybean nodules, leaves, stems and fruit.

Although it is known that many plant species exude substantial
quantities of carbonaceous materials from their roots, little
work has been done on the effect of root exudates on the growth
of adjacent species (1, 2). Root exudates of host plants have
been shown to affect nitrogen fixing bacteria, but little study
has been made of the activity of root exudates of adjacent species on nodulation of host species (3). Therefore, we chose
eight common soybean weeds and soybean to investigate the effect
of their root exudates on soybean germination nodulation and
growth.

Materials and Methods

Ten 10 cm diameter by 305 cm long polyvinyl chloride pipes
were supported on wooden frames in the greenhouse. Fifty seven
3.5 cm diameter holes were drilled in a line along the length
of each pipe. One end of each pipe was fitted with a
fiberglass screen; the other end was fitted with an elbow

joint. The elbow end of each pipe was elevated 20 cm above the
screened end to facilitate drainage. Each pipe was filled with
a mixture of 12 liters each of coarse vermiculite and perlite,
0.5 liter of peat moss and 1.8 grams of FTE 555 , a fritted
trace element mixture. Each pipe was planted with seeds of one
of the following species: common cocklebur, ivyleaf morningglory, red root pigweed, bermudagrass, smooth crabgrass,
sicklepod, hemp sesbania or soybean 'Centennial'. Soybean were
planted one week later than the other species in order to sche-
dule emergence of all species as closely as possible. The
soybean seed were coated with Nitragin , a commercial preparation of Rhizobium japonicum. Johnsongrass rhizomes were
planted in one pipe. One pipe, without plants, served as the
check.

Soybean dry matter production. One week after planting
in the pipes, two-hundred-forty 0.7 liter styrofoam cups with
four holes in the lower sides of each were filled with a 2 cm
layer of pea gravel, then filled to within 1 cm of the top with
a 1/1/1 v/v/v mix of perlite, coarse sand and a sandy loam
soil. Each cup was planted with three soybean seed
('Centennial' or 'Coker 136') previously coated with Nitragin.
The seed were covered with 1 cm of soil mix and watered with
tap water.
Two weeks after planting in the pipes, the plants were
thinned to 35 pipe per pipe each and the cups to one plant
each, and the treatments begun. Each first, third and fifth
day of the week for twelve weeks the pipes were flushed with
three liters of tap water poured in the elbow end. The water
flowed past the plant root systems and drained out the screened
end of the pipes into a flask. One hundred milliliter aliquots
of this water ('root exudate') were used to water the soybean
plants in the cups three times weekly. After each flushing,
two liters of a low nitrogen (50 ppm N) complete nutrient solution (Peter's Hydro-sol) were added to each pipe. The soybean
plants in cups were watered as needed at other times with tap
water. On alternate weeks the soybean plants were fertilized
with the complete nutrient solution. At 4, 8 and 12 weeks
after the root exudate treatments started eighty soybean plants
(10 treatments x 2 soybean varieties x 4 blocks) were randomly
chosen for analysis. The soil was washed free of the plant
roots and each soybean plant was divided into roots, nodules,
stems, leaves and fruits. The plant parts were dried at 105°C
for four days and weighed.

Soybean germination and root growth bioassay. Ten ml aliquots of each root exudate were pipetted onto six layers of
Anchor germination paper in 10 cm by 10 cm by 1.5 cm petri
dishes. A 2×10^{-3}M solution of p-hydroxybenzaldehyde was used
as a standard treatment. Nine soybean seed (Centennial or Coker
136) were placed in a 3 x 3 array in each dish. The dishes were
incubated at 20°C for 168 hours. The number of seed germinated
was counted at 48, 72, 96 and 168 hours. At 168 hours, the
seedlings were frozen to stop growth. Seedling root length was
measured after thawing for five minutes.

Aliquots of each treatment (root exudate) were dried at 105°C to determine total dissolved solids. The pH of each treatment was measured.

The experimental design was a randomized complete block with eleven treatments, two soybean varieties and five blocks (reps). The experiment was conducted six times at two week intervals, starting four weeks after the weeds were planted in the pipes.

Soybean nodulation. Styrofoam cups (180 ml) with four holes in the lower sides were filled with coarse vermiculite mixed with FTE 555 (2 g/1) and Nitragin (2.5 g/1). A germinated seed (1-5 mm root length) of one of four soybean cultivars ('Centennial', 'Coker 136', 'Dare', 'Essex') was planted in each cup and watered three times weekly with 50 ml of one of the ten root exudates. The cups were watered as needed at other times with tap water or Hydro-sol. Three weeks after planting the soybeans were harvested, dried for four days at 105°C, and the plant parts weighed. The experimental design was a randomized complete block with ten treatments (root exudates), four soybean cultivars and six replications.

Results and Discussion

Soybean germination and root growth. The results of this bioassay are shown in Table I. Germination at 48 hours was the only time period showing significant differences; the treatment effects on germination at 72, 96, and 168 hours were not significantly different from the control. Germination averaged about 20% at 48 hours and was essentially complete at 72 hours; few seed germinated after 72 hours. The treatment effects on germination were similar for both soybean cultivars tested except Amaranthus and Ipomoea root exudates significantly reduced 'Coker 136' germination while not affecting 'Centennial' germination. Both Cynodon root exudate and 2×10^{-3} M p-hydroxybenzaldehyde stimulated germination; Cassia root exudate inhibited germination of both cultivars. No root exudate had a significant effect on root length; the p-hydroxybenzaldehyde slightly reduced 168 hour root length. The pH of the root exudates ranged between 6.5 and 7.3. The total dissolved solids (TDS) in the root exudates ranged between 300 to 350 ppm by weight. Regression analysis of the pH and TDS versus the treatment effects of the root exudates did not indicate a significant relationship. The pH and TDS of the p-hydroxybenzaldehyde solution was 5.6 and 244 ppm, respectively; these factors may have been involved in activity of this solution versus the control treatment.

Soybean dry matter production. The results of this bioassay are shown in Table II. The relative relationships of the different treatments were essentially the same for weeks four, eight, and twelve, so the results shown are mean values over all three harvest times. Values shown for fruit are means of weeks eight and twelve, since no fruit was produced at week four. Xanthium significantly reduced both nodule dry weight and stem dry weight. However, neither fruit dry weight nor the total dry

Table I.   Effect of weed root exudates on soybean germination
           and root growth

Root Exudate	Germination – 48 hours [a,b]	Root Length – 168 hours [a,b]
Control	100	100
Xanthium	58	102
Ipomoea	64	100
Amaranthus	58	101
Cynodon	231 b	106
Digitaria	116	100
Sorghum	100	99
Cassia	49 b	102
Glycine	67	101
Sesbania	76	101
p-hydroxybenzaldehyde	304 b	92 b

a   Germination or root length as a percentage of the control.

b   Figures followed by b are significantly different from
    the control at the 0.05 confidence level by Waller–Duncan
    K-ratio T-test.

Table II. Effect of root exudates on soybean dry matter production [a]

Root Exudate	Nodules	Root	Leaves	Stem	Fruit	Total
Control	100	100	100	100	100	100
Xanthium	82 b	104	87	84 b	108	92
Ipomoea	88	96	93	90	97	93
Amaranthus	92	104	96	95	112	100
Cynodon	92	112	97	108	104	103
Digitaria	87	103	100	94	92	97
Sorghum	93	103	100	91	104	98
Cassia	86	98	90	86	103	92
Glycine	73 b	85 b	78 b	85 b	86	83 b
Sesbania	69 b	94	80 b	79 b	80 b	81 b

[a] Dry weight as a percentage of the control treatment; each figure is mean of four, eight and twelve week harvest except for fruit, which the mean of the eight and twelve week harvest.

[b] Figures followed by b are significantly different from the control at the 0.05 confidence level by Waller-Duncan K-ratio T test.

weight was significantly reduced compared to the control. *Glycine* root exudates significantly reduced dry weight of all plant parts except fruit; fruit dry weight was significantly reduced at the 0.07 confidence level. *Sesbania* root exudates significantly reduced dry weight of all plant parts except roots. Since the soybean bioassay plants were fertilized regularly with a complete nutrient solution, it was not thought likely that variation in the nutrient supply to the bioassay plants could explain the results. As early as 1910 Lipman (4) noted that legumes released soluble nitrogenous compounds into the soil; this could explain the reduction in nodule dry weight since nodulation is generally inversely proportional to the supply of nitrogen obtained from the soil (5). Both *Glycine* and *Sesbania* growing in the pipes were heavily nodulated, and dark green in appearance. However, one would expect the rest of the bioassay plant to be heavier than the control since the large quantities of photosynthate usually diverted to supply energy for nitrogen fixation would be available for fruit and other dry matter production. Since total dry weight of soybean plants treated with *Glycine* or *Sesbania* root exudates was reduced, it is likely that the root exudates reduced nodulation of soybeans due to factors other than nitrogen supply. Dry weight of the rest of the soybean plants may have been reduced indirectly due to lack of nitrogen through inhibition of nodulation, or directly through activity of the root exudates against other growth processes of soybeans.

Soybean nodulation. The results of this bioassay, shown in Table III, are essentially the same as the dry matter production bioassay. *Xanthium* root exudates reduced soybean nodule dry weight; *Glycine* and *Sesbania* root exudates reduced nodule, leaf and total dry weight. The percentage reduction is not quite as great as in the dry matter production bioassay. The effect apparently begins early in the life of the soybean plant, and either the root exudates continue to reduce growth or the soybean plant is not able to overcome the early effects later on. The data are consistent with the hypothesis that soybean nodulation or dinitrogen fixing ability is reduced by *Sesbania* and *Glycine* root exudates.

Table III. Effect of root exudates on soybean nodulation [a]

Root Exudate	Nodules	Root	Leaves	Stem	Total
Control	100	100	100	100	100
Xanthium	81 b	100	90	97	93
Ipomoea	102	94	89	94	92
Amaranthus	93	101	98	101	99
Cynodon	85	98	89	93	92
Digitaria	92	95	95	101	97
Sorghum	106	100	98	109	102
Cassia	87	100	93	98	95
Glycine	79 b	90	85 b	93	88 b
Sesbania	76 b	92	84 b	97	89 b

a Dry weight as a percentage of the control treatment; each figure is the mean of twenty-four plants.

b Figure followed by b are significantly different from the control at the 0.05 confidence level by the Waller-Duncan K-ratio T test.

Literature Cited

1. Rovira, A. D. Bot. Rev. 1969, 35, 35-57.
2. Rice, E. L. "Allelopathy"; Academic Press. 1974.
3. Gitte, R. G.; Rai, P. V.; Patil, R. B. Plant and Soil, 1978, 50, 553-566.
4. Lipman, J. G. J. Agri. Sci. 1910, 3, 297-300.
5. Johnson, J. W.; Welch, L. F.; Kurtz, L. T. J. Environ. Qual. 1975, 4, 303-306.

RECEIVED August 21, 1984

Rye (*Secale cereale* L.) and Wheat (*Triticum aestivum* L.) Mulch: The Suppression of Certain Broadleaved Weeds and the Isolation and Identification of Phytotoxins

DONN G. SHILLING[1], REX A. LIEBL, and A. DOUGLAS WORSHAM

Crop Science Department, Weed Science Center, North Carolina State University, Raleigh, NC 27695-7627

Research over the years has indicated improved control of certain annual broadleaf weeds in no-till cropping systems when mulches are left on the soil surface. Therefore, we initiated studies in no-till corn, soybeans, sunflower, and tobacco to evaluate and separate the effects of small grain mulches and tillage on broadleaf weeds. Planting corn no-till into a desiccated green wheat cover crop reduced Ipomoea spp. biomass 79% compared to a non-mulched, tilled treatment. Elimination of tillage in non-mulched treatments was as effective in reducing weed biomass as replacing mulch on tilled soil. In double-crop soybeans, there was no mulch effect; tillage, however, greatly increased Ipomoea spp. biomass. In no-till tobacco, elimination of tillage and presence of a rye mulch reduced biomass of Amaranthus retroflexus L., Chenopodium album L., and Ambrosia artemisiifolia L. by 51, 41, and 73%, respectively. Rye mulch reduced C. album biomass in both the tilled and the non-tilled systems by 60%. In full-season soybeans and sunflowers planted into desiccated green rye, the elimination of tillage and the presence of rye mulch reduced aboveground biomass of C. album, A. artemisiifolia, and A. retroflexus 99, 92, and 96%, respectively. Rye mulch was as effective as elimination of tillage in reducing C. album and A. retroflexus biomass. A. artemisiifolia was not significantly affected by rye mulch in any of these cropping

[1]Current address: Monsanto Agricultural Products Co., 800 N. Lindbergh Blvd., St. Louis, MO 63167

0097-6156/85/0268-0243$08.25/0

systems. Allelopathic suppression of several broadleaf
weeds by the mulches is implicated in these studies.
In further studies to examine possible causes of weed
suppression in no-till cropping systems utilizing killed
rye (Secale cereale L.) and wheat (Triticum aestivum L.)
as a mulch, two phytotoxic compounds not previously
implicated in allelopathy, β-phenyllactic acid (βPLA)
and β-hydroxybutyric acid (βHBA), were identified from
aqueous extracts of field-grown rye. The βPLA and βHBA
at 8 mM inhibited Chenopodium album L. hypocotyl growth
68 and 30%, respectively in laboratory bioassays. Both
acids inhibited C. album root growth 20% at 2 mM.
Amaranthus retroflexus L. hypocotyl growth was inhibited
17% by βPLA at 0.8 mM and 100% at 8 mM with βHBA giving
27% inhibition at 8 mM. A. retroflexus root growth was
inhibited 59 and 39% at 2 mM by βPLA and βHBA, respectively. The compound identified in alkali extracts from
wheat having greatest inhibitory effects on Ipomoea
lacunosa L. seed germination was identified as ferulic
acid (4-hydroxy-3-methoxy cinnamic acid). At 5 mM,
germination and root length was inhibited 23 and 82%,
respectively. Sida spinosa L. germination and root
length, with carpels present on the seed, was inhibited
85 and 82%, respectively. Ferulic acid was decarboxy-
lated by a bacterium living on the carpels of S.
spinosa seed to a more phytotoxic styrene derivative,
4-hydroxy-3-methoxystyrene. These compounds plus other
unidentified phytotoxic chemicals could, in part, explain
suppression of certain weeds by rye and wheat mulches in
no-till crops.

Cultivation of soil has and will continue to be an important means
of controlling weeds (1). However, extensive soil cultivation leads
to various problems such as losses of soil, soil moisture and
nutrients. This results in water pollution by both the soil itself
and pesticides and nutrients associated with it (2, 3, 4). Minimum
or no-till cropping systems can reduce these problems because
various crop residues (i.e., mulch) are left on the soil surface
with a minimum of soil disturbance in planting the crop.
    The presence of crop residues has been reported to both
increase (5, 6) and decrease crop yields (7) and not tilling to
increase certain difficult to control weeds (8). However, other
reports indicate that the presence of certain mulches can reduce the
biomass of certain weeds (9-15) and allow for higher crop yields (5,
6). Thus, under certain conditions, mulches can suppress certain
weed species, but determining the reason(s) presents many logis-
tical problems, especially under field conditions. To determine the
cause(s), the physical and chemical (i.e., allelopathy) effects of
the mulch and the role of soil disturbance (or the lack of, as would
be the case in a no-till system) must be separated.
    Under no-till conditions, many parameters are affected in such
a manner as to alter germination of weed seeds. Putnam et al. (16)
stated that "eliminating tillage reduces densities of many annual
species presumably because numerous seeds are isolated from

favorable sites for germination." One major result of soil disturbance is that it provides the seed with light which is required in
many instances for germination (17, 18, 19). The elimination of
tillage alone (without any mulch present) has been shown to reduce
weed populations (9, 15, 16). No-till systems are also known to
affect soil moisture, temperature and pH, all of which could affect
the emergence and growth of plants--both crops and weeds (4, 20).

Even with these problems, attempts have been made to demonstrate
that mulches suppress weeds allelopathically. Putnam and DeFrank
(12) and Barnes and Putnam (39) used Populus wood shavings to
separate chemical and physical effects of mulches. Their work
indicated that certain mulches do possess allelopathic potential.
Liebl and Worsham (9) and Shilling and Worsham (14) placed mulch on
tilled soil, after tilling, in an attempt to provide the weeds with
an exposure to light. Their work also indicated that at least part
of the suppression of weeds by wheat and rye mulch is allelopathic.
Thus, research to date indicates that both mulch and the lack of
soil tillage contributes to the suppression of weeds in no-till
cropping systems.

Allelopathic effects are probably due to inhibitory compounds
being released directly from plants (or their residues) or microbial
metabolites. In an early study, Guenzi and McCalla (21) showed that
most crop residues contain water-soluble substances that can depress
the growth of corn (Zea mays L.), wheat (Triticum aestivum L.) and
sorghum (Sorghum bicolor L.). Guenzi and McCalla (22) identified
five phenolic acids from a number of plant residues and showed that
all inhibited the growth of wheat. Chou and Patrick (23) isolated
and identified a number of phytotoxic compounds from soil amended
with rye (23). Other studies have also indicated that compounds
from plant residues can adversely affect plant growth (11, 23, 24,
25, 26). Shettel and Balke (27) have further demonstrated that
allelopathic compounds can selectively suppress certain weed species.

The implication that "living crops" can allelopathically
suppress weeds has been made by Putnam and Duke (28) and Leather
(29). These studies demonstrated that the potential for suppression
of weeds could be enhanced by crop selection. Putnam and Duke (28)
suggested that there is potential for breeding crops to better
suppress weeds by utilizing and improving allelopathic characteristics. Fay and Duke (30) evaluated the amount of scopoletin (6-
methoxy-7-hydroxy coumarin) exuded from 3,000 accessions of Avena
sp. They suggested that the "wild types" of crop varieties could
have once possessed allelopathic potential, but this character has
been inadvertently selected against over time.

The possibility that microorganisms play a pivotal role in the
production and persistence of phytotoxic compounds has also been
demonstrated (7, 23, 26, 31, 32). Guenzi et al. (7) found that
aqueous extracts of weathered corn and sorghum residues were most
phytotoxic to wheat growth after 4 and 16 weeks of decomposition,
respectively. The greatest phytotoxicity exhibited by aqueous
extracts of wheat and oat (Avena sativa L.) straw occurred at or near
harvest time, with essentially all toxicity gone after four weeks of
decomposition. Kimber (33) found that aqueous extracts of several
grasses and legumes that had been rotting for periods of up to 21
days were inhibitory to the growth of wheat. He also showed that
straws cut while still green produced a higher level of toxicity

than those cut when fully mature. Toussoun et al. (34) found that
the production of water-soluble phytotoxins of barley (Hordeum
vulgare L.) residue in soil required a soil moisture content above
30% of the soil and residue dry weight. Extracts became toxic 7-10
days after incorporation into soil with phytotoxic activity reaching
a maximum in 3 weeks. Thus, microorganism can enhance and/or reduce
the phytotoxicity of plant residues.

Since the early 1960s, increased emphasis has been placed on the
role of soil microorganisms in the production of phytotoxic substances
from plant residues. Norstadt and McCalla (35) postulated that the
inhibitory effects of crop residues might be due to a combination of
toxins from plant residues and from microorganisms that are more
prolific where plant residues are present. A number of fungi have
been isolated from soil obtained from no-till plots. Many of the
fungi produced substances toxic to higher plants (36). Patrick and
Koch (37) reported that aqueous extracts of various plant residues
had no effect on the respiration of tobacco (Nicotiana tabacum L.)
seedlings. They did, however, demonstrate that substances formed
during the decomposition of plant residues in the soil markedly
inhibited the respiration of tobacco seedlings. In later studies,
Patrick (38) and Chou and Patrick (23) isolated and identified a
wide range of toxic compounds from decomposing rye (Secale cereale
L.) and corn residues. In another study, Cochran et al. (32) found
that plant residues produced wheat seedling root inhibitors only
after conditions became favorable for microbial growth.

The purposes of the studies reported here were to: (1) characterize the effects of rye and wheat mulch and soil disturbance on
certain weed species, (2) determine if allelopathy was involved,
(3) determine if rye and wheat straw extracts were phytotoxic to
certain weed species, (4) isolate, characterize, and identify
inhibitory compounds, and (5) determine quantitatively the phytotoxicity of inhibitory compounds that were identified.

MATERIALS AND METHODS

Rye Mulch Studies

Field Research.   In 1982 and 1983, experiments were conducted to
evaluate the effects of primary tillage (i.e., soil tilled at time
of planting) and above- and below-ground rye residue [i.e., shoot
(mulch) versus root residue] on weed populations in three cropping
systems (flue-cured tobacco, soybeans and sunflowers). No residual
herbicides were used in any of the following experiments. These
tests were conducted at the Central Crops Research Station near
Clayton, NC (Johns sandy loam soil type) and Lower Coastal Plain
Research Station near Kinston, NC (Goldsboro sandy loam soil type).
In October of 1981 and 1982, fields were disked and rye (Secale
cereale L. 'Abuzzi') planted at 188 kg/ha. Two no-rye treatments
were also established at this time and maintained plant free by
applications of paraquat (1,1'-dimethyl-4,4'-bipyridinium ion) at
0.6 kg/ha as needed. Each test was a completely randomized block
design replicated four times.

In April of 1982 (flue-cured tobacco) and 1983 (soybean and
sunflower), treatments were set up as follows: (1) mulch removed;

(2) mulch removed and soil tilled; (3) mulch cut; (4) mulch removed, soil tilled and mulch replaced; (5) mulch desiccated with 3.36 kg/ha of glyphosate; (6) mulch desiccated with 0.6 kg/ha of paraquat; (7) no rye, soil tilled; and (8) no rye, no-till. Once these treatments were established, the tests were treated with 3.36 kg/ha of glyphosate [N-(phosphonomethyl) glycine] to desiccate weeds and living rye. In 1982, tobacco (<u>Nicotiana tabacum</u> L. 'McNair 944') was transplanted approximately 10 days after treatment and in 1983 soybean (<u>Glycine max</u> L. 'Essex') and sunflower (<u>Helianthus annuus</u> L.) were planted on the same day. During the growing season data were collected to determine the effects of soil tillage and rye mulch on % weed control, weed biomass, weed density, soil temperature and incident light. Weed ratings were determined by visually comparing treatment plots to the no rye, tilled plot (i.e., this treatment represented 0% weed control). A completely randomized design was used with four replications.

<u>Growth Chamber Experiment</u>. The experiment was conducted to evaluate the allelopathic potential of rye. This was accomplished by watering common lambsquarters with an aqueous extract of rye. Field-grown rye was collected in April of 1983 (at anthesis) and air dried for seven days. The plant material was extracted with distilled water for six hours with shaking. The extract was obtained by filtration. The extract was then diluted to obtain the following concentrations: 1 g/20 ml, 1 g/30 ml, 1 g/60 ml. All extracts were adjusted to pH 5.5 with 1N KOH. Osmotic potential of each extract was determined and appropriate controls established using polyethylene glycol. Common lambsquarters (0.052 g) was seeded into 1 L pots containing quartz sand and watered with nutrient solution. Starting on the second day of the experiment, pots were watered with 20 ml of the appropriate solution. A total of six applications of the extract were applied. The plants were grown for 28 days at 25°C with a 12 hr day length. At this time the number of plants per pot was determined. A completely randomized block design was used with four replications.

Data were subjected to analysis of variance and regression analysis using the general linear model procedure of the Statistical Analysis System (<u>40</u>). Means were compared using Waller-Duncan procedure with a K ratio of 100. Polynomial equations were best fitted to the data based on significance level of the terms of the equations and R^2 values.

Wheat Mulch Studies

<u>Field Studies</u>. At all locations, the area planted in small grains was divided into two parts. One part was planted in corn the following spring, and the other part planted in soybeans ('Ransom') following small grain harvest.

   <u>Study 1</u>. Wheat ('McNair 1813') was planted at the Central Crops Research Station near Clayton (Lynchburg sandy loam) and the Tidewater Research Station near Plymouth (Bayboro loam), North Carolina, at a rate of 101 kg/ha in October of 1980. The following spring or early summer, plots were set up in which a green wheat cover crop or wheat straw and stubble remaining after wheat harvest was: (1) left

intact, (2) mowed with a sickle bar mower and wheat mulch or straw
removed, (3) mowed, removed and soil tilled with a disk, or (4) mowed,
removed, tilled and the wheat mulch or straw replaced following
tillage. Following either residue removal or tilling, but prior to
replacement, corn or soybeans were planted using a John Deere 2-row
Max-Emerge no-till planter. Planter adjustments were made to obtain
a uniform planting depth in the various treatments. Paraquat at
0.56 kg ai/ha and alachlor at 2.2 kg ai/ha were applied following
planting or mulch/straw replacement. Herbicides were applied as a
tank mix using a CO_2 back sprayer with a carrier volume of 171.2
L/ha. Treatments within each crop were arranged in a randomized
block design with four replications. Plots consisted of four 96-cm-
width rows 15 m long.
        At both locations, morningglory was the predominant weed species.
During the growing season, morningglory plant counts and biomass data
were obtained to evaluate the effects of wheat residues and tillage
on morningglory growth.

        Study 2. This study was conducted at the Upper Coastal Plain
Research Station near Rocky Mount (Norfolk loamy sand) and the Lower
Coastal Plain Research Station near Kinston (Goldsboro loamy sand),
North Carolina, to investigate the effects of various small grain
residues on weed growth in no-till corn and soybeans.
        Wheat, oats (Avena sativa L. 'Brooks'), barley (Hordeum vulgare
L. 'Keowee') and rye ('Abruzzi') were planted at rates of 101, 72,
108, and 94 kg/ha, respectively, in October of 1980. Two treatments
were also included which were not planted in small grains. Bare
soil plots were kept weed-free during the winter and early spring
months with applications of paraquat (0.56 kg ai/ha). The following
growing season, small grains were either killed with paraquat in
early spring and planted no-till in corn or harvested in June and
soybeans planted no-till in stubble and residue remaining. Prior to
corn or soybean planting, one of the bare soil treatments was tilled
with a disk to simulate a conventional seedbed. The other bare
treatment remained undisturbed and was planted no-till. Herbicide
treatment, statistical design and plot size were identical to those
of Study 1. Throughout the growing season, visual weed ratings were
obtained to evaluate the effects of the four small grain residues
and tillage on weed growth.

Isolation and Identification of Phytotoxins

Rye Mulch Studies.

        Extraction Procedure. A flow chart of the isolation and
identification procedure is presented in Figure 1. Field-grown rye
('Abruzzi', harvested at early flowering stage on March 24, 1983,
from the Central Crops Research Station, Clayton, NC) was air-dried
for 7 days. The tissue (150 g) was extracted with 3 L of distilled
water for 10 hr with agitation. The extract was filtered through
cheesecloth and then centrifuged at 28,000 x g for 20 min. The
supernatant was reduced in volume to 300 ml in vacuo at 50°C. Sixty
ml of the concentrated aqueous extract was dried in vacuo, the
residue extracted with 20 ml of methanol and filtered. The methanolic extract was stored at 0°C until use.

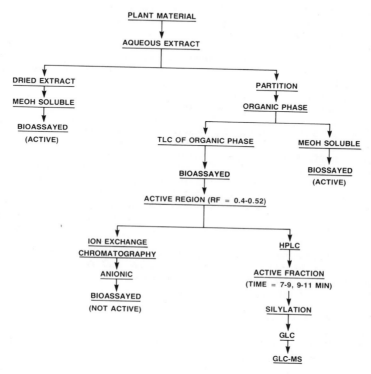

Figure 1. Flow diagram of separation technique.

The remaining 240 ml of concentrated aqueous extract was
adjusted to pH 2.5 with 2 N HCl and partitioned three times with
240 ml of diethyl ether and then 240 ml of ethyl acetate. The
organic fractions were combined and dried for 2 hr over $MgSO_4$ and
then filtered, dried in vacuo and reconstituted with 80 ml of diethyl
ether:ethyl acetate (1:1).

The components in the organic phase were separated first using
thin-layer chromatography (TLC). The organic phase (80 ml) was
streaked onto 20 plates of 1000 μm silica GF (Analtech, Inc.,
Newark, DE) and developed using toluene-ethyl formate-formic acid
[(5:4:1), 41]. Compounds were then located with ultra-violet light
(UV) at 250 nm. The compounds were eluted from the silica gel with
ethyl acetate-methanol (2:1) followed by filtration with Whatman #1
filter paper. The filtrate was dried in vacuo and the residue
dissolved in 80 ml of methanol.

Thirty ml of the above methanol solution which contained
biologically active compounds (R_f = 0.4-0.52) was filtered (0.4 μm
millipore filter) and dried in vacuo. This material was then
dissolved in 15 ml of 35% aqueous methanol and components separated
using high performance liquid chromatography (HPLC). The instrument
used was a Waters Associates HPLC (Milford, MA; Model 6000A pumping
system, Model U6K injector, Model 480 Lambda-max LC spectrophotometer) with a μBonddapak C_{18} column (7.8 mm x 30 cm). The solvent
(35% aqueous methanol) was delivered isocratically at a flow rate of
1.5 ml/min and the column effluent monitored at 350 nm (0.5 aufs for
6-30 min). Effluent was collected every 2 min with a total run time
of 30 min (Figure 1). Each of the 15 fractions was dried in vacuo
and residue dissolved in 15 ml of methanol and stored at 0°C.

An extract blank (i.e., distilled water only) was also included
throughout the extraction procedure to ensure that all isolated
compounds were of biological origin.

Qualitative Analysis. Extract samples (200 μl) which had been
purified by TLC (R_f = 0.4-0.52) and HPLC (7-11 min) were dried at
40°C under N_2 gas. Trimethylsilyl (TMS) derivatives of samples were
formed by adding 25 μl of N,O-bis(trimethylsilyl)trifluoroacetamide
(BSTFA, Pierce Chemical Co., Rockford, IL) and 25 μl pyridine to a
1-ml reaction vial and heating to 70°C for 30 min (42). Deuterated
trimethylsilyl derivatives (d-TMS) of samples were prepared by reaction with Deutero-regisil (Regis Chemical Co., Morton Grove, IL).
The derivatized samples (1-4 μl) were injected (splitless) into a
30-m (0-35-mm i.d.) DB-5 fused silica capillary column [95% dimethyl-
(5%)-diphenyl poly-siloxane, J and W Sci., Inc., Rancho Cordova, CA].
A Hewlett Packard GC/MS 5985B was used to obtain retention times
(t_r), electron impact (EI) and chemical ionization (CI) mass spectra
of the derivatized samples and standard compounds. Helium was used
as the carrier gas (ca. 1 ml/min) and methane as the makeup gas
(15 ml/min) for CI runs. Chromatographic conditions were as follows:
injection port temperature, 250°C; detector temperature, 270°C;
column temperature, 50°C for 2 min and then 8°C/min programmed to
280°C. All mass spectra were acquired at 70 eV. A Varian GC 3700
(Varian Associates, Sunnyvale, CA) was used to corroborate identifications by co-injection of samples and standard TMS derivatives.

Bioassays. Biological activity of various isolated compounds
was determined using Chenopodium album L. seed collected in North

Carolina in 1981. Each sample (1.4 ml of methanolic sample) was
placed into a 3-cm petri dish and the solvent evaporated under a
laminar flow hood at room temperature. Seventy seeds (0.035 g) were
then placed into the petri dishes and 1.4 ml of sterilized (0.2 μm-
filter) 15 mM Mes [2-(N-morpholino)ethanesulfonic acid; Sigma
Chemical Co.] buffer adjusted to pH 5.5 was added. The dishes were
kept in the dark at $25°C_2$ for 84 hr, exposed to 12-hr fluorescent
light (250 μ einsteins/m^2/sec), and then placed back in the dark for
an additional 4 days (17). Percent germination, root and hypocotyl
lengths were then determined.

The following chemicals were obtained commercially (Sigma
Chemical Co.) and bioassayed with C. album and Amaranthus retroflexus
L. (seeds collected in North Carolina in 1980) following identification: DL-β-hydroxybutyric acid (DL-3-hydroxy-butyric acid as a Na
salt) and L-β-phenyllactic acid (L-2-hydroxy-3-phenyl-propanoic
acid).

Data were subjected to analysis of variance and regression
analysis using the general linear model procedure of the Statistical
Analysis System (40). Polynomial equations were best fitted to the
data based on significance level of the terms of the equations and
R^2 value.

Wheat Mulch Studies

Extraction Procedure. Wheat ('McNair 1813') plant material used
in this study was harvested from the field in early spring after the
wheat had tillered but before heading. The plant material was dried
at 50°C for 48 hr.

Five-g samples of wheat plant material were soaked in 150 ml
of water for 10 hr at room temperature. The mixture was filtered and
the filtrate used to germinate pitted morningglory (Ipomoea lacunosa
L.) and ragweed (Ambrosia artemisiifolia L.) seed. The effect of
the extract on germination was tested in the presence and absence of
light.

Extractions were done by the procedure of Guenzi and McCalla
(22). Five-g wheat samples were hydrolyzed with 2 N NaOH for 4 hr
at room temperature. The alkaline extract was filtered and the
filtrate acidified to pH 2 with concentrated HCl and extracted with
diethyl ether. The concentrated ether fraction was used to spot
thin-layer chromatography (TLC) plates. Silica gel G obtained from
Sigma Chemical Company was used. The plates were developed in a
benzene-methanol-acetic acid system (80:10:5 v/v/v). Compounds
separated by one-dimensional TLC were detected by exposing the
developed plates to long-wave (3360 A) UV light. The isolated
compounds noted under UV light were individually scraped from the
TLC plates and then bioassayed for morningglory seed germination
inhibition.

Identification. One compound isolated by TLC was found to be
very inhibitory to morningglory seed germination and was identified
using mass spectrometry. A LKB 2091 GC-MS was used for GC-MS
analysis. In addition to GC-MS, the sample was also analyzed by
direct probe.

Bioassay for Toxicity of Inhibitory Compound(s). Since the
identity of the inhibitory compound was determined to be ferulic acid
(4-hydroxy-3-methoxycinnamic acid), ferulic acid obtained from Sigma
Chemical Company was used in germination bioassays.

Ferulic acid solutions of 5.0, 1.0, and 0.5 mM buffered to pH
6.5 with sodium phosphate were used to germinate morningglory,
prickly sida, ragweed, crabgrass (Digitaria sanguinalis L.), corn,
and soybean (Glycine max L. Merr.). The weed and crop seed were
germinated in 9-cm Petri dishes. To each dish, 10 ml of test solution was added. At the termination of the bioassay (4 days),
percent germination and root length data were taken.

RESULTS AND DISCUSSION

Rye Mulch Studies

Field Research. Rye mulch (aboveground herbage) caused a significant increase in the percent control of grass species even when the
soil was tilled (cut mulch, glyphosate and paraquat desiccation and
till and replace mulch versus no rye till or no-till (Table I). Rye
root residue also caused a significant improvement in grass control
although not up to the level provided by the addition of rye mulch.
Soil disturbance did not significantly affect the level of grass
control (no rye, till versus no-till) which is different than the
response of broadleaf weeds.

All three broadleaf weeds, redroot pigweed, common lambsquarters
and Ambrosia artemisiifolia (common ragweed) responded to tillage in
the same manner--tillage caused an increase in density and biomass
(Table II) and a decrease in percent weed control [Table I (no rye,
no-till versus till)]. Many weed species are known to respond
positively to soil tillage (9, 15, 16). At least part of the reason
is that tillage provides a light stimulus which is a requirement for
many species (18, 19). Common lambsquarters seed collected in North
Carolina were shown to respond positively to light both qualitatively
and quantitatively (data not shown). The requirement of light for
the germination of common lambsquarters has been reported previously
(17).

The effect of tillage (i.e., increase in biomass, density and
decrease in percent weed control) was eliminated by placing rye
mulch on tilled soil--no rye till versus till and mulch replaced
(Table II). The reduction in biomass of common lambsquarters and
redroot pigweed was greater than for common ragweed; 96, 78, and 39%,
respectively. Also note that the density of only redroot pigweed
was significantly reduced by replacement of the rye mulch. In other
words, the mulch caused a tremendous decrease in the size of common
lambsquarters and ragweed but not a reduction in numbers.

The broadleaf weeds responded to tillage in the same general
manner when rye root residue was present, however, it does appear
that the growth of all the weeds was inhibited by the presence of
rye root residue. This can be seen in Table II by comparing results
from the tilled treatment with and without rye root residue. In the
case of common ragweed, biomass was reduced 34% (178.3 g no rye,
till versus 118.3 remove mulch till), but the density increased by
69% (140 plants/2.2 m^2 versus 280 plants/2.2 m^2) when rye root

Table I. The Effect of Tillage and Rye Residue (Above- and Belowground) on Weed Control Averaged Across Three Cropping Systems at Clayton, NC, in 1982 and 1983

Treatment	% Weed Control[a]			
	Grass[b]	Redroot[c,d] Pigweed	Common[c] Lambsquarters	Common[c] Ragweed
Remove mulch, no till	60c	55ab	49d	68a
Remove mulch, till	71bc	58ab	60cd	33c
Cut mulch, no-till	91a	69a	84a	73a
Remove mulch, till, replace mulch	82ab	63ab	70bc	43bc
Glyphosate, no-till	84ab	74a	65bc	80a
Paraquat, no-till	81ab	79a	77ab	73a
No rye, till	28d	8c	10f	10d
No rye, no-till	33d	41b	32e	55b

[a]Means within a column followed by the same letter are not significantly different as determined by Waller-Duncan T-test (K ratio = 100).

[b]Large crabgrass and fall panicum; rated 31 days after planting.

[c]Rated 49 days after planting.

[d]Sunflower and soybeans only.

Table II. The Effect of Tillage and Rye Residue (Above- and Belowground) on Weed Biomass and Density Averaged Across Three Cropping Systems at Clayton, NC, in 1982 and 1983[a]

Treatments	Redroot Pigweed		Common Lambsquarters		Common Ragweed	
	Density[b] (plants/2.2m^2)	AGB[c] (g)	Density[b] (plants/2.2m^2)	AGB[c] (g)	Density[b] (plants/2.2m^2)	AGB[c] (g)
Remove mulch, no-till	8abc	2.4bc	25b	65.6cd	21c	34.0c
Remove mulch, till	4bc	0.8bc	132a	117.8bc	236a	118.3b
Cut mulch, no-till	2c	0.4c	7b	32.4d	17c	29.9c
Remove mulch, till, replace mulch	3bc	0.8bc	120a	44.8d	186ab	108.2b
Glyphosate, no-till	2c	0.4c	5b	44.3d	9c	5.9c
Paraquat, no-till	1c	0.2c	5b	25.9d	16c	12.3c
No rye, till	26a	22.1a	140a	199.9a	140b	178.3a
No rye, no-till	21ab	12.0ab	41b	124.3b	15c	34.7c

[a]Means within a column followed by the same letter are not significantly different as determined by Waller-Duncan T-test (K ratio = 100).

[b]Sunflower and soybeans only.

[c]AGB: Aboveground biomass. AGB and densities determined 49 days after test initiation.

residue was present in the tilled soil. Common lambsquarters
responded to tillage in the presence of rye root residue in the same
manner as common ragweed, in that biomass was decreased but the
density was unaffected. Redroot pigweed appears to be the most
inhibited by rye root residue in that both density and biomass were
reduced in the tilled treatment.

The growth of common lambsquarters and redroot pigweed was
significantly reduced by rye root residue in the no-till system as
well (no rye, no-till versus remove mulch, no-till). Thus, both
common lambsquarters and redroot pigweed are affected by both rye
mulch and root residue, but common ragweed appears to be only
moderately affected by rye residues. However, all weed species
observed responded to soil disturbance. Overall, the elimination of
tillage and the presence of rye (cut mulch, no-till versus no rye,
till) caused a decrease in the biomass of redroot pigweed, common
lambsquarters and common ragweed of 96, 84, and 83%, respectively.
The majority of the reduction in common ragweed was due to the
elimination of tillage and not the presence of rye. This is
contrary to the findings of Putnam et al. (13) who reported that
common ragweed was inhibited by rye mulch.

Growth Chamber Tests. The extraction of field grown rye with water
was an initial attempt to determine the allelopathic potential of
rye simulating a natural release mechanism (i.e., rain). Common
lambsquarters was used because it seemed to be the most sensitive
weed species of the indigenous weeds observed. As shown in Figure 2,
common lambsquarters emergence (density) was significantly inhibited
by aqueous rye extracts at the higher concentration. Low concentrations of known allelopathic compounds are known to cause stimulation
of weed species (26). The rye extract caused a similar response
from common lambsquarters. Another explanation for the stimulation
at low concentrations could be that the extract provided additional
nitrogen for the common lambsquarters.

It should be pointed out that various measurements were taken
in the field during the two years of research to determine which
growth parameters (light and soil temperatures) were affected by
these treatments. The degree to which each parameter was affected
was then used to establish laboratory limits, within which their
possible role in explaining the observed effects was evaluated. Of
those tested, light and allelopathy appear to be the most probable
causes of the observed effects. The implications of these studies
are first that agro-ecosystems could be manipulated to biologically
reduce certain weed pressures. And, second, chemicals present in
mulch and/or root exudates could be of practical significance in
terms of new herbicide chemistry.

Wheat Mulch Studies

Field Studies.

Study 1. Morningglory plant populations and biomass were found
to be greatly affected by both primary tillage and wheat mulch. In
general, tillage greatly increased densities of morningglory while
the presence of a wheat mulch had the tendency to reduce morningglory growth. Morningglory populations consisted of a mixture of
pitted (I. lacunosa L.) and tall (I. purpurea L.) species.

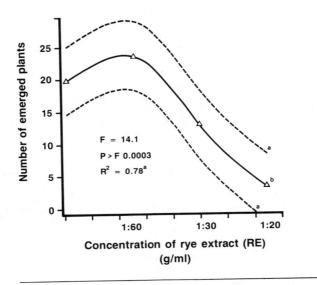

[a]95% mean confidence intervals

[b]Emergence = 19.75 + 974.52 (RE) - 54,295.51 (RE)^2 + 562,102.0 (RE)^3

Figure 2.   The effect of an aqueous extract of rye on the emergence of Chenopodium album.

In the corn test (Table III), the non-mulched tilled treatment accumulated an average of 302% more morningglory biomass than non-mulched non-tilled plots. The tillage effect, however, was essentially eliminated when tillage was followed by replacement of the wheat mulch. Morningglory growth in the tilled and replaced plots was similar to that observed in the non-tilled treatments. In the non-tilled treatments, there was little effect of the wheat mulch on morningglory biomass. Enhanced soil moisture under the mulch may have compensated for the reduced numbers of morningglory plants initially observed in the non-tilled mulch treatment as compared to the non-tilled non-mulched plots.

In the soybean test (Table IV), there was no effect of the wheat mulch on morningglory growth. Age differences of the mulch between the two tests may have accounted for the lack of weed control. Wheat at the time of treatment was fully mature in the soybean test and green in the corn test. Researchers have shown that straw cut while still green produced a higher level of phytotoxicity than those cut when fully mature. In addition, rainfall required to leach toxins from the mulch into the soil may have been lacking when soybeans were planted since the growing season in North Carolina in 1981 was generally dry. Although a mulch effect on morningglory growth in the soybean test was not observed, tillage increased morningglory biomass 269% over non-tilled treatments at the Clayton location (Table IV). In addition to morningglory, similar trends were observed for cocklebur (Xanthium pensylvanicum Wallr.). Higher cocklebur populations were observed in tilled plots than in non-tilled plots (data not shown).

Table III. Effect of Tillage and Wheat Mulch on Morningglory Populations and Biomass in Corn [a]

| | Morningglory | | | |
| | Plants/7.4m[2b] | | Dry wt. gm/7.4[2c] | |
	Plymouth	Clayton	Plymouth	Clayton
Remove mulch, no-till	4.0b	245.7ab	13.7b	73.0b
Remove mulch, till	72.4a	346.8a	157.0a	192.2a
Remove mulch, till, replace	3.2b	86.4b	30.7b	63.1b
No-till into mulch	3.4b	31.7b	37.5b	34.6b

[a]Within each column, values sharing the same letter are not significantly different at the 0.05 level, according to Duncan's multiple range test.

[b]Counts made 25 days after planting.

[c]Biomass obtained 45 days after planting.

No deleterious effects of either the wheat mulch or tillage were observed on soybeans (Table V) or corn (data not presented). In general, crop growth was better in mulched or no-till plots. Enhanced soil moisture in the mulched treatments and reduced morningglory densities in the non-tilled treatments may have contributed to improved crop growth.

Table IV.   Effect of Tillage and Wheat Mulch on Morningglory
            populations and biomass in soybeans [a]

	Morningglory			
	Plants/7.4m[2b]		Dry wt. gm/7.4[2c]	
	Plymouth	Clayton	Plymouth	Clayton
Remove mulch, no-till	15.8ab	1.0c	253.4a	58.9b
Remove mulch, till	24.0ab	71.5a	194.1a	253.7a
Remove mulch, till, replace	9.0b	44.0b	202.5a	207.1a
No-till into mulch	28.0a	1.5c	262.9a	65.8a

[a]Within each column, values sharing the same letter are not
significantly different at the 0.05 level, according to Duncan's
multiple range test.

[b]Counts made 30 days after planting.

[c]Biomass obtained 90 days after planting.

Table V.   Soybean biomass as influenced by tillage and
           wheat mulch [a]

	Soybean fresh wt. Kg/6.0m[2b]	
	Plymouth	Clayton
Remove mulch, no-till	7.82b	11.93ab
Remove mulch, till	7.91b	10.57b
Remove mulch, till, replace	9.98ab	10.57b
No-till into mulch	10.21a	12.93a

[a]Within each column, values sharing the same letter are not
significantly different at the 0.05 level, according to Duncan's
multiple range test.

[b]Biomass obtained 90 days after planting.

Study 2.   The effects of wheat, oats, barley and rye mulches on
three broadleaf weed species and crabgrass (Digitaria spp.) are
shown in Table VI.   Weed control data for the corn test at Kinston
are not presented because of poor cover crop kill by the paraquat
treatment.
     In both the corn and soybean tests, there were no appreciable
differences in the broadleaf weed control among the four small grain
mulches.   In addition, age of the small grain plant material at
planting time had essentially no effect on weed control.   The reduction in weed control by the older wheat straw in the soybean test of
Study 1 was not observed here.   Broadleaf weed control in the small
grain stubble and residue of the soybean test was similar to the
control observed with the younger mulches present in the corn test.
The reasons for the observed differences between the two studies are
unclear.   Perhaps it is because weed densities are so greatly reduced by eliminating tillage the presence of a mulch would have little
influence.   This is probably why in the corn test of Study 1 a
reduction in weed growth by the mulch was only observed in plots
that were tilled followed by replacement of the wheat mulch.

Table VI. Effect of Small Grain Mulch and Tillage on Weed Control in Corn and Soybeans[a]

Small grain Mulch	Soybean[b]					Corn[c,d]		
	Rocky Mount			Kinston		Rocky Mount		
	Morning-glory	Crab-grass	Prickly Sida	Sickle-pod	Crab-grass	Morning-glory	Crab-grass	Prickly Sida
Wheat	78a	76a	83a	68b-e	71a-d	93a	35de	85ab
Oats	74a	69ab	81a	60c-e	85a-c	81a-d	41c-e	84ab
Barley	77a	68ab	81a	53de	85a-c	88ab	65a-b	83a-c
Rye	82a	88a	89a	84a-c	90ab	88ab	61a-d	79a-c
No cover, no-till	39c	35c	28cd	50de	80a-c	68a-d	41c-e	66a-d
No cover, tilled	36c	74a	3d	41e	96a	49b-e	90ab	20e

[a]Within each crop, and within each location, values sharing the same letter are not significantly different at the 0.05 level, according to Duncan's multiple range test.

[b]Ratings conducted 35 days after planting.

[c]Ratings conducted 95 days after planting.

[d]Corn test at Kinston abandoned due to poor cover crop kill with paraquat.

When broadleaf weed control ratings were averaged over location, crop, and weed species [morningglory, prickly sida (Sida spinosa L.), and sicklepod (Cassia obtusifolia L.)], weed control in the no-till mulched plots was 37 and 62% better than the control achieved in the non-mulched non-tilled and the non-mulched tilled treatments, respectively (Table VII). Of the three broadleaf weed species studied, the growth of prickly sida was influenced the most by mulch and tillage treatments. The finding that ferulic acid leached from small grain straws is converted to a more phytotoxic styrene derivative by bacteria on the carpels of prickly sida seed may account for the greater reduction of this weed by small grain residues (26).

Crabgrass control in the mulched plots was generally lower than the control of broadleaf species, despite the alachlor treatment at planting time (Table VI). Poor crabgrass control in the non-tilled treatments may have been due to the low early season rainfall which reduced the effectiveness of alachlor.

Table VII. Mean broadleaf weed control of mulched and unmulched treatments averaged over location, crop, and weed species[a]

Small grain mulch	Weed control[b]
Wheat	81.0a
Oats	76.0a
Barley	76.1a
Rye	84.1a
No cover, no-till	50.0b
No cover, tilled	29.8c

[a]Weed species include: morningglory, prickly sida and sicklepod.

[b]Values sharing the same letter are not significantly different at the 0.05 level, according to Duncan's multiple range test.

As in Study 1, no adverse effects of either mulch or tillage were observed on corn or soybean growth.

Small grain mulches studied here did suppress weed growth, but equally important is the role of tillage. Eliminating tillage reduced densities of morningglory, prickly sida and sicklepod. When a soil is tilled, environmental factors will change which may enhance weed seed germination. Of these factors, exposure of weed seed to light may be the most important (19), even though morningglory germination studied here was not influenced by light (data not shown).

For the most part, weed control is achieved mainly through cultural, mechanical, or chemical practices. Because weed control options are limited with the adoption of minimum tillage cropping practices, the difficulty of controlling weeds is often cited as a problem in no-till crop production (8). However, environmental and ecological differences which distinguish no-till from conventional tillage may benefit growers by enhancing control of certain weed species in no-till cropping systems. With the proper choice and management of cover crops and plant residues, it may be possible to supplement if not reduce the number and amount of herbicides used in these cropping systems by eliminating tillage which restricts weed seeds to poor germination sites and by utilizing natural phytotoxic substances leaching from plant residues.

Isolation and Identification of Phytotoxins

Rye Mulch Studies. Extraction of the dried aqueous extract (Figure 1) with methanol gave a preparation which showed the greatest activity, giving 58, 80 and 86% inhibition of C. album hypocotyl length, root length and germination, respectively (data not shown). Therefore, methanol was used for all subsequent transfers.

The activity in the aqueous extract was also effectively partitioned into a diethyl ether organic phase. At this step, relative inhibition increased compared to the methanol extract. The organic phase extract caused 96, 93, and 89% inhibition of C. album hypocotyl length, root length, and germination, respectively (data not shown).

"Semi-preparative HPLC" was then used to further purify the extract. Instrumentation and techniques were the same as described previously. Active compounds were contained in the 7-11 min fraction (data not shown).

β-phenyllactic acid (βPLA) and β-hydroxybutyric acid (βHPA) were identified in the HPLC active fraction by comparing retention time (t_r) and mass spectra with those of standard compounds. Chemical ionization (CI) of the TMS derivatives and electron impact (EI) of the TMS and d-TMS derivatives indicated a MW of 248 for one of the unknown compounds and 310 for another unknown (Table VIII). The absence of molecular ions (M+) in TMS derivatives of hydroxy acids has been previously reported (43). By comparing the EI spectra of TMS and d-TMS derivatives, the number of TMS groups was also determined. In the case of βPLA, the difference in mass between the TMS derivative (295) and the d-TMS (310) was 15. If the M+ was present in the spectra, the mass difference (TMS vs. d-TMS) would have been 18 (Table VIII). Thus, the CI and EI spectra give the M+ and the number of silylated functions (Table VIII). Once this was determined, EI spectra of TMS derivatives were compared to standard spectra (44, 45) in the literature. Standard spectra were then generated for βPLA and βHBA. In so doing, t_r of the unknown compounds was also matched to standard compounds. Also, co-injection of TMS derivatives of standards and samples was used to further substantiate the identifications.

Biological Activity of Identified Compounds. Both βHBA and βPLA inhibited C. album hypocotyl length (HL) at 8 mM by 30 and 68%, respectively (Figure 3). It is probable that total seedling growth would be more sensitive. Only βPLA significantly decreased HL at lower concentrations (17% at 4 mM). Both compounds also inhibited C. album root length (RL) [Figure 4]. There was not a significant difference in effect on RL between the two compounds, although βPLA caused significant inhibition at a lower concentration (20% at 2 mM) as compared to the control.

Hypocotyl length of A. retroflexus was more sensitive to βPLA inhibition than C. album (Figures 3 and 6). Hypocotyl length of C. album was inhibited by 17% at 4 mM. With A. retroflexus, 0.8 mM caused a 17% inhibition of HL. Thus, βPLA was 5X more inhibitory on hypocotyl growth of A. retroflexus. Also, βPLA completely

Table VIII. Tabulation of Important Ions Present in Mass Spectra, GC Retention Times (t_r), Molecular Weights of Unknown and Standard Compounds and Number of Trimethylsilyl (TMS) Groups per Molecule

Chemical	EI				CI (TMS)			MW	t_r	No. of TMS groups
	M^+		M^+-15							
	TMS	d-TMS	TMS	d-TMS	M+1	M+29	M+41			
(1) β-hydroxybutyric acid	—[a]	—[a]	233	248	249	277	289	104	16.1[b]	2
(2) β-Phenyllactic acid	—[a]	—[a]	295	310	311	339	351	166	24.3[b]	2

[a]Not present in mass spectra.

[b]Compared to corresponding standards.

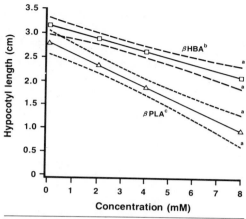

[a] 95% confidence intervals.

[b] HL = 3.15-0.13 (βHBA), F = 50.41 P>F 0.0001 R^2 = 0.84

[c] HL = 2.80-0.23 (βPLA), F = 73.51 P>F 0.0001 R^2 = 0.85

Figure 3. The effect of DL-β-hydroxybutyric acid (βHBA) and L-β-phenyllactic acid (βPLA) on C. album hypocotyl length (HL).

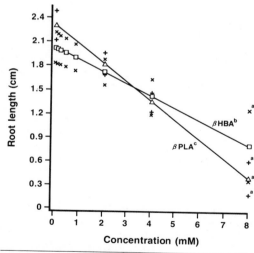

[a] 95% mean confidence intervals.

[b] RL = 2.02-0.15 (βHBA), F = 22.31 P>F 0.0001 R^2 = 0.51

[c] RL = 2.30-0.24 (βPLA), F = 179.75 P>F 0.0001 R^2 = 0.95

Figure 4. The effect of DL-β-hydroxybutyric acid (βHBA) and L-β-phenyllactic acid (βPLA) on C. album root length (RL).

inhibited hypocotyl length of A. retroflexus at 8 mM (Figure 5).
There was little difference in response between the two weed species
to βHBA (a 30% inhibition at 8 mM for C. album versus 27% for A.
retroflexus). There was no significant interaction effect on HL
between the two compounds (F = 0.97, P > F 0.50). The difference in
sensitivity of the weed species to the two compounds could have been
due to the duration of the experiments (8 days for C. album versus
60 hr for A. retroflexus). The longer duration may have resulted in
a lower concentration of the compound because of microbial degradation.

Amaranthus retroflexus RL was significantly inhibited by both
compounds (Figure 6). There was no significant difference in
inhibition of C. album root growth. This again indicates species
specificity in response to the two compounds. The effect on A.
retroflexus RL indicated a significant interaction between the two
compounds. This interaction was antagonistic, as the combined
inhibition by both compounds was less than the sum of the individual
inhibition. At 2 mM βPLA alone reduced RL by 59% and βHBA alone
reduced RL by 33%. However, in combination the two compounds reduced
RL by 62%, as opposed to an expected inhibition of 92% (i.e., 59 +
33 = 92%) had the interaction been additive.

Aliphatic acids such as butyric acid have been previously
implicated as being allelopathic compounds (46, 47, 23). Chou and
Patrick (23) isolated butyric acid from soil amended with rye and
showed that it was phytotoxic. Hydroxy acids have also been shown to
possess phytotoxic properties (48) but have not been implicated in
any allelopathic associations. Since βHBA is a stereo isomer, and
the enantiomer was not identified because of impurity, all bioassays
were run using a racemic mixture. The D-(-) stereo isomer of βHBA
has been isolated from both microorganisms and root nodules of
legumes and is suspected to be a metabolic intermediate in these
systems (49). It is likely that only one enantiomer was present in
the extract; therefore, the true phytotoxic potential of this compound awaits clarification of the phytotoxicity of the individual
enantiomers.

Most of the simple aromatic acids which have been implicated in
allelopathic associations are derived from either cinnamic acid (in
the case of phenolic acids) or benzoic acid (47). Therefore, βPLA
is unique among allelopathic aromatic acids in that it is believed
to be an intermediate (although of minor importance) in the shikimic
acid pathway and not an end product (50, 51, 52, 53). Although βPLA
is a stereo isomer and the exact enantiomer was not identified in
this study, Tamura and Chang (54) and Kimura and Tamura (55) isolated
and identified L-βPLA from two fungal species and demonstrated plant
growth regulator activity. These workers also showed that L-βPLA
affected hypocotyl and root growth of lettuce, although in both
cases growth was stimulated in this species. They also reported that
D-βPLA did not affect growth as much as the L form. This could
explain why DL-βPLA did not affect the growth of the bioassay species
in this study (data not shown). It is possible that both βHBA and
βPLA were of microbial origin as most of the reports on these compounds deal with plant microorganism associations.

No definite conclusions can be drawn as to the potential use of
these results in explaining allelopathic associations under field

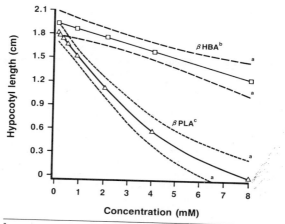

^{a}95% mean confidence intervals.

^{b}HL = 1.93-0.09 (βHBA), F = 24.26 P > F 0.0003 R^2 = 0.65

^{c}HL = 1.82-0.39 (βPLA) + 0.02 (βPLA)^2, F = 124.09 P>F 0.0001

Figure 5.   The effect of DL-β-hydroxybutyric acid (βHBA) and L-β-phenyllactic acid (βPLA) on <u>A. retroflexus</u> hypocotyl length (HL).

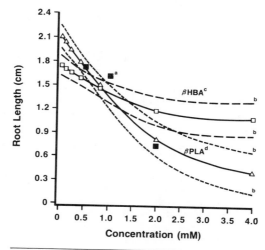

^{a}Equal molar concentration of both compounds: RL = 1.99-0.34 (βHBA) -0.14 (βPLA) -0.23 (βPLA)^2 + 0.16 (βHBA $\times$ βPLA), F = 47.78 P>F 0.0001 R^2 = 0.82

^{b}95% mean confidence intervals.

^{c}RL = 1.75-0.40 (βHBA)-0.29 (βHBA)^2, F = 19.36 P>F 0.0001 R^2 = 0.68

^{d}RL = 2.10-0.85 (βPLA) + 0.11, F = 83.07 P>F 0.0001 R^2 = 0.90.

Figure 6.   The effect of DL-β-hydroxybutyric acid (βHBA) and L-β-phenyllactic acid (βPLA) individually and in combination on <u>A. retroflexus</u> root length (RL).

conditions. However, both βPLA and βHBA are easily extractable by
water and thus could be readily leached from rye mulch to inhibit
weed growth.

Wheat Mulch Studies. The effect of the aqueous wheat extract on
morningglory and ragweed germination is shown in Table IX. In the
dark, the presence of the wheat extract reduced morningglory germination and root length 27 and 66%, respectively. The wheat extract
did not, however, significantly reduce ragweed germination in the
dark, but did reduce ragweed root length 86%. Ragweed was chosen in
this test because levels of this weed have not been observed to be
reduced by no-till cropping practices. Greater phytotoxicity was
observed when the bioassays were conducted in the presence of light.
The enhanced phytotoxicity due to light was particularly apparent on
ragweed. Ragweed seed wetted with the wheat extract and germinated
in the presence of light did not germinate, yet light alone had a
slight stimulatory effect on ragweed germination. Morningglory
germination and root length were reduced 65 and 62%, respectively,
when seed wetted with the wheat extract were germinated in the light.

Table IX. Effect of Light on Morningglory and Ragweed Germination
           with Aqueous Wheat Extract

Treatment	Morningglory		Ragweed	
	Germination (%)	Root Length (cm)	Germination (%)	Root Length (cm)
Dark + extract	53 b[a]	1.5 c	29 b	1.1 c
Dark check	73 a	4.4 a	34 b	7.6 a
Light + extract	30 c	1.0 c	0 c	0.0 d
Light check	85 a	2.6 b	45 a	3.0 b

[a]Values within columns sharing the same letter are not significantly different according to the Waller-Duncan procedure, assuming
a K ratio of 100.

Reprinted with permission of Plenum Publishing Corp. (28).

Isolation of Inhibitory Compound. The hydrolyzed extract of
wheat straw yielded many compounds when subjected to TLC separation.
When individual spots were removed and bioassayed for morningglory
seed germination, only two of the spots significantly inhibited germination (Table X). Of the two inhibitory compounds, the compound
with R_f 0.5 was the more inhibitory. This compound reduced morningglory germination and root length 94 and 89%, respectively. Although
the substance at R_f 0.5 was found to be the most inhibitory, concentration differences of the various components in the ether extract
may have been partially responsible for the greater inhibition of
the test solution containing the compound at R_f 0.5.

Identification of Inhibitory Compound. Of the two compounds
isolated by TLC that had activity on morningglory germination, only
the one at R_f 0.5 on the TLC plate was analyzed by GC-MS. From the

Table X. Effect of Compounds Isolated from Wheat
by TLC on Morningglory Seedling Growth

R_f value	Germination (%)	Root length (cm)
0.5	6 c[a]	0.3 c
0.95	67 b	1.5 b
Check	94 a	2.7 a

[a]Values within columns sharing the same letter are not significantly different according to the Waller-Duncan procedure assuming a
K ratio of 100.

Reprinted with permission of Plenum Publishing Corp. (28).

mass spectral data, the identity of the compound at R_f 0.5 was
determined to be ferulic acid.

<u>Biological Activity of Identified Compound.</u> The effect of
ferulic acid on the germination and root length of morningglory,
ragweed, prickly sida, and crabgrass is shown in Table XI. Significant reductions in germination or root length of morningglory, crabgrass, and prickly sida seed enclosed in carpels were observed only
at the highest concentration of ferulic acid. Ferulic acid, however,
had no effect on either ragweed or prickly sida seed with carpels
removed. Of the four weed species bioassayed with ferulic acid,
crabgrass appeared to be the most sensitive. No crabgrass germination was observed at the high concentration of ferulic acid. None
of the concentrations of ferulic acid used had any effect on the
germination of corn or soybean (Table XII). Ferulic acid did, however, inhibit the root growth of both corn and soybean at the highest concentration.

Table XII. Influence of Ferulic Acid
on Germination of Corn and Soybean Seed

Ferulic acid (M)	Corn		Soybean	
	Germ. (%)	Root length (cm)	Germ. (%)	Root length (cm)
5×10^{-3}	68	0.9	63	2.7
1×10^{-3}	83	3.7	65	2.7
5×10^{-4}	85	3.3	70	3.7
Check	83	2.9	82	4.4
LSD (0.05)	NS	1.2	NS	1.2

Reprinted with permission of Plenum Publishing Corp. (28).

Of the many types of phytotoxic compounds released from decaying plant material by microbial activity or leaching, the phenolic
acids are probably the most common (47, 56). Ferulic acid, as well
as other phenolic acids, is produced from intermediates of

Table XI. Effect of Ferulic Acid on Morningglory, Ragweed, Crabgrass, and Prickly Sida Seed Germination

| Ferulic Acid (M) | Morningglory | | Ragweed | | Crabgrass | | Prickly sida | | | |
| | | | | | | | Without Carpel | | With Carpel | |
	Germ. (5)	Root Length (cm)	Germ. (%)	Root Length (cm)	Germ. (%)	Root Length (cm)	Germ. (%)	Root Length (cm)	Germ. (%)	Root Length (cm)
5×10^{-3}	58	0.3	26	1.5	0	0	16	1.4	3	0.3
1×10^{-3}	68	1.4	29	1.6	29	0.9	18	1.6	13	1.7
5×10^{-4}	65	1.8	27	1.6	34	1.0	15	1.8	21	2.1
Check	75	1.7	27	1.6	30	1.1	19	1.6	20	1.7
LSD (0.05)	15	0.6	NS	NS	19	0.3	NS	NS	9	0.3

Reprinted with permission of Plenum Publishing Corp. (28).

respiratory metabolism via the shikimic acid pathway. Ferulic acid has been found in a variety of crop residues by a number of researchers (22, 57, 58). The reported inhibitory effects of ferulic acid on germination and seedling growth have varied widely. Borner (59) found that a ferulic acid concentration as low as 10 ppm inhibited the growth of wheat and rye roots. Guenzi and McCalla (22), however, reported that a 2500 ppm solution of ferulic acid had no effect on the germination of wheat and reduced the growth of wheat roots by only about 50%. In work with soybean seedlings, Patterson (25) showed that soybean total dry weight, leaf area, plant height, and number of leaves were all significantly reduced when soybean plants were grown in solution culture containing 194 ppm of ferulic acid. Ferulic acid had no effect on soybean growth when the ferulic acid concentration was reduced to 19.4 ppm. Lodhi (57) reported that a ferulic acid concentration of 194 ppm was very inhibitory to the seed germination and radicle growth of radish (Raphanus sativus L.).

With the extraction procedure we employed (22), ferulic acid was isolated as the most inhibitory component in wheat straw. There could also be other unknown compounds in the straw which would not be evident with this procedure. In addition, we ignored the possible influence of toxin-producing microorganisms. Microorganisms may have influenced the phytotoxicity exhibited by the aqueous wheat extract in Table IX. Although the present study was not concerned with the phytotoxic effects of microbially decomposed wheat straw, an influence of microbial activity on ferulic acid phytotoxicity was observed. From the results shown in Table XI, it appears that the presence of the prickly sida seed carpel enhanced the inhibitory effects of ferulic acid. In addition to ferulic acid in test solutions containing prickly sida seeds with carpels, a second compound, 4-hydroxy-3-methoxy styrene, was also found to be present. This compound is formed by the decarboxylation of ferulic acid and was produced by a bacterium present on the carpel of prickly sida seed. The decarboxylation of ferulic acid was detected in aqueous solutions of ferulic acid inoculated with the bacterium isolated from the carpels of prickly sida seed. No conversion occurred when the bacterium was not present.

It seems most likely that the presence of the styrene compound was at least partially responsible for the inhibition of prickly sida germination and root length, since ferulic acid alone (prickly sida seed without carpels plus ferulic acid) had no effect on prickly sida germination or root length (Table XI). The decarboxylation of phenolic acids to corresponding styrenes is known from studies on fungi and bacteria (60, 61). However, in a number of studies directly concerned with the microbial decomposition of ferulic acid, as well as other phenolic acids, no mention is made of any styrene compounds produced as a result of phenolic acid decarboxylation (62, 63, 64, 65).

It is unlikely that any one particular compound could be responsible for reduced weed growth in no-till. Higher plants and microorganisms produce a myriad of phytotoxic substances. If these substances are present in the right combination and concentration, phytotoxic effects may be observed. With the proper choice and management of various cover crops and plant residues, it may be possible to supplement if not reduce the number and amount of herbicides used in no-till cropping systems by utilizing natural phytotoxic substances.

Acknowledgments

[1]Paper No. 9340 of the Journal Series of the North Carolina Agricultural Research Service, Raleigh, NC 27695-7627. The work reported here was supported in part from the Consortium of Integrated Pest Management Grant jointly funded by EPA (Agreement No. 806277-03) and USDA (Agreement No. 71-59-2481-1-2-039-1) and the North Carolina Tobacco Foundation, Inc.
[2]Current address: Monsanto Agricultural Products Co., 800 N. Lindbergh Blvd., St. Louis, MO 63167.

Literature Cited

1. Anderson, W.P. "Weed Science: Principles"; West Publishing Co.: New York; 1977; 598 pp.
2. Edwards, W.M. Proc. Nat'l. No-tillage Systems Symposium, February, 1972, p. 30-41.
3. Oschwald, W.R. (ed.). "Crop Residue Management Systems." Amer. Soc. of Agron., Crop Sci. Soc. Amer. and Soil Sci. Amer., Inc., 1979; Madison, Wisconsin, 248 pp.
4. Phillips, R.E.; Blevins, R.L.; Thomas, G.W.; Frye, W.W.; Phillips, S.H. Science 1980, 208, 1109-13.
5. Gallaher, R.N. Proc. South. Weed Sci. Soc., 1978, 31, 127-33.
6. Putnam, A.R.; DeFrank, J. Abs. Weed Sci. Soc. Amer., 1980, p. 35.
7. Guenzi, W.D.; McCalla, T.M.; Norstadt, F.A. Agr. J. 1967, 59, 163-4.
8. Worsham, A.D. Proc. 35th Annu. Corn and Sorghum Res. Conf., 1980, 35, 146-63.
9. Liebl, R.A.; Worsham, A.D. Proc. South. Weed Sci. Soc., 1983, 36, 405-14.
10. Forney, D.R.; Foy, C.L.; Wolf, D.C. Proc. South. Weed Sci. Soc., 1983, 36, 358.
11. Overland, L. Amer. J. Bot. 1966, 53, 423-32.
12. Putnam, A.R.; DeFrank, J. Crop Protection. 1983, 2(2), 173-81.
13. Putnam, A.R.; DeFrank, J.; Barnes, J.P. J. Chem. Ecol. 1983, 9 (8), 1001-10.
14. Shilling, D.G.; Worsham, A.D. Abs. Weed Sci. Soc. Amer. 1984, p. 56.
15. Shilling, D.G.; Worsham, A.D. Proc. South. Weed Sci. Soc., 1983, 36, 404.
16. Putnam, A.R.; DeFrank, J. Abs. Weed Sci. Soc. Amer. 1980, p. 35.
17. Karssen, C.M. Acta Bot. Neerl. 1970, 19 (3), 297-312.
18. Sauer, J. and Struik, G. Ecology 1982, 45, 884-86.
19. Wesson, G.; Wareing, P.F. J. Exp. Bot. 1969, 20, 402-13.
20. Lal, R. Plant and Soil. 1974, 40, 321-31.
21. Guenzi, W.D.; McCalla, T.M. Soil Sci. Soc. Am. Proc., 1962, 26, 456-58.
22. Guenzi, W.D.; McCalla, T.M. Agron. J. 1966, 58, 303-04.
23. Chou, C-H.; Patrick, Z.A. J. Chem. Ecol. 1976, 2, 369-87.
24. Guenzi, W.D.; McCalla, T.M. Soil Sci. Soc. Am. Proc., 1966, 30, 214-16.
25. Patterson, D.T. Weed Sci. 1981, 29, 53-59.

26. Liebl, R.A.; Worsham, A.D. J. Chem. Ecol. 1983, 9 (8), 1027-43.
27. Shettel, N.L.; Balke, N.E. Weed Sci. 1983, 31, 293-98.
28. Putnam, A.R.; Duke, W.B. Science 1974, 185, 370-72.
29. Leather, G.R. Weed Sci. 1983, 31, 37-42.
30. Fay, P.K.; Duke, W.B. Weed Sci. 1977, 25, 224-28.
31. Norstadt, F.A.; McCalla, T.M. Soil Sci. Soc. Am. Proc., 1968, 32, 241-45.
32. Cockran, V.L.; Elliot, L.F.; Papendic, R.I. J. Soil Sci. Soc. Am. 1977, 41, 903-08.
33. Kimber, R.W.L. Plant Soil 1973, 38, 347-61.
34. Toussoun, T.A.; Weinhold, A.R.; Linderman, R.G.; Patrick, Z.A. Phytopathology 1968, 58, 41-54.
35. Norstadt, F.A.; McCalla, T.M. Science 1963, 140, 410-11.
36. McCalla, T.M.; Haskins, F.A. Bacteriol. Rev. 1964, 28, 181-207.
37. Patrick, Z.A.; Koch, L.W. Can. J. Bot. 1985, 36, 621-47.
38. Patrick, Z.A. Soil Sci. 1971, 111, 13-18.
39. Barnes, J.P.; Putnam, A.R. J. Chem. Ecol. 1983, 9, 1045-57.
40. Helwig, J.T.; Council, K.A. (ed.) "SAS User's Guide"; SAS Institute, Inc.: Cary, N.C., 1979, p. 1-101.
41. Van Sumere, C.F.; Wolf, G.; Tenchy, H.; Kint, J.J. Chromatog. 1965, 20, 48-60.
42. Pierce, A.E. "Silylation of Organic Compounds." Pierce Chemical Company: Rockford, IL, 1983.
43. McLafferty, F.W. "Interpretation of Mass Spectra"; University Science Books: Mill Valley, CA, 1980.
44. Stenhagen, E.; Abrahamsson, S.; McLafferty, F.W. "Registry of Mass Spectral Data"; John Wiley and Sons: New York, 1974.
45. Markey, S.P.; Urban, W.G.; Levine, S.P. (eds.). "Mass Spectra of Compounds of Biological Interest"; Technical Information Division, U.S. Atomic Energy Commission (26553 P1-P3), 1975, Vol. I-II, Oak Ridge, TN.
46. Rao, O.N.; Mikkelsen, D.S. Agr. J. 1977, 69, 923-28.
47. Rice, E.L. "Allelopathy." Academic Press: New York, 1974.
48. Gross, D. Phytochem. 1975, 14, 2105-12.
49. Wong, P.P.; Evans, H.J. Plant Physiol. 1971, 47, 750-55.
50. Pridham, J.B. Ann. Rev. Plant Physiol. 1965, 16, 13-36.
51. Gamborg, O.L. Can. J. Biochem. 1966, 44, 791-99.
52. Robinson, T. "The Organic Constituents of Higher Plants"; Cordus Press: North Amherst, MA, 1980.
53. Conn, E.E. (ed.). "The Biochemistry of Plants: Secondary Plant Products"; Academic Press: New York, 1981.
54. Tamura, S.; Chang, C-F. Agr. Biol. Chem. 1965, 29, 1061-62.
55. Kimura, S.; Tanura, S. Agr. Biol. Chem. 1973, 37, 2925.
56. Whittaker, R.H. In "Chemical Ecology"; Sondheimer, E.; Simone, J.B., Eds.; Academic Press: New York, 1970; pp. 43-70.
57. Lodhi, M.A.K. J. Chem. Ecol. 1979, 5, 429-37.
58. Wang, T.S.C.; Yang, T.K.; Chuang, T.T. Soil Sci. 1967, 103, 239-46.
59. Borner, H. Bot. Rev. 1960, 26, 393-424.
60. Finkle, B.J.; Lewis, J.C.; Corse, J.W.; Lundin, R.E. J. Biol. Chem. 1962, 237, 2926-31.
61. Indahl, S.R.; Scheline, R.R. Appl. Microbiol. 1968, 16, 667.
62. DiMenna, M.E. Gen. Microbiol. 1959, 20, 13-20.
63. Henderson, M.E.K. Pure Appl. Chem. 1963, 7, 589-602.
64. Henderson, M.E.K.; Farmer, V.C. J. Gen. Microbiol. 1955, 12, 37-46.
65. Turner, J.A.; Rice, E.L. J. Chem. Ecol. 1975, 1, 41-58.

RECEIVED August 21, 1984

Allelopathy in Tall Fescue

ELROY J. PETERS[1] and K. T. LUU[2]

[1]Agricultural Research Service, U.S. Department of Agriculture and University of Missouri, Columbia, MO 65211
[2]Department of Plant and Soil Science, University of Tennessee, Knoxville, TN 37901–1071

In bio-assay procedures, aqueous leachates of tall fescue (Festuca arundinacea Schreb.) leaves and roots were shown to inhibit seedling growth of rape (Brassica nigra L.), birdsoot trefoil (Lotus corniculatus L.) and red clover (Trifolium pratense L.). Genotypes of tall fescue varied in intensity of inhibition of other species, indicating that production of phytotoxic compounds may be genetically controlled. Tall fescue extracts were separated into anion, cation and neutral fractions. Inhibition was apparent mainly in the anion fraction. Substances in the anion fraction were identified with gas-liquid and paper chromatography as lactic, succinic, malic, citric, shikimic, glyceric, fumaric and quinic acids and several unknowns.

Our research on allelopathy in tall fescue (Festuca arundinacea Schreb.) was begun to determine if growth inhibitors were present, if tall fescue had an inhibitory effect on plants growing in association with it, and finally to identify the chemicals responsible for inhibition.

Osvald (1) in 1949, found that alcohol extracts of soil, on which red fescue (Festuca rubra L.) had been growing inhibited the germination of rape (Brassica nigra L.) seed. Hoveland (2) in 1964, found that water extracts of dry ground roots of tall fescue had little effect on germination and seedling vigor of white (Trifolium repens L.), ball (T. nigrescens Viv.), crimson (T. incarnatum L.), and arrow leaf (T. vesiculosum Savi.) clovers.

We began to explore the possibility of growth inhibitors in tall fescue because we noted that fields of tall fescue frequently had no other plant species growing with them.

Toxicity of Tall Fescue Leachates to Rape (Brassica nigra L.) and Birdsfoot trefoil (Lotus corniculatus L.)

Plants of Kentucky tall fescue were dug from the field in July, and the soil was washed from the roots. Tops and roots were separated and cut into 7- to 9-cm segments. Ten g of fresh roots or tops were soaked in 100 ml of distilled water for 24 hours and the plant material was then filtered from the leachate. Twenty five rape and 25 birdsfoot trefoil seeds were placed on filter paper in 10 by 1.5 cm petri dishes and wet with 5 and 10 ml of leachate. A control group of petri dishes received 10 ml of distilled water. The experiment was replicated four times. Seeds were germinated in the dark at 20°C for 5 days. Seedlings were counted, and those with normal appearing radicles were separated from those that appeared abnormal.

The leachates did not affect total percent germination of rape and birdsfoot trefoil seeds (Table I), and all trefoil seedlings had

Table I. Germination of rape and birdsfoot trefoil seeds
          in petri dishes containing fescue extracts *

Treatment[†]	Rape % germination		Birdsfoot trefoil % germination	
	Normal	Total	Normal	Total
Root extract				
Water	77 a	84 a	43 a	56 a
5 ml	34 b	60 ab	41 a	56 a
10 ml	20 b	42 b	40 a	50 a
Leaf extract				
Water	73 a	90 a	39 a	59 a
5 ml	31 b	71 a	35 a	49 a
10 ml	5 c	83 a	27 a	47 a

*Percentages within a column followed by the same letter are not significantly different at the 5% level according to Duncan's Multiple Range Test
†ml=Volume of extract put in petri dishes.
Reproduced with permission of the Crop Science Society of America, Table 1 (3). Copyright 1968, Crop Science Society of America.

normal radicles. However, both root and leaf extracts of tall fescue caused radicles of rape to be shorter than those of the distilled water check.

A second experiment similar to the one above was set up except that 3 ml of extract was used from leachates prepared by using ratios of 5, 10 and 15 g of tops and roots to 100 ml of distilled water.

The leachate from fescue leaves significantly reduced germination of rape seeds, but the root leachate had no effect (Table II).

Table II.   Germination and root growth of rape and
birdsfoot trefoil seedlings exposed for 5 days to
various concentrations of fescue extracts *

Treatment[†]	Rape		Birdsfoot trefoil	
	Root length	Germination	Root length	Germination
	mm	%	mm	%
Root extract				
0 g (H₂O)	28.0 a	77 a	10.0 a	66 a
5 g	7.5 b	80 a	4.0 a	55 b
10 g	5.8 b	84 a	2.8 a	46 bc
15 g	4.0 cd	77 a	2.0 a	32 cd
Leaf extract				
0 g (H₂O)	28.0 a	77 a	10.0 a	66 a
5 g	8.8 b	57 b	4.0 a	48 bc
10 g	7.0 b	57 b	3.5 a	23 d
15 g	3.0 d	41 c	1.0 a	1 e

*Numbers followed by the same letter in each column are not significantly different at the 5% level according to Duncan's Multiple Range Test.
[†] g=Grams of plant material soaked in 100 ml of water to make up the concentration.
Reproduced with permission of the Crop Science Society of America, Table 2 (3). Copyright 1968, Crop Science Society of America.

Both root and leaf leachates reduced germination of birdsfoot trefoil seeds. The root lengths of rape seedlings were reduced by all concentrations of leachate from fescue roots and leaves. As the concentrations of fescue leachate increased, root growth of rape was reduced. The leachate had no significant effect on the root length of birdsfoot trefoil roots.

Influence of Ethanol Extracts of Soil and Fescue Plants

In order to determine if inhibiting substances were present in soil in which tall fescue was growing, soil was dug from the field on March 20. Fescue roots were separated from the soil and saved along with leaves. Two hundred grams (fresh weight) each of fescue leaves and roots were leached overnight in 1,000 ml of ethanol and then filtered. Four hundred grams of the fescue soil was extracted overnight with 200 ml of ethanol and then filtered. The ethanol extracts were evaporated to dryness and the residues taken up in sufficient Hoagland's (4) solution to equal the original volumes of extract. Thirty five ml of each solution was added to 100 ml beakers containing 150 g of silica sand. The beakers were wrapped in black paper to exclude light. Rape and birdsfoot trefoil seeds were germinated on moist filter paper and five seedlings were transplanted to each beaker. Seedlings were also planted in beakers containing Hoagland's solution without extracts. Distilled water was added periodically as water loss occurred. Iron chelate was added with the distilled water to prevent iron deficiency. The beakers containing plants were placed in the greenhouse on March 23.

After 15 days, the plants were harvested and the length of tops and roots were measured. Each treatment was replicated four times. Rape and birdsfoot trefoil seedlings growing in nutrient solution containing extracts of fescue leaves and roots had shorter shoots and roots than when grown without these extracts (Table III). Ex-

Table III. Length of shoots and roots
of rape and birdsfoot trefoil seedlings grown in
nutrient solutions containing ethanol extracts *

Extracts	Rape		Birdsfoot trefoil	
	Shoots	Roots	Shoots	Roots
	------------ Length in mm ------------			
Leaves	18.5 d	14.5 b	4.5 c	6.5 c
Roots	21.0 cd	17.0 b	6.5 bc	15.5 b
Soil with roots removed	27.2 ab	30.7 a	10.2 a	37.0 a
Control	27.5 a	27.2 a	9.5 a	31.8 a

*Numbers followed by the same letter within a column are not significantly different at the 5% level according to Duncan's Multiple Range Test.
Reproduced with permission of the Crop Science Society of America, Table 2 (3). Copyright 1968, Crop Science Society of America.

tracts of soil had no adverse affect on growth of rape and birdsfoot trefoil. These data suggest that if an organic inhibitor is present in soil, it is strongly adsorbed on the soil particles, is rapidly broken down or the concentration is too low to cause inhibiting effects.

Growth of Rape Seedlings When Competition for Mineral Nutrients is Eliminated

In order to eliminate competition for mineral nutrients as a possible cause of retarded growth of rape plants growing with fescue, the following system was set up. Tap roots of week-old rape seedlings were split with a razor blade and grown for 10 days in Hoagland's solution until two root systems had developed on each plant. Tall fescue plants were dug from the field and the fibrous root system was combed into two separate root systems. The tall fescue plants were grown in washed silica sand containing Hoagland's solution until they were used. Each plant was placed on the edge of two adjoining 9.2 x 9.2 cm square plastic containers so that one "root system" of a plant grew in one container, and the second "root system" grew in an adjacent container, Figure 1. With this arrangement, one "root system" of each plant had a source of nutrient solution while the second "root system" grew in distilled water in competition with the "root system" of the other plant. The plants were grown together for 6 weeks. Solutions were flushed from the pots and replaced at 10-day intervals.

Rape roots grown with fescue in distilled water weighed significantly less than those grown alone in distilled water and less than those grown in nutrient solution with other rape roots (Table IV).

Figure 1. Plants arranged on the edges of containers so that half of each root grew in adjacent containers. Nutrient solution in first and third containers from the left and distilled water in the second and fourth. Reproduced with permission from Ref. 3. Copyright 1968, Crop Science Society of America.

Table IV. Weight of rape tops and roots when grown
with part of the root system in distilled water with
fescue roots or with rape roots and part of the
root systems in an independent source of nutrients *

| | Weight in mg | |
Placement of rape roots	Rape tops	Rape roots
Grown with fescue:	17.2 a	-
Half of root with fescue in water	-	42 a
Half of root with rape in nutrient solution	-	109 b
Grown with rape:	22.1 a	-
Half of root with rape in nutrient solution	-	124 b
Half of root in distilled water	-	92 b

*Numbers followed by the same letter within a column are not signi-
ficantly different at the 5% level according to Duncan's Multiple
Range Test.
Reproduced with permission of the Crop Science Society of America,
(3). Copyright 1968, Crop Science Society of America.

    Weights of rape roots grown with rape either in distilled water
or nutrient solution were similar. These data indicate that factors
other than competition for nutrients may be responsible for the
depressing effect of fescue on rape.   Thus, it appears that fescue
roots may excrete an inhibiting substance.

Genetic Influences on Allelopathy

A grass breeding nursery in the field at Columbia, Missouri contained tall fescue genotypes collected from various sources around
the world.   It was noted that infestations of large crabgrass
(Digitaria sanguinalis (L.) Scop.) varied from fescue genotype to
genotype.   Differences in crabgrass infestations could not be explained by competition because even some thin stands of fescue had
little or no crabgrass.   Soil was collected from the top 2.5 cm of
each plot in October 1972 and put in pots in the greenhouse.   The
soil was permitted to dry until February 25, 1973 and then watered
until March 13 when crabgrass plants were counted.   The pots were
divided into a weedy group and a clean group depending on the
crabgrass infestation observed in the field.
    The crabgrass infestations in the greenhouse showed differences
according to genotype of tall fescue and followed the observations
in the field with one exception (Table V).   The genotype from "Fawn"

Table V.   Number of crabgrass seedlings per pot in soil collected
from field plots containing tall fescue genotypes *

Origin of genotype	No. of crabgrass plants per pot
Weedy in field	
Alta	87 a
Kentucky 31	56 ab
Fawn	22 bc
Clean in field	
Switzerland	20 cd
Oregon experimental	17 cd
Holland	3 d
Netherlands	21 cd

*Numbers followed by the same letter within a column are not signi-
ficantly different at the 5% level according to Duncan's Multiple
Range Test.
Reproduced with permission of the American Society of Agronomy (5).
Copyright 1981, American Society of Agronomy.

had a high weed population in the field but had a low population in
the greenhouse.   Further investigations were done by selecting 13
genotypes that had been used in the development of the variety "MO-
96".   Leaves from these plants were cut from the roots and soaked in
distilled water for 24 hr and filtered.   A solution prepared in a
ratio of 10 g of fresh plant tissue per 100 ml of water was put in
each petri dish containing filter paper.   Control petri dishes
contained distilled water.   Twenty five seeds each of red clover or
birdsfoot trefoil were placed in the petri dishes.   The seeds were
germinated and observed for 2 weeks.   Distilled water was added
periodically to keep the seeds moist.   The data in Table VI shows

Table VI.   Percent germination of birdsfoot trefoil
and red clover seeds exposed to leaf extracts
of tall fescue genotypes *

Fescue genotype	Birdsfoot trefoil	Red clover
196:207	53 ab	41 c
237	52 ab	53 bc
302	40 bc	56 b
360	29 c	50 bc
446	48 b	54 bc
698	50 ab	52 bc
Water control	66 a	82 a

*Numbers followed by the same letter within a column are not signi-
ficantly different at the 5% level according to Duncan's Multiple
Range Test.

that genotypes of tall fescue differed in inhibition of germination
of legumes.
An additional study was done to determine the effects of tall
fescue leachates on the growth of birdsfoot trefoil.   The genotypes
237, 360 and 698 appeared to have the greatest inhibition on germination in the previous test so leaves of these plants were cut from
the roots and soaked in distilled water for 24 hours.   Ratios of 3,
5 and 10 g of leaves per 100 ml of distilled water were used.

Legume seedlings were germinated in distilled water on filter paper and five seedlings were transplanted to 180 ml plastic cups containing half strength Hoaglands solution in silica sand. After a few days when seedlings were growing well, the tall fescue extracts were added to the containers to replace water lost by evaporation. Seventy ml of extract were used for each beaker and distilled water was added subsequently as water use occurred. Seedlings were harvested 2 weeks after the leachates had been added.

Dry weights of birdsfoot trefoil roots and shoots were retarded by the 10 g concentration of genotypes 360 and 698 as compared with the control (Table VII). Other treatments had no significant effect on root weight.

Table VII. Dry weights of roots and shoots of
birdsfoot trefoil and red clover seedlings grown in
water extracts of leaves and of tall fescue genotypes

Tall fescue genotypes	Extract	Dry weight			
		Birdsfoot trefoil		Red clover	
		Roots	Shoots	Roots	Shoots
	g/100 ml	------------------- mg -------------------			
Check (water)		98 a*	220 a	300 a	490 a
I96-237	3	87 ab	190 ab	180 b	410 b
	5	62 ab	170 abc	180 b	380 bc
	10	55 ab	170 abc	170 bc	320 cd
I96-360	3	94 a	190 ab	190 bc	410 b
	5	70 ab	180 ab	130 bc	290 de
	10	50 b	130 c	90 c	250 e
I96-698	3	95 a	200 ab	180 b	400 bc
	5	57 ab	180 ab	170 bc	370 bc
	10	50 b	150 bc	150 bc	370 bc

*Numbers followed by the same letter within a column are not significantly different at the 5% level according to Duncan's Multiple Range Test.
Reproduced with permission of the American Society of Agronomy (5).
Copyright 1981, American Society of Agronomy.

All solutions containing extracts of tall fescue significantly inhibited growth of red clover shoots and roots. Genotype 360 was significantly more inhibitory of shoot growth than the other genotypes at the 10 g concentration. From the data presented, it appears that allelopathy may be genetically controlled, and it may be possible to select fescue with a high amount of allelopathy to exclude other plants from the stand or to select fescue with less allelopathy so that certain species could be grown with tall fescue.

Identification of Inhibitory Substances

Extract of Kentucky 31 tall fescue was prepared by soaking 75 g fresh fescue leaves for 24 hours in 500 ml of distilled water. The extract was filtered and fractionated through cation and anion exchange columns according to the procedure described by Williams (6), Boland (7) and Boland and Garner (8). The extract was first passed through a column of Dowex MSC-1[1]. The column was then washed with a 2 bed volume of distilled water and combined with the filtrate prior to passing through an Amberlite IRA-78[1]. The column was washed with deionized distilled water for complete removal of the neutral fraction. The cation fraction was eluted from the Dowex MSC-1 column with 6N NH$_4$OH, and the anion fraction was eluted from the amberlite IRA-78 column with 1N HCOOH. The solutions were then dried in vacuum and diluted with distilled water to the volume of the original extract (9). Aliquots of these fractions were adjusted to various pH and used in bioassays with birdsfoot trefoil seeds. Seeds were germinated on filter paper in petri dishes with the test solutions and the percent germination and lengths of hypocotyls and roots were measured after 5 days.

The anion fraction, with unadjusted pH, showed significantly greater inhibition than the cation and neutral fractions (Table VIII).

Table VIII. The effect of three fractions separated
from extracts of tall fescue (pH 5.5) *

Extract	Germinations	Root length	Hypocotyl length
	%	mm	mm
Anion	15.3 b	1.5 c	9.0 b
Cation	66.4 a	11.9 a	29.5 a
Neutral	65.4 a	7.4 b	26.4 a
Water control	65.5 a	16.0 a	29.8 a

*Numbers followed by the same letter within a column are not significantly different at the 5% level according to Duncan's Multiple Range Test.

When the fractions were adjusted to pH 7, seedlings in the anion fraction had significantly less germination and less root and hypocotyl length than those in the cation and neutral fraction (Table IX).

Table IX. The effects of three fractions separated from
tall fescue when the fractions were adjusted to pH 7 *

Extract	Germination	Root length	Hypocotyl length
	%	mm	mm
Anion	58.2 bc	4.4 d	14.0 b
Cation	64.2 a	8.4 c	18.8 a
Neutral	64.9 a	8.4 c	19.3 a
Water control	60.0 abc	14.0 a	21.1 a

*Numbers followed by the same letter within a column are not significantly different at the 5% level according to Duncan's Multiple Range Test.

---

[1]Mention of a trademark or proprietary product does not constitute a guarantee or warranty of the product by the U.S. Dept. of Agric., and does not imply its approval to the exclusion of other products that may also be acceptable.

A 200 ml sample of the anion fraction was evaporated to dryness and dissolved in 10 ml of 80% ethanol. The extract was streaked onto Whatman #17 Chromatography paper, dried and then the chromatographs were developed in a descending manner with ethyl acetate: acetic acid:formic acid:water (18:3:1:4) as an irrigant. A spraying reagent was prepared according to Stahl (9).

Seven individual spots that appeared on the chromatogram were cut from the paper and soaked in 80% ethanol for 2 hours. The extract was evaporated to dryness. To insure removal of formic acid, 3 to 5 ml of 80% ethanol was added and repeatedly evaporated to dryness. After final drying, 16 ml of distilled water was added to each fraction and used for the germination test with birdsfoot trefoil.

The samples used for identification with gas-liquid chromatography were lyophilized as suggested by Thimann and Bonner (10) to minimize loss of organic acids. After lyophilization, the samples were used in a derivatization procedure that makes organic acids become volatile in the gas-liquid chromotography column. This was a modification of the procedure developed by Boland and Garner (8) as described by Luu (11). After identification with gas-liquid chromatography, standards of the identified compounds were put on paper chromatograms to confirm their identification by checking retention times.

Lactic, succinic, glyceric, glycolic, shikimic, malic, citric and quinic acids were identified and several unknowns were present. To determine the effect of organic acids on birdsfoot trefoil, solutions of organic acids were made up in concentrations from 0.001N to 0.1N and were bioassayed with birdsfoot trefoil grown in petri dishes using the technique described previously (Table X).

Table X. Percent germination and root hypocotyl length of birdsfoot trefoil as affected by organic acids (concentration 10^{-1}N and pH 4) *

Organic acid	Germination	Root length	Hypocotyl length
Lactic	18.2 b	1.2 c	1.4 c
Succinic	50.2 b	2.2 c	3.7 c
Malic	43.8 b	1.6 c	2.5 c
Citric	46.1 c	1.1 c	4.1 c
Quinic	41.2 c	1.0 c	2.6 c
Glycolic	00.0 b	0.0 c	0.0 c
Glyceric	48.7 b	2.0 c	2.5 b
Water control	63.1 a	9.2 a	12.0 a

*Numbers followed by the same letter within a column are not significantly different at the 5% level according to Duncan's Multiple Range Test.

All organic acids at concentrations of 10^{-1}N retarded germination and growth of birdsfoot trefoil seedlings, but no growth retardation occurred at 10^{-3}.

General Discussion

If tall fescue has inhibiting effects on other species growing in associations, the inhibiting substances must be excreted by the fescue plants. Plant material used in our experiments was fresh green material and intact except for cut edges. Leaching was for a 24-hour period. Therefore, we assumed that the leachate was somewhat similar to what may be present in the field. The data show that inhibition does occur from leachate. Tall fescue had an inhibitory effect on other plants grown in association indicating that inhibitory substances must be excreted from roots of tall fescue. Evidence for allelopathy in tall fescue has been shown by other researchers (11, 12). Walters and Gilmore (13), in one instance, showed growth inhibition from decomposing tissue of tall fescue; and therefore, decomposition products may be an additional factor.

Our data show that organic acid concentrations in the anion fraction of tall fescue is about 0.0357N, and significant inhibition occurred at 0.01N. Ulbright et al. (14) found that 3 to 8 C straight-chain carboxylic acids inhibit monocot and dicot root growth when applied at concentrations of 5×10^{-4} M and toxicity occurred at 5×10^{-3} M. Other authors have implicated organic acids as being involved in inhibition (15, 16).

Literature Cited

1. Osvald, H. Ann. R. Agr. Coll. Sweden. 1949, 16, 789-796.
2. Hoveland, C. S. Crop Sci. 1964, 4, 211-213.
3. Peters, E. J. Crop Sci. 1968, 8, 650-653.
4. Hoagland, D. R.; Snyder, W. C. Amer. Soc. Hort. Sci. 1933, 30, 288-294.
5. Peters, E. J.; Mohammed Zam, A. H. B. Agron. J. 1981, 73, 56-58.
6. Williams, M. Ph.D. Thesis, University of Missouri, Columbia, 1976.
7. Boland, R. L. Ph.D. Thesis, University of Missouri, Columbia, 1971.
8. Boland, R. L.; Garner, G. B. J. Agric. Food Chem. 1973, 21, 661.
9. Stahl, E. G. "A Laboratory Handbook", Acad. Press: New York, 1965.
10. Thimann, K. V.; Bonner, W. D. Ann. Rev. of Plant Phys. 1950, 1, 75-108.
11. Luu, K. T. Ph.D. Thesis, University of Missouri, Columbia, 1980.
12. Gilmore, A. R. Ill. Res. 1977, 19, 8-9
13. Walters, D. T.; Gilmore, A. R. J. Chem. Ecol. 1976, 2, 469-479.
14. Ulbright, C.; Pichard, B.; Varner, J. Plant Phys. 1979, 63(5), 80.
15. Patrick, M. Soil Sci. 1971, 111, 13-18.
16. Rice, E. L. "Allelopathy", Acad. Press: New York, 1974; p. 247.

RECEIVED June 12, 1984

Germination Regulation by *Amaranthus palmeri* and *Ambrosia artemisiifolia*

JUDITH M. BRADOW

Southern Regional Research Center, Agricultural Research Service, U.S. Department of Agriculture, New Orleans, LA 70179

An apparent allelopathic interference having been observed in which residues of Palmer amaranth, Amaranthus palmeri S. Wats., inhibited the growth of onion, carrot and Palmer amaranth itself, aqueous and organic solvent extracts of amaranth plant parts were made and used to investigate the growth regulatory activity of the chemical constituents of this amaranth species. The resultant crude extracts, as well as isolated and identified chemical compounds, were assayed in vitro as germination regulators, by using both crop and weed seeds. Some crude organic solvent extracts proved inhibitory, but inhibitions of onion, Palmer amaranth, and carrot germination did not result from the same extracts or compounds. Four compounds have been isolated from Palmer amaranth: 2,6-dimethoxy-benzoquinone, vanillin, phytol, and chondrillasterol. All showed biological activity, to varying degrees, in seed germination bioassays.
Parallel studies of extracts and mixtures of compounds isolated from ragweed, Ambrosia artemisiifolia L., were also made. Water extracts of ragweed proved highly inhibitory of seed germination. Mixed sesquiterpenes isolated from ragweed were very inhibitory of onion, oat, ryegrass, and Palmer amaranth germination when applied in a dichloromethane solution pretreatment.

Amaranthus Palmeri S. Wats., Palmer amaranth, is a coarse, weedy, drought-resistant, dioecious member of the Amaranthaceae and is related to A. retroflexus L., redroot pigweed, and the ornamentals A. tricolor L., summer poinsettia, and A. caudatus L., love-lies-bleeding. Palmer amaranth was used by natives of the North American desert as a protein source and cereal to supplement maize and beans. Another member of the same family, A. spinosus L., spiny amaranth, has been reported to exhibit allelopathic activity toward coffee

(1); and Dr. R. M. Menges, USDA, Weslaco, TX, has observed that Palmer amaranth residues plowed into fields later inhibit the growth of onion, carrot, and Palmer amaranth itself (see Menges, this publication). The reports from Texas prompted the formation of an interdisciplinary (and multi-site) group to investigate the apparent allelopathic activity of Palmer amaranth. Dr. N. H. Fischer, and his group at Louisiana State University, Baton Rouge, LA, were already investigating the secondary metabolites of Ambrosia artemisiifolia L., common ragweed. Although the group itself had no field observations to indicate allelopathic interference by A. artemisiifolia, a tropical ragweed species, A. cumanensis HBK, has been shown to have allelopathic potential (2); and western ragweed, A. psilostachya DC, has been found to be inhibitory in laboratory seed germination tests (3) and appears to have a role in old-fields succession (4). Common ragweed itself is reported to exert allelopathic effects in the first steps of secondary succession (5).

Field observations indicated that the inhibitions result from Palmer amaranth residues, rather than the growing plant itself, and it was felt that competitive interference at the higher plant level could be ignored in favor of the study of the apparent chemical interactions. Both "phytoinhibitins" and "saproinhibitins" (6) were to be considered, beginning with chemicals of plant origin (phytoinhibitins). This paper, a section of a three-part presentation, describes the in vitro seed germination assays performed with crude and pure substances isolated from the various parts of ragweed and Palmer amaranth.

Materials and Methods

The Palmer amaranth plants used as raw material for crude extract preparation and compound isolation were supplied by Dr. Menges or grown at Southern Regional Research Center (SRRC) and harvested after seed dispersal (water-extracted samples only). Ragweed plants were collected in Baton Rouge in June or in Kenner, LA, just before flower anthesis. The details of organic solvent extract preparation and compound isolation and identification are discussed elsewhere (see Fischer et al., this publication). Water extracts of air-dried leaves of ragweed (1:10, w/w) and Palmer amaranth (1:16, w/w) were made by blending the leaves (and amaranth thyrses) in deionized water, pH 6.8, and agitating the slurry at room temperature for 3 h before filtration through 4 layers of cheesecloth. The filtrate was centrifuged 20 min at 13,000 x g (avg.) and the pellet discarded. These crude water extracts were used in the seed germination bioassay described below.

The crude organic solvent preparations were produced by sequential extractions of the plant material with petroleum ether (PE), followed by dichloromethane (DCM), and finally methanol (MeOH). Some Palmer amaranth samples prepared at SRRC were first extracted with hexane (HX), rather than PE. The organic solvent extracts were supplied for bioassay in the form of oils or solids remaining after solvent removal in vacuo. After each extraction step aliquots of the crude extracts from the roots, stems, or leaves (and thyrses) were evaluated for seed germination regulatory activity.

The germination bioassay consisted of germinating the seeds of a number of crop and weed species (Table I) for 72 h in the dark at

Table I. Seed Species Used to Determine <u>In Vitro</u> Germination
Regulatory Activity of Palmer Amaranth and Ragweed Extracts

Seed species	No. seed/ replicate	Class	Common name
<u>Allium cepa</u> L.,			
cv. Texas Early Grano	20	Monocot	Onion
<u>Amaranthus palmeri</u> S. Wats.	25	Dicot	Palmer Amaranth
<u>Amaranthus retroflexus</u> L.	25	Dicot	Redroot Pigweed
<u>Avena sativa</u> L.	15	Monocot	Oat
<u>Bromus secalinus</u> L.	20	Monocot	Cheatgrass*
<u>Cucumis sativus</u> L.,			
cv. Marketmore	10	Dicot	Cucumber
<u>Daucus carota</u> L.,			
Danvers Half-long	20	Dicot	Carrot
<u>Echinochloa crus-galli</u>			
(L.) Beauvois	20	Monocot	Barnyard Grass*
<u>Eragrostis curvula</u>			
(Schrader) Nees	20	Monocot	Lovegrass*
<u>Lactuca sativa</u> L.			
cv. Grand Rapids	20	Dicot	Lettuce
cv. Great Lakes	20		
<u>Lepidium sativum</u> L.,			
Curly Cress	20	Dicot	Garden Cress
<u>Lolium</u> spp.	20	Monocot	Ryegrass
<u>Lycopersicon esculentum</u> Mill.,			
cv. Homestead	20	Dicot	Tomato
<u>Portulaca oleracea</u> L.	20	Dicot	Purslane*
<u>Sorghum bicolor</u> (L.) Moench	15	Monocot	Sorghum
<u>Trifolium incarnatum</u> L.	20	Dicot	Red Clover
<u>Triticum aestivum</u> L.	15	Monocot	Wheat

*Not used in studies of Palmer amaranth extracts.

25 C in 10 cm plastic petri dishes containing one 9 cm sheet of
Whatman #1 filter paper which had been saturated with 3 ml of a test
or control solution. The number of seeds per replicate was chosen
to minimize contact between seedlings at 72 h and to avoid water-
stress during germination in the larger seeds. There were 8
replicates of each treatment and of parallel controls consisting of
either deionized water, pH 6.8, or 0.1% (v/v) dimethylsulfoxide
(DMSO) in deionized water. The DMSO was necessary to aid water
solvation of the extracts made with the less polar organic solvents.
In a few instances where DMSO was ineffective and the test material
was either a pure, identified compound or more fully characterized
than a crude extract, seeds were pretreated with DCM solutions of
the substance (0.1 mM where possible) and germinated as above with
parallel controls pretreated with DCM alone. This pretreatment
technique was utilized in a time-course germination study of the
effects of chondrillasterol (5α-stigmasta-7,22-dien-3β-ol) isolated
from Palmer amaranth, using the following five amaranth species:
<u>Amaranthus caudatus</u> L., cv. Love-lies-bleeding; <u>A. palmeri</u>; <u>A.
retroflexus</u>; <u>A. tricolor</u> L., cv. Tampala; and <u>A. salicifolius</u> L.,

cv. Flaming Fountain. The seeds were soaked in 3 ml 0.1 mM chondril-lasterol in DCM for 1.5 h; the solvent was allowed to evaporate, and the seeds were air-dried for an additional 24 h before use. The seeds were germinated under the conditions described above, and germination counts were made every 24 h for 7 days. All seed manipulations were carried out under a green safelight.

The chondrillasterol was also assayed for activity in mung bean hypocotyl and radicle elongation tests where mung beans, Vigna radiata L. Wilczek, cv. Golden Gram, were soaked 3 in 5 ml 0.1 mM chondrillasterol in DCM. Controls treated with DCM alone were included. The solvent was allowed to evaporate, and the seeds were planted, 20 seeds per 10 cm plastic petri dish, on a single Whatman #1 filter paper disk saturated with 3 ml deionized water, pH 6.8. Untreated seeds were similarly planted, using deionized water (control), 0.4% dimethylformamide (DMF control) or 0.05 mM chondril-lasterol in 0.4% DMF (4 replicates each). Incubation conditions were the same as above and germination and growth evaluations were made at 24, 48, 72, and 96 h.

Some crude extracts from ragweed leaves and roots were tested before the multi-seed species assays were regularly in use at SRRC. These extracts, PE, DCM, MeOH, and H_2O were tested, using two lettuce seed cultivars, Grand Rapids (GR) and Great Lakes (GL), germinated at 28 C, a thermal-stress temperature, and radish (Raphanus sativus L., cv. Early Scarlet Globe) germinated 64 h at 25 C, plus carrot germinated 96 h at 25 C.

Since the germination data obtained for the controls, regard-less of solvent or pretreatment, fit normal distributions with different means and variances according to seed species, statistical analyses were made using the standard normal deviates for each species. To allow comparisons between species the germination data are shown as percent of the germination observed in the appropriate control.

Results and Discussion

In vitro germination assays such as those used to obtain the results reported here are not intended as close approximations of the field conditions where the apparent allelopathic interactions occur. The small quantities of suspected allelochemicals which are available for use in bioassays and the complexity of the field situations dictate that some sort of screening assay be developed which is simple and rapid to perform, easily controlled with respect to experimental conditions, and reproducible. Since we anticipated testing crude extracts, isolated compounds, and mixtures of related chemicals we chose a multi-species assay where total germination and pre-emergent seedling development were the experimental parameters. The seeds varied considerably in size and germination requirements. The mean control germination values for each species are a result of more than 30 experiments containing 8 replicates each and control germina-tion data for each species fit a normal distribution. Using Student's paired t-test, the presence of 0.1% DMSO was found to have no significant effect (P >= 0.05), except in the case of wheat where DMSO decreased germination 5.8%.

Concentration was another confounding factor. When using crude extracts determination of the concentrations of individual components is impossible. In dealing with crude extracts the most useful measure of application rate proved to be weight percent (wt.%). The crude extract concentrations used ranged from 0.015 to 0.4 wt.%. These levels are within the concentration limits used in exogenously-applied growth regulator studies where mixtures of gibberellic acids (0.1 mM = 0.035 wt.%) and abscisic acid isomers (0.05 mM = 0.0003 wt.%) have been found effective (7). When tested with species used in our bioassays, the allelochemical, coumarin, inhibited carrot at 0.0005 wt.%, onion at 0.0019 wt.%, lettuce at 0.0033 wt.%, and wheat at 0.0219-0.438 wt.% (8). The wt.% concentrations of the crude extracts tested are indicated in the following tables showing the effects of Palmer amaranth and ragweed on seed germination.

Palmer Amaranth Extracts. The aqueous extracts made from senescent Palmer amaranth leaves and thyrses after seed development and dispersal had little effect on any of the seeds tested (Table II).
    The most concentrated (1:16, w/w) extract inhibited only lettuce, tomato, and ryegrass. The osmolarity of the solution was 40.3 mOsm, insufficient to affect germination of most seeds (9). The responses of the test seed to increased external osmolarity were determined by germinating the seeds in a polyethylene glycol (PEG) solution of 40.3 mOsm (10). The only test seeds significantly affected by this treatment were those of tomato where PEG-treated seed germination was 69.3% of that observed in deionized water. Dilution of the aqueous extract to 1:32 (w/w) removed all inhibitory action, but the diluted solution increased wheat germination.
    Crude organic solvent extracts from the Palmer amaranth leaves and thyrses produced a mixture of promotions and inhibitions of germination, depending on seed species and concentration of the extract. The two most nonpolar solvents used, PE and HX, extracted mixtures inhibitory of carrot (HX) and promotive of sorghum (PE). Further extractions with other organic solvents of Palmer amaranth aerial parts, excluding stems, produced mixtures that were all promotive of cucumber germination. The inhibition of carrot by both a DCM and a MeOH extract prompted further, incomplete investigations of these extracts. No extracts of the leaves and thyrses affected onion, and only a second DCM extraction of leaf tissue and MeOH extraction of the roots produced inhibitors of Palmer amaranth. Inhibitors of onion germination were found in MeOH extracts of both the roots and the stems. The stem extract using MeOH after extraction with HX and DCM contained an onion germination inhibitor which was effective at a very low concentration (0.025 wt.%). The carrot inhibitor in the DCM leaf extract following initial HX extraction was also very active. Cucumber was promoted by root and stem DCM extracts, as was wheat by the HX/DCM extract of the stems.
    These bioassays of crude extracts were intended only as screening tests for the guidance for the natural products chemists in their isolation work, but a few generalities were observed in the results. Although all the seeds in Table I were used in these assays, only three species were promoted: wheat, sorghum, and primarily cucumber which was promoted by DCM extracts from most of the plant parts. Five species were inhibited by the crude Palmer amaranth extracts, including carrot, onion, and Palmer amaranth itself, as seen in the field tests. Ryegrass and sorghum were

Table II. Effects of Palmer Amaranth (<u>Amaranthus</u> <u>Palmeri</u>)
Extracts on Seed Germination

Extract (Solvent used)	Concen.	Seed	% of Germination in Control
Leaves and Thyrses			
Aqueous	1:16 (w/w)	Lettuce	70.7*
		Tomato	74.5*
		Ryegrass	86.0
	1:32 (w/w)	Wheat	120.5
PE	0.380 wt.%	Sorghum	114.4*
HX	0.015	Carrot	79.3
PE/DCM	0.391	Sorghum	85.6*
		Cucumber	107.4*
		Carrot	84.7
	0.330	Palmer amaranth	62.8*
HX/DCM	0.016	Cucumber	114.8*
PE/DCM/MeOH	0.423	Cucumber	110.1*
		Carrot	74.7
HX/DCM/MeOH	0.031	Cucumber	113.8*
Roots			
PE/DCM	0.385 wt.%	Cucumber	107.8*
PE/DCM/MeOH	0.495	Sorghum	86.7*
		Ryegrass	82.8*
		Palmer amaranth	83.7
		Onion	93.1
Stems			
HX/DCM	0.017 wt.%	Wheat	117.5*
		Cucumber	107.5*
		Carrot	84.2
HX/DCM/MeOH	0.025	Onion	83.8*

*$P \leq 0.01$, for all other percentages shown, $P \leq 0.05$.

inhibited by the MeOH root extract, and ryegrass was sensitive to
the aqueous leaf extract as well. Palmer amaranth and carrot
inhibition coincided in the case of the PE/DCM extract of the
leaves, but onion and carrot inhibition did not coincide, an indica-
tion that the two plants are sensitive to different allelochemicals.
Onion seed is inhibited by the more polar (MeOH-extracted) components of the root and stem; while carrot is inhibited by DCM extracts
of the leaves and stem, plus the MeOH extract of the leaves. None
of the inhibitors from the crude organic solvent extracts were
present at effective concentrations in the aqueous extract of the
leaves, since the species affected by the aqueous extract were different from the species affected by the organic extracts of the same
plant parts.

Allelochemicals from Palmer Amaranth. It is fairly easy to extract metabolic inhibitors and promoters from plant material; but isolating and identifying even major components of the crude extracts is time-consuming and frustrating, particularly when the compound identified turns out to be well-known and relatively inactive in the bioassays. To date, only four compounds have been isolated from Palmer amaranth and identified. One is 2,6-dimethoxybenzoquinone which has previously been isolated from fermented wheatgerm (11) and Rauwolfia vomitoria (12). The 2,6-dimethoxybenzoquinone and its 2,5-isomer exhibit cytotoxicity when administered with sodium ascorbate in a murine Ehrlich ascites assay, apparently forming long-lived semiquinone radicals (13). Cytotoxicity may arise from changes in electron charge transfer reactions and electronic delocalization effects in biological systems such as the oxidation-reduction reactions of respiration. Indeed, another quinone, juglone (5-hydroxy-1,4-naphthoquinone) causes 90% reduction in corn root respiration after 1 h exposure (14). The effects of the quinone on seed germination are shown in Table III. At 0.1 mM (0.0017 wt. %) promotion was

Table III. Effects of Isolated and Identified Chemical
Constituents of Palmer Amaranth on Seed Germination

Compound	Concen. mM	Seed	% of Control Germination
2,6-dimethoxybenzoquinone (Deionized Water Control)	0.1	Sorghum	108.6
		Oat	112.1
		Ryegrass	110.4
		Cucumber	94.6
		Tomato	91.1
Vanillin (0.1% DMSO Control)	0.1	Ryegrass	116.0*
		Wheat	111.8
		Cucumber	95.3
Phytol (aqueous) (0.1% DMSO Control)	0.1	Onion	112.8
		Sorghum	89.4*
		Wheat	78.6
		Carrot	84.4
Chondrillasterol (DCM Pretreat. DCM Control)	0.1	Onion	88.0
		Wheat	88.9
		Carrot	118.1
		Palmer amaranth	147.4*

*P <= 0.01; for all other percentages shown, P <= 0.05.

observed in 3 monocots, sorghum, oat, and ryegrass; while both cucumber and tomato were slightly inhibited. Adding 0.16 mM ascorbate as an electron donor inhibited oat germination 13%, compared to the 12% promotion by the quinone alone. Ascorbate alone (0.16 mM) increased oat germination 11%. Further investigations of this quinone/ascorbate system are planned.

Another phenolic compound isolated from Palmer amaranth was vanillin (4-hydroxy-3-methoxybenzaldehyde). This compound has been previously reported in extracts of soils, plant roots and leaf

litter, (15). Vanillin is reported to be toxic to ladino clover
(Trifolium repens L.) seedlings at a minimal level of 1000 ppm, or
6.6 mM (16). At 0.1 mM (0.0015 wt. %) vanillin promoted germination
in ryegrass and wheat while slightly inhibiting cucumber.

Vanillin occurs in the roots of many species and in the soil
surrounding the growing plants, but at low pH it remains tightly
bound in both roots and soils (15, 17). The analogous acid,
vanillic, appears to be much more biologically active (18, 19), and
no further studies of vanillin are planned at this time.

Another very common phytochemical isolated was phytol (3,7,11,
15-tetramethyl-2-hexadecen-1-ol), the diterpenoid alcohol which
forms the "tail" of chlorophyll. This compound is nearly insoluble
in water, but when applied as a 0.1 mM (0.003 wt. %) solution in
0.1% DMSO it increased onion germination while decreasing germination in sorghum, wheat, and carrot. The germination effects are
significant at only the 95% level, but pretreatment studies are
planned as part of other terpene chemistry studies at SRRC and LSU.

The fourth isolated and identified compound from Palmer amaranth
is chondrillasterol (5α-stigmasta-7,22-dien-3β-ol), a sterol closely
related structurally to the major plant sterols, stigmasterol and
sitosterol. This compound, isolated as the free sterol, is not
soluble in water or 0.1% DMSO, and germination bioassays required
pretreatment of the test seed with a 0.1 mM solution of the sterol
in DCM.

Soaking seeds in DCM and allowing the solvent to evaporate
before germination assays significantly affected germination (P <=
0.05) as shown in Table IV. Three of the monocot species, sorghum,

Table IV. Effects of 1 Hour Pretreatment with Dichloromethane
           on Seed Germination Percentages

Seed	% Change in Germination Percentage	
Onion	-5.0	NS
Sorghum	-24.1	**
Oat	-16.5	**
Ryegrass	-4.4	NS
Wheat	-21.6	**
Lovegrass	+0.7	NS
Cheatgrass	-1.3	NS
Barnyard grass	-0.7	NS
Lettuce	-5.3	NS
Red clover	-6.5	*
Carrot	-4.2	NS
Cress	-0.3	NS
Cucumber	-0.4	NS
Palmer amaranth	-2.6	NS
Tomato	+8.9	*
Redroot pigweed	+6.5	NS

*P <= 0.05; ** P <= 0.01; NS Nonsignificant.

oat, and wheat, were greatly inhibited by this pretreatment; while tomato germination was increased. When the germination percentages obtained after pretreatment with chondrillasterol in DCM solution are compared with those for controls pretreated with DCM alone, only two cases of inhibition and two of promotion are observed (Table III). Onion and wheat were both inhibited, but carrot and Palmer amaranth, the source of the chondrillasterol, were promoted. The promotion of Palmer amaranth germination was so marked, 1.5-2 fold in repetitions of the experiment, that a time-course study of the effect of chondrillasterol on amaranth germination was made, using the five amaranth species listed in Materials and Methods.

While the DCM pretreatment itself had no significant effect on the germination of the amaranth seed over 7 days, the chondrillasterol produced the effects observed in Table V. The effects of

Table V.   Time-Course Study of Effects of Chondrillasterol on Seed Germination in Amaranthus Species

Species	% of DCM-treated Control Germination at t Hours						
	t=24	48	72	96	120	144	168
A. Caudatus	98.3	99.1	117.5*	104.3	102.3	105.1	100.5
A. Palmeri	97.2	113.6	147.4*	118.0*	118.0*	140.2*	115.4
A. Retroflexus	102.1	91.2	106.7	107.2	110.8	126.2*	119.2
A. Tricolor	92.3	98.9	97.8	97.6	97.6	97.6	98.1
A. Salicifolius	104.8	88.2*	93.5	96.5	99.8	99.3	99.3

* $P \leq 0.05$.

chondrillasterol are both species and time-dependent. Amaranthus tricolor germinated so rapidly and completely that chondrillasterol had no measurable effect. Germination of A. retroflexus and A. salicifolius was somewhat slower; and the sterol initially appeared to retard germination slightly in both species. This retardation disappeared with time; and A. retroflexus was ultimately promoted by treatment with chondrillasterol. Amaranthus palmeri and A. caudatus controls were slow to germinate and chondrillasterol induced significant increases in the germination rate of both species at 72 h. Untreated A. caudatus seed eventually germinated as completely as the sterol-treated seed, but chondrillasterol produced a second increase in A. palmeri germination at 144 h. At the conclusion of the experiment the three cultivated amaranths had germinated more than 93%; while untreated A. palmeri had only achieved 59% germination and A. retroflexus 65%. Total germination of both weed species was significantly increased by 0.1 mM chondrillasterol at 144 h before development slowed in the more advanced, etiolated seedlings.

No appreciable development of betacyanin pigment in the dark was observed in any of the weed seedlings during the course of the experiment. The development of betacyanin in the cotyledons of dark-grown A. retroflexus seedlings is reported to be induced by the plant growth regulator, kinetin [20], but not by brassinolide, a steroidal lactone [21].

Both radicle and hypocotyl elongation in germinating mung beans have been used as measurements of steroidal activity in plant growth (22). Two methods were used to introduce chondrillasterol into mung beans: a pretreatment in 0.1 mM chondrillasterol in DCM and germination directly in 0.05 mM chondrillasterol in dimethylformamide (DMF). The lower sterol concentration in the DMF studies was due to by the lower solubility of chondrillasterol in DMF. Dichloromethane pretreatment of mung beans initially delayed germination (84% of the H_2O control germination at 24 h, increasing to 95% at 48 h, Table VI); and the presence of 0.4% DMF reduced germination to 55% that

Table VI. Time-course Study of the Effects of Chondrillasterol on Germination and Hypocotyl and Radicle Elongation in Mung Beans

	Germination Period in Hours			
Treatment	24	48	72	96
	Percent Germination			
Sterol in DCM	30.0	87.5	93.0	100.0
DCM Control	56.3	90.0	96.0	100.0
Sterol in DMF	48.7	100.0	100.0	100.0
DMF Control	37.5	95.0	100.0	100.0
Water Control	67.5	100.0	100.0	100.0
	Mean Hypocotyl Length (mm)			
Sterol in DCM	0	1.7	2.7	4.2
DCM Control	0	1.8	2.6	3.8
Sterol in DMF	0	1.6	3.8	3.8
DMF Control	0	2.7	3.0	3.8
Water	0	3.6	3.8	3.8
	Mean Radicle Length (mm)			
Sterol in DCM	1.4	8.3	14.2	26.7
DCM Control	3.5	7.1	12.8	20.6
Sterol in DMF	2.5	9.8	21.7	27.3
DMF Control	0.6	13.0	17.4	25.4
Water Control	4.7	12.3	20.9	26.0

observed in the H_2O control at 24 h. By 48 h seeds in the DMF had reached 95% of the germination in the H_2O control, and at 96 h all seeds had germinated.

Hypocotyl elongation was not linear with time, except in the seeds pretreated in DCM alone (r = 0.99). Initially the pretreatment and DMF retarded hypocotyl elongation, compared to the water control. Chondrillasterol in DCM pretreatment relieved the inhibition somewhat by 96 h. After 72 h chondrillasterol in DMF increased the degree of hypocotyl elongation to that observed in the water control, exceeding that in the DMF control. All differences vanished by 96 h, except a 10.5% increase in mean hypocotyl length in those seeds pretreated with the sterol in DCM.

Radicle elongation was quite linear with time (r = 0.98, minimum). Pretreating the seed with DCM reduced early radicle growth, but by 96 h the sterol/DCM treated seed achieved radicle lengths equal to that of the H_2O control and 30% more than those of DCM controls. The experiment was terminated at 96 h when space limitations in the petri dishes and near-anaerobic conditions during incubation began to interfere with seedling development. There were no secondary roots visible after 96 h.

Sterols have not been reported to have effects on seed germination per se. Mung beans germinated in 0.04 mM hydrocortisone are reported to show stimulation of both radicle and hypocotyl elongation (22, 23, 24); while germination itself is unaffected. The glucosides of stigmasterol, sitosterol, and campesterol did not affect cress germination but did increase both root and shoot growth in cress seedlings (25). The steroidal lactone, brassinolide, applied to barley seeds in DCM before planting, accelerated plant growth (24). Chondrillasterol appears to function as a growth regulator like these other sterols, increasing radicle elongation in the mung bean bioassay; but it also increases the germination rate of the two weedy amaranth species. A mixture of stigmasterol and sitosterol (0.018 wt.%) isolated from ragweed did not affect any of the test seeds. Further tests are in progress with commercially available samples of these sterols.

Ragweed Extracts. More effort has been expended on studies of Palmer amaranth than of ragweed, but some screening germination tests have been completed, including examination of aqueous and several organic solvent extracts. At a 1:10 (w/w) concentration an aqueous extract of air-dried ragweed leaves proved highly inhibitory in the seed germination bioassay (Table VII). All decreases in germination are highly significant, but we have been unable to rule out osmotic inhibition by the ragweed extracts. Even after diluting the extract to 1:640 and filtering through a 5 μm membrane filter, preseeding in the solution prevented determination of the osmolarity using a freezing point depression apparatus.

Dilution to 1:20 eliminated inhibition in onion, oat, ryegrass, red clover, and cheatgrass. Further dilution to 1:40 limited inhibition to lettuce, carrot, and Palmer amaranth. At this concentration lovegrass germination was promoted. The addition of 0.1% DMSO as a carrier in the 1:20 extract increased inhibition in onion (-13.1%, compared to the extract without DMSO), and, similarly, in oat (-22.7%), lettuce (-19.2%), cress (-33.8%), and tomato (-46.0%). The inhibition of cress by the aqueous ragweed extracts was the only significant effect observed in our assays using cress seed which is considered sensitive enough for use in cumulative seed germination assays (26, 27). In these ragweed extract assays the seed appeared to hydrate normally, but no radicle protrusion was observed in the 1:10 extract. After the first dilution the extract still reduced germination of cress. Seeds which did germinate in the 1:20 solution developed into normal seedlings.

Table VII. Effect of Ragweed (<u>Ambrosia</u> <u>artemisiifolia</u>) Aqueous
            Leaf Extract on Seed Germination

Seed	% of Germination in Deionized H_2O		
	1:10 (w/w)	1:20 (w/w)	1:40 (w/w)
Onion	72.5*	---	---
Oat	52.9*	---	---
Ryegrass	79.5*	---	---
Lovegrass	46.6*	89.9	112.6
Cheatgrass	16.0*	---	---
Barnyard grass	92.1*	94.7	---
Lettuce	16.2*	85.5	85.7
Red clover	71.2*	---	---
Carrot	14.2*	48.3*	75.2
Cress	0.0*	83.1*	---
Cucumber	90.9*	91.3*	---
Tomato	2.3*	62.8*	---
Palmer amaranth	14.6*	45.1*	59.6*
Redroot pigweed	27.1*	38.3*	---

*P <= 0.01, all other numbers shown are significant at P <=
0.05.

     Some crude extracts from ragweed leaves and roots were supplied
by LSU cooperators before the multi-seed species assays were in use
at SRRC. These extracts, PE, DCM, MeOH, and H_2O, were tested, using
two lettuce seed cultivars, Grand Rapids (GR) and Great Lakes (GL),
germinated at 28 C, a thermal-stress temperature. The PE extract
(Table VIII) was highly inhibitory of both lettuce cultivars, even
after dilution of the assay solution to half strength. The MeOH
extract of the leaves was also very inhibitory, followed in activity
by the DCM extract, and finally the water extract. Root extracts
inhibited GR germination somewhat, but had no effect on GL. These
same solutions had no effect on radish germination and only the PE,
DCM, and water extracts of the leaves affected carrot germination.
The half-strength solutions were not tested with carrot or radish
seed.
     A mixture of sesquiterpenes, all MW 204, has been isolated from
ragweed. This mixture appears to contain β-bisabolene, bergamotene,
patchoulin, bulnesene, and guayene, and when it was applied in DCM
pretreatment at the wt.% shown in Table IX, the effects were mainly
inhibitory, although cucumber germination was promoted. Ryegrass,
oat, and onion were very sensitive to the sesquiterpenes, as was
Palmer amaranth. When this same mixture was applied in 0.1% DMSO
solution, oat, ryegrass, cucumber, and Palmer amaranth were all
inhibited. Further characterization of this mixture, using updated
GC-MS procedures, is planned, as are additional germination assays
using commercially available sesquiterpenes.
     The results presented and discussed above show some of the dif-
ficulties inherent in the bioassay portions of allelopathy studies.
It is relatively simple to extract inhibitory metabolic products
from suspected allelopathic plants and even to demonstrate <u>in</u> <u>vitro</u>

Table VIII. Effect of Ragweed Organic Solvent Extracts on Lettuce and Carrot Germination

Extract	Seed	% of Control Germination	
		0.227 wt.%	0.113 wt.%
LEAF			
PE	GR Lettuce	1.9*	20.3*
	GL Lettuce	2.0*	5.5*
	Carrot	72.2*	NA
PE/DCM	GR Lettuce	30.6*	65.6*
	GL Lettuce	25.9*	25.4*
	Carrot	75.0*	NA
PE/DCM/MeOH	GR Lettuce	4.5*	36.3*
	GL Lettuce	0.5*	10.7*
	Carrot	98.3	NA
PE/DCM/MeOH/H$_2$O	GR Lettuce	71.9*	87.3
	GL Lettuce	32.0*	73.6*
	Carrot	76.0*	NA
ROOT			
PE/DCM	GR Lettuce	75.8*	103.2
	GL Lettuce	106.0	105.6
	Carrot	98.3	NA
PE/DCM/MeOH	GR Lettuce	79.0*	90.4
	GL Lettuce	100.0	109.6
	Carrot	86.4	NA

*Highly significant (P <= 0.01).

effects for these compounds. However, these effects often disappear under field conditions (28). Most of the compounds isolated and identified so far in our Palmer amaranth studies are too water-insoluble to be transported in soil or even penetrate seeds under normal germination conditions unless some natural solubilizing and transport processes exist. The laboratory experiments did not include the soil microflora which play major solubilizing and conversion roles in ecological systems in the field, lessening the effect of phytoinhibitins or converting innocuous plant excretions and decay products into toxic saproinhibitins (16, 29, 30).

Some inhibitory mixtures of metabolites have been extracted from Palmer amaranth and ragweed, and several isolated compounds have shown marked bioactivity in the germination assays. Currently, work is continuing on the identified compounds and root exudates from Palmer amaranth, as well as the constituents of soil containing Palmer amaranth residues.

Table IX. Effects of Sesquiterpenes (MW 204) Isolated from
Ragweed on Seed Germination

Treatment	Seed	% of Control Germination
DCM Pretreatment 0.2488 wt.% (DCM Control)	Onion	10.7*
	Sorghum	85.6*
	Oat	77.7*
	Ryegrass	17.7*
	Wheat	82.9*
	Lettuce	86.5*
	Red clover	89.6*
	Carrot	83.8*
	Cucumber	105.3*
	Palmer amaranth	41.5
	Tomato	85.0*
Aqueous 0.0629 wt.% (0.1% DMSO Control)	Oat	88.7
	Ryegrass	92.4
	Cucumber	93.7
	Palmer amaranth	41.5*

*P <= 0.01; all other percentages significant (P <= 0.05).

Acknowledgments

Many thanks to Mary Palazzolo whose technical assistance and patience make the replicated assays possible. Thanks also to Joel Carpenter, the LSU student, who supplied the crude ragweed extracts discussed here.

Names of companies or commercial products are given solely for the purpose of providing specific information; their mention does not imply recommendation or endorsement by the U.S. Department of Agriculture over others not mentioned.

Literature Cited

1. Van der Veen, R. Arch. Koffiecult. 1935, 3, 65.
2. Anaya, H. L.; Del Amo, S. J. Chem. Ecol. 1978, 4, 289.
3. Dalrymple, R. L.; Rogers, J. L. J. Chem. Ecol. 1983, 9, 1073.
4. Weill, R. L.; Rice, E. L. Amer. Midl. Nat. 1971, 86, 344.
5. Jackson, J. R.; Willemsen, R. W. Amer. J. Bot. 1976, 63, 1015.
6. Fuerst, E. P.; Putnam, A. R. J. Chem. Ecol. 1983, 9, 937.
7. Khan, A. A.; Samimy, C. In "The Physiology and Biochemistry of Seed Development, Dormancy and Germination"; Khan, A. A., Ed.; Elsevier: Amsterdam, 1982; p. 213.
8. Mayer, A. M.; Poljakoff-Mayber, A. "The Germination of Seeds, 3rd Edition", Pergamon: Oxford, 1982, p. 69.

9. Hadas, H. In "The Physiology and Biochemistry of Seed Development, Dormancy and Germination"; Khan, A. A., Ed.; Elsevier: Amsterdam, 1982; p. 514.
10. Michel, B. E. Plant Physiol. 1983, 72, 66.
11. Cosgrove, D. J.; Daniels, D. G. H.; Whitehead, J. K.; Goulden, J. D. S. J. Chem. Soc. 1952, 4821.
12. Kupchan, S. M.; Obasi, M. E. J. Am. Pharm. Assoc. 1960, 49, 257.
13. Pethig, R.; Gascoyne, P. R. C.; McLaughlin, J. A.; Szent-Gyorgyi, A. Proc. Natl. Acad. Sci. USA. 1983, 80, 129.
14. Koeppe, D. E. Physiol. Plant. 1972, 27, 89.
15. Whitehead, D. C.; Dibb, H.; Hartley, R. D. Soil Biol. Biochem. 1983, 2, 133.
16. McCalla, T. M.; Haskins, F. A. Bacteriol. Rev. 1964, 28, 181.
17. Whitehead, D. C.; Dibb, H.; Hartley, R. D. J. Appl. Ecol. 1982, 19, 579.
18. Williams, R. D.; Hoagland, R. E. Weed Sci. 1982, 30, 206.
19. Wang, T. S. C.; Yang, T.; Chuang, T. Soil Sci. 1967, 103, 239.
20. Bamberger, E.; Mayer, A. M. Science 1960, 131, 1094.
21. Mandava, N. B.; Sasse, J. M.; Yopp, J. H. Physiol. Plant. 1981, 53, 453.
22. Guens, J. M. C. Z. Pflanzenphysiol. 1974, 74, 42.
23. Guens, J. M. C. In "Aspects and Prospects of Plant Growth Regulators", British Plant Growth Regulator Group. Monograph 6; Jeffcoat, B., Ed.; Wantage, 1981, pp. 209-17.
24. Gregory, L. E. Amer. J. Bot. 1981, 68, 586.
25. Tietz, A.; Kimura, Y.; Tamura, S. Z. Pflanzenphysiol. 1977, 81, 57.
26. Lehle, F. R.; Putnam, A. R. Plant Physiol. 1982, 69, 1212.
27. Lehle, F. R.; Putnam, A. R. J. Chem. Ecol. 1983, 9, 1223.
28. Etherington, J. R. "Environment and Plant Ecology"; Wiley: New York, 1975, pp. 244-96.
29. "Principles of Plant and Animal Pest Control, Vol. 2, Weed Control"; National Academy of Sciences, 1975, pp. 6-35.
30. Norstadt, F. A.; McCalla, T. M. Science 1963, 140, 410.

RECEIVED June 12, 1984

The Influence of Secondary Plant Compounds on the Associations of Soil Microorganisms and Plant Roots

R. E. HOAGLAND[1] and R. D. WILLIAMS[2]

[1]Southern Weed Science Laboratory, Agricultural Research Service, U.S. Department of Agriculture, Stoneville, MS 38776
[2]Southern Plains Watershed and Water Quality Laboratory, Durant, OK 74701

This review summarizes aspects concerning the complexity and importance of associations of soil microorganisms and plant roots in the environment. Emphasis is not on compiling a comprehensive review, but rather on problems and potential for research in this area. Allelochemical sources, synthesis, metabolism, degradation, binding in soils, and mode of action are briefly presented and discussed with regard to root-microbe interactions. Data on these areas is accessed with recommendations and suggestions for further investigation.

Allelopathy includes any direct or indirect harmful effect by one plant (including microorganisms) on another through the production of chemicals that escape into the environment. Generally in allelopathic interactions among plants, a secondary compound is released from one plant through exudation, leaching or decomposition of various plant parts which directly inhibits the growth and development of another plant. Such cases have been reported in the literature and have been implicated in successional changes in the plant community and/or spatial patterning of vegetation (1, 2). However, these effects may be indirect rather than direct due to the dynamic nature of the soil-root interactions. One well-documented example is the inhibition of nitrogen-fixing and nitrifying microorganisms by various plants during old field succession (1, 2).

Ecological succession is the orderly process of community change and is the sequence of communities which replace one another in a given area (3). Generally, the driving force behind succession has been attributed to changes of physical factors in the habitat, availability of essential minerals, differences in seed production and dispersal, competition, or a combination of these. In addition, Rice and co-workers (as cited in 1) have indicated that

allelopathy may be responsible for some successional changes in
grassland communities. These studies indicated that certain
species from the weedy stage of succession are inhibitory to other
species in this stage, autotoxic, and/or inhibitory to
nitrogen-fixing and nitrifying organisms. Furthermore, since the
majority of plants absorb nitrogen through their roots (usually
ammonium and nitrate ions) often assisted by bacteria and
mycorrhizal fungi, possible allelopathic effects on these processes
could have a significant impact on plant growth. Recent
compilations of aspects of nitrogen in soils (4) and nitrogen
fixation in plants (5) demonstrate the importance and complexity of
nitrogen cycling. The objective of this paper is to examine the
relationship between the root and its' natural microflora, as well
as to discuss possible mediation of the potential allelopathic
interactions by the microflora.

Rhizosphere

As the root system develops in the soil, the soil micro-environment
is altered, and organic and inorganic compounds exuded from roots
stimulate growth and activity of various soil microorganisms –
particularly bacteria and fungi. The zone of soil in which the
microflora is influenced by the plant root is termed the
rhizosphere. Microbial activity is greater in rhizosphere than
non-rhizosphere soils and the microflora in the rhizosphere differs
quantitatively and qualitatively from that of the non-rhizospheric
soil. Bacteria and fungi associated with the rhizosphere may be
either attached to the surface of the root (rhizoplane) or present
in the soil surrounding the root. Mycorrhizal fungi not only grow
on the root surface but also penetrate the root cortex and develop
a symbiotic relationship with the host plant. Papavizas and Davey
(6) demonstrated that bacterial and fungal numbers decreased with
increased distance from blue lupine roots and that some fungi,
particularly Cylindrocarpon radicicola, were associated only with
the root surface. For example, the number of bacteria per gram of
soil at the root surface, 3-6 mm from the root surface and in a
control soil was 1.59×10^8, 3.8×10^7, and 2.7×10^7,
respectively. Transmission electron micrographs indicate that the
rhizosphere is generally in the range of 1-2 mm thick (7).

Although some authors exclude the mycorrhizal associations from
general discussions of the rhizosphere, we have used the term in
its broadest sense. Since we will be unable to cover all the
literature, we refer the reader to several general references on
the rhizosphere (8-12) and on mycorrhizae (13-18).

Rhizosphere initiation

During seed imbibition, germination and radicle elongation, various organic compounds such as sugars and amino acids are exuded. These compounds stimulate the germination and growth of bacterial and fungal spores. Initiation of the rhizosphere begins as early as seed germination. Further rhizosphere development takes place as the root elongates, but not all areas of the root stimulate microfloral development equally. The area just behind the root cap is the most active part of the root and contributes significantly to mycorrhizal formation.

Early rhizosphere establishment is demonstrated in 2-3 day-old wheat plants when there is a shift towards a population of amino acid requiring bacteria (19). Maximum activity and numbers of rhizosphere microorganisms correlated with maximum vegetative plant development (20-22). Once established, the rhizosphere remains qualitatively similar, but quantitatively increases from seedling stage to maturity (23). After maturity the bacterial population reverts to a population similar to that in non-rhizospheric soils.

Rhizospheric organisms

Bacteria and fungi are stimulated by root growth and are found in the rhizosphere. Generally amino acid-requiring bacteria comprise a higher proportion of the rhizosphere microflora than the general soil microflora, and the rhizosphere contains a higher proportion of bacteria with simple nutritional requirements than soil distant from the plant root (24). Various physiological bacterial groups (motile forms, chromagenic forms, ammonifiers, denitrifiers, gelatin liquifiers and aerobic cellulose-decomposing forms) are present in greater numbers in the rhizosphere than the general soil, whereas other groups (nitrifying forms, anaerobic cellulose-decomposing forms and nitrogen-fixing anaerobes) are fewer in number (12). Rhizobium, Azotobacter and Psuedomonas are bacterial genera common to the rhizosphere.

Although numerous genera of fungi may be associated with plant roots, only a restricted number appear to be isolated from apparently healthy roots with high frequency, i.e. Fusarium, Cylindrocarpon, Rhizoctonia, Glicladium and Mortierella (12).

In nature, most plant roots are invaded by fungi and transformed into mycorrhizae or "fungus roots" (25). The host plant and fungus form a symbiotic relationship whereby nutrients absorbed from the soil by the fungus are released into the host cell and the mycorrhizal fungus obtains nutrients from the host. Mycorrhiza formation is complex and depends on the dynamic interaction of the host plant, fungus and soil. Once formed, mycorrhizae have a profound influence on growth and development of the host plant (26-28).

There are three distinct types of mycorrhizae, but the vesicular-arbuscular mycorrhiza is found on more plant species than any other type (29). It occurs in the bryophytes, pteridophytes and spermatophytes, and is of particular interest because it is found on many economically important agronomic and horticultural species. Vesicular-arbuscular mycorrhizae have been reported to enhance: absorption of nutrients such a phosphorus (30-37), sulfur (38, 39), zinc (40), silicon (32), potassium (41), calcium (42) and copper (42); salt (43) and drought tolerance (43); and nodulation (44-46). The increase in nodulation may be due to an increase in the nutritional level of the plant (particularly with regard to phosphorus) and to overall better growth dynamics of mycorrhizal versus non-mycorrhizal plants (47).

Factors Influencing Rhizosphere Formation

Some factors which influence the development of rhizosphere organisms are: plant species, plant age, root exudates, nutrient pool, soil type, soil moisture, soil temperature and pH. Basically any condition which alters root exudation (quantitatively or qualitatively) of the host plant can alter the nature of the rhizosphere. For example high light intensity and temperature can increase root exudation, particularly during the first few weeks of growth (48). Increases in total bacterial number, glucose fermenting forms, and ammonifiers were detected on wheat roots grown under high light intensity (49). In a similar study fungi development was not affected by light intensity (50). These studies suggest that increased root exudation (caused by increased light intensity) promotes bacterial growth to a greater degree than fungal growth. Furthermore, this indicates how one environmental parameter significantly influences the relationship between host exudation patterns and microbial interactions.

Nature of Exudate and Factors Influencing Exudation

One major factor influencing rhizosphere development is the root exudate. Rovira and Davey (8) compiled a list of compounds occurring in wheat root exudates which are typically found in exudates in general. Several sugars (glucose, fructose, ribose, etc.), amino acids (most of the commonly occurring amino acids), organic acids (oxalic, citric, glycolic, etc.), nucleotides, flavonones, and enzymes (invertase, amylase, protease) have been identified (8). Several phenolic acids such as caffeic, ferulic and cinnamic acids have also been reported to exude from plant roots and have been implicated as allelochemicals (1). Most of the compounds exuded are simple organic compounds of relatively low molecular weight.

The nature of the compounds in the exudate can be altered by biotic and abiotic factors experienced by the plant. Factors involved in quantitative or qualitative changes in exudates are:

plant species, plant age, plant nutrition, light quality and quantity, temperature, soil moisture, soil type, soil pH, foliar sprays (agricultural chemicals) and microorganisms (51, 52). These are generally the same factors that influence rhizosphere microorganisms.

Exudates may be grouped according to their mobility through the soil as diffusible-volatile, diffusible-water soluble and nondiffusible compounds (8). Most of the techniques used to study root exudates yield information only on the diffusible-water soluble compounds. However, ^{14}C-labeling techniques indicate that for every unit of carbon exuded as water-soluble material, 3 to 5 units are released as non-water soluble components (mucilage and root cap cells) and 8 to 10 units as volatile material (8). Based on this, Rovira and Davey (8) stress that (a) under natural conditions many of the simple compounds will be quickly adsorbed or modified by microflora, and (b) techniques used largely ignore the volatile materials and water insoluble materials that may far exceed the soluble compounds under natural conditions. The majority of the allelochemicals identified are water soluble, or partially water soluble, and of relatively low molecular weight. Also only a few volatile compounds have been suggested as inhibitors in allelopathy studies (see for example 53-56).

Influence of Rhizosphere Organisms on Plants

The nature of exudates plays a key role in rhizosphere development and in turn rhizosphere organisms have either a direct or indirect effect on the host plant's growth and development. Rhizosphere organisms can: alter root morphology; change phase equilibria of soil; enhance nutrient availability to the plant; change the chemical composition of soil participating in symbiotic processes; and physically block root surfaces (52). Rhizosphere organisms can affect exudation by: altering root cell permeability or root metabolism; preferentially utilizing certain exudate components; or, excreting toxins (52). It is clear that the process(es) involved in root-rhizosphere interactions are both complex and dynamic. All nutrients and/or compounds a plant obtains from the soil must pass through the rhizosphere and be subjected to biological and chemical transformation in this zone. Studies of toxic compounds in the soil should include processes of the rhizosphere. An allelochemical exuded by a plant could be altered (detoxified or enhanced) by the rhizosphere of the exuding or target plant. Alternatively, allelochemicals may have indirect effects on the target species by influencing its rhizosphere.

Modification of Exudates by the Rhizosphere

Modification of chemicals by the rhizosphere has been followed in

studies using $^{14}CO_2$. For example, Martin (57) studied
water-soluble compounds exuded by roots of wheat, clover and
ryegrass grown in sandy soil, after labeling the shoots with $^{14}CO_2$.
Differences in the amounts of ^{14}C-labeled material in the leachate
occurred among species, but in this near natural nonsterile
environment, exudates were mainly compounds of high molecular
weight, i.e. 45% had molecular weights above 10,000 and 70% were
above 1,000. Under sterile conditions, the bulk of soluble
exudates released by roots consists of low molecular weight
compounds. This indicates that these low molecular weight
compounds are readily utilized by microorganisms in the
rhizosphere, and are subsequently transformed into more complex
forms.

Synthesis and Sources of Allelochemicals from Plants

The term allelochemical applies to phytotoxic substances that cause
growth inhibition (58). These compounds are generally composed of
by-products of main plant biochemical pathways, i.e., secondary
plant compounds (59). Rice (1) has designated 14 allelochemical
categories based on chemical properties and metabolic pathways.
Moreland et al. (60) and Moje (61) have reviewed major
allelochemicals and their roles in plant interactions. The most
thoroughly studied and perhaps the most important allelochemicals
are those derived from the shikimate biosynthetic pathway (Fig. 1)
(62). The most important chemical groups in this regard are:
phenolic acids and their derivatives, terpenoids and steroids,
coumarins, flavonoids, alkaloids and cyanohydrins, and tannins (1).
These compounds are generally water soluble and can be leached from
living and decaying plant tissues; exuded from roots; and some
compounds are volatile and can diffuse from roots and leaves.
There are numerous reports of allelopathy in the literature, but
often the identity of the allelochemical(s) is unknown. There are,
however, many cases where specific compounds or groups of compounds
have been implicated as allelopathic agents. Table 1 summarizes
some examples of sources and identities of allelochemicals that
directly inhibit plant growth. These secondary compounds have been
implicated as a driving force in ecological succession (1).

Allelochemicals Implicated in Affects on Rhizosphere and/or
Root-microbial Association

The bulk of the allelopathy literature has dealt with direct toxic
effects on other plants. However, as developed in this review, it
is obvious that allelochemicals may have a major impact on plant
root-microbial interactions. Such interactions could lead to
growth inhibition in the microorganisms (or in roots) and affect
other factors of the root-microbe association resulting in effects

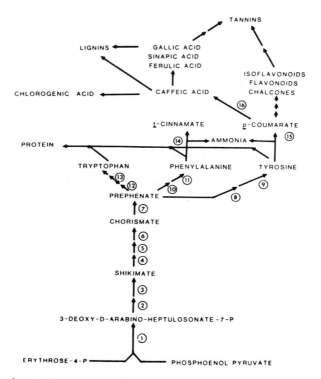

Figure 1.   Schematic outline of various products and associated enzymes from the shikimate and phenolic pathways in plants (and some microorganisms).   Enzymes: (1) 3-deoxy-2-oxo-D-arabino-heptulosate-7-phosphate synthase; (2) 5-dehydroquinate synthase; (3) shikimate dehydrogenase; (4) shikimate kinase; (5) 5-enol-pyruvylshikimate-3-phosphate synthase; (6) chorismate synthase; (7) chorismate mutase; (8) prephenate dehydrogenase; (9) tyrosine aminotransferase; (10) prephenate dehydratase; (11) phenylalanine aminotransferase; (12) anthranilate synthase; (13) tryptophan synthase; (14) phenylalanine ammonia-lyase; (15) tyrosine ammonia-lyase; and (16) polyphenol oxidase. (From ACS Symposium Series No. 181, 1982) (62).

Table I. Sources and Identity of Allelochemicals from Higher Plants

Method of Isolation	Species and Tissue	Chemical Class	Reference
Extraction – ethanol	Cyperus esculentus, tubers and leaves	Ferulic acid and other phenolics	63, 64
	Cyperus rotundus, tubers and leaves	Ferulic acid and other phenolics	64, 65
– water	Abutilon theophrasti, leaves	Phenolics	66
	Agropyron smithii, litter	Phenolics	68
	Ambrosia artemisifolia, shoots	Chlorogenic and caffeic acids	69
	Aster pilosus, shoots	Chlorogenic and caffeic acids	69
	Avena fatua, dead litter	Ferulic and coumaric acids	70, 71
	Bidens pilosa, leaves	Phenylheptatriyne	72, 73
	Brachiaria mutica, dried leaves	Vanillic, p-hydroxybenzoic, o-hydroxyphenylacetic acids	74
	Chloris gayana	Phenolics - ferulic acid	75
	Cynodon dactylon, dried leaves	Phenolics - p-coumaric acid	74
	Digitaria decumbens, dried leaves	Ferulic acid and other phenolics	74
	Digitaria sanguinalis, whole plant	Chlorogenic, isochlorogenic and sulfosalicylic acids	76, 77
	Erica arborea, leaves	Salicylic acid, scopoletin, p-hydroxybenzaldehyde	78
	Helianthus annuus, root	Chlorogenic, isochlorogenic acids; scopoletin	79-81
	Kochia scoparia, leaves	Ferulic acid, myricetin, quercitin	82
	Panicum maximum, dried leaves	o-Hydroxyphenylacetic acid	74
	Paspalum plicatulum, dried leaves	Ferulic acid, other phenolics	74
	Polygonum orientale, roots, stems, leaves	Flavone glycosides	83
	Rumex crispus, leaves	Phenolics	84
	Setaria sphacelata, dried leaves	Ferulic acid, other phenolics	74
	Sorghum halepense, leaves, rhizomes	Chlorogenic, p-coumaric acids, p-hydroxybenzaldehyde	85, 86
	Tripsacum laxum, dried leaves	Ferulic acid, other phenolics	74

		Species	Compounds	Ref.
Exudates	– root	Ambrosia artemisifolia Chenopodium album Helianthus annuus	Chlorogenic and caffeic acids Oxalic acid Chlorogenic, isochlorogenic acids; scopoletin	69 87 79–81
		Hemarthria altissima	Cinnamic and benzoic acids and derivatives	88
		Polygonum aviculare	Phenolic glucosides, fatty acids	75, 89
	– root and rhizome	Sorghum halepense,	Chlorogenic, p-coumaric acids; p-hydroxybenzaldehyde	85, 86
Leachates	– leaves	Cyperus esculentus Datura stramonium Polygonum aviculare	Ferulic acid, other phenolics Scopolamine, hyoscyamine Phenolic glucosides, fatty acids	63, 64 90 75, 89
	– fronds	Pteredium aquilinum Salsola kali	Phenolics Quercitin, ferulic acid, others	91–93 94
	– fruits	Ammi majus	Xanthotoxin, (furanocoumarin)	95
	– seeds	Abutilon theophrasti Datura stramonium	Phenolics Scopolamine, hyoscyamine	96 90
	– roots and rhizomes	Agropyron repens Cyperus esculentus	Acetic, butyric acids Ferulic acid, other phenolics	97–101 63, 64
Volatiles	– leaves	Artemisia tridentata Salvia leucophylla	Terpenes, camphor, pinene, Various volatiles	102 103

interpreted as direct allelopathic effects. Most of the
allelopathic literature disregards the rhizosphere interaction with
plant roots even though some studies suggest the importance of
direct effects of allelochemicals on soil microorganisms (Table 2).
Many varied plant sources contain components that affect
rhizosphere bacteria and fungi. In contrast to numerous examples
of direct effects of allelopathy (of which Table 1 is only a small
fraction) there are limited examples of indirect effects (presented
in Table 2, which comprises most of the major reports in this
area). Here most of the work is from Rice's laboratory and deals
with interactions of allelochemicals and plant growth with regard
to growth of nitrogen-fixing and nitrifying bacteria, nodulation
(nodule mass and number), and hemoglobin content of nodules. Also
presented are other important examples of interactions, i.e. growth
of plants in the presence of free-living bacteria (104), effects of
rhizosphere fungi and bacteria on Rhizobium growth (106, 133, 134),
effects of mycelial extracts of fungi on Rhizobium growth and on
nodulation and growth of soybean (121), and root exudate effects on
mycorrhizal fungi growth (122, 132). Elkan (135) found that root
extracts of a hybrid between a nodulating and non-nodulating
soybean caused reduced nodule weight, total dry weight, and total
nitrogen per plant in the nodulating variety of soybean, and did
not inhibit R. japonicum growth. Azotobacter cultures caused
increased growth in several plant species, probably due to
production of plant growth regulators, i.e. IAA, GA, and cytokinins
(136). Schenck and Stotzky (137) showed that unidentified
volatiles from several germinating seeds promoted growth of soil
bacteria and fungi. Menzies and Gilbert (138) showed that exposure
of soils to alfalfa, corn and other plants caused an increase in
respiration rate of the soil microflora, followed by increased
bacterial and fungal growth. Studies of effects of rhizosphere
fungi on seed germination and root growth of leguminous weed
seedlings indicated that culture filtrates of some rhizosphere
fungi inhibited germination in some species, but increased
germination of others (139).

As stated earlier, mycorrhizae enhance nutrient absorption.
Greater soil exploitation by mycorrhizal roots as a means of
increasing phosphate uptake is well established. The normal
phosphate depletion zone around non-mycorrhizal roots is 1-2 mm,
but an endomycorrhizal root symbiont increased this zone to 7 cm
(140). This ability to increase the nutritional level
(particularly with regard to phosphorus), and subsequently the
overall better growth dynamics of the mycorrhizal plant has been
suggested as the reason for the salt (43) and drought (44-46)
tolerance and increased nodulation (47) observed in mycorrhizal
associations. Another interesting aspect of this enhanced nutrient
uptake is the possible effect of mycorrhizae on competitive ability
between two plant species. Under some conditions, mycorrhizal

Table II. Sources and Action of Secondary Plant Compounds on Root – Rhizosphere Interactions

Species	Source Tissue-isolation	Identity	Effect	Reference
Camelina sativa	H₂O-leaf wash	?	stim'd. radicle elongation in presence of N₂-fix. bacteria	104
Cassia fistula, C. occiden-talis, Leucaena leucocephala, Trifolium alexandrinum	non-nodulating roots and root extracts	?	inhib'd. R. japonicum growth	105
	fungi and bacteria of T. alexandrinum	?	antagonistic and stim. to R. japonicum	106
Hyparrhenia fillipendula, Cynodon dactylon, Rhynchelytrum repens, Sporobolus pyramidalis, Eragrostis curvula, Themeda triandra, Pennisetum purpureum	root extracts	?	inhib'd. growth of NO_3^- & NH_4^+ oxidizers	107, 108
Populus balsamifera	H₂O-extracts of all plant parts	?	red'd. growth, nodulation & acetylene red'n in Alnus crispa	109
Abries balsamea, Populus balsamifera	extracts & leachates of leaves and buds	?	inhib'd. nitrification	110
Pinus ponderosa	extracts—needles, bark	caffeic, chlorogenic acids, tannins	red'd nitrification	111
Aristida adscensionis	extracts & leachates - roots, shoots, litter	?	inhib'd. Rhizobium & Azotobacter nodulation	112–114
Atriplex confertifolia, Eurotia lanata, Artemisia tridentata	H₂O-extract, leaves	?	inhib'd. N₂-fixation	115

Table II. Continued on next page

Table II. (Continued)

Species	Source	Identity	Effect	Reference
Various plant species	volatiles fr. plant residues	?	inhib'd. Rhizoctonia growth, inc'd pigmentation in mycelium, dec'd saprophytic activity	116
Pseudomonas spp.	growing w. Azotobacter	acid end-products	inhib'd. A. chroococcum	117
Trachypogon plumosus	H₂O-extract, roots	?	inhib'd. E. coli, Bacillus subtilis; Staph. aureus, Strep. haemolyticus	118
Oryza sativa	decomposing straw	phenolics	toxic in lettuce & rice seed bioassays; mungbean root assay	119
Soils	extracts of soil	humic & fulvic acids	stim'd plant growth & nodule mass; dec'd nodule no.	120
Trichoderma viride, Rhizopus nigricans, Mucor vesiculosis	mycelial exudates	?	inhib'd. R. japonicum; T. veridi inhib'd. nodulation & M. vesiculosis; inc'd nodule no.	121
Calluna vulgaris	root leachate	?	inhib'd. mycorrhizal fungi growth	122
Various plant species, ex. Ambrosia elatir, Euphorbia corrollater, Helianthus annuus	extracts, exudates, leachates of plants & soil	sugar-phenolic complexes, tannins	inhib'd. Nitrosomonas, Nitrobacter; red'd nodule size & no.; red'd hemoglobin in nodules	123-131
Populus tremula	H₂O-extracts, leaves	benzoic acid, catechol	inhib'd. mycorrhizal fungal growth	132

plants have been more competitive than non-mycorrhizal plants (141, 142).

Although there is no doubt as to the importance of mycorrhizae in nutrient absorption, reviews on ion uptake have generally not considered it. Hatling et al. (143) made this same point more than 10 years ago. In addition, although phenolic acids inhibit phosphate (144, 145) and potassium (146) uptake, no work has examined the effects of these compounds on nutrient absorption of mycorrhizal associations. Since soil microorganisms produce the bulk of the volatile compounds emitted from soil, which are known to inhibit or stimulate fungal development (147-148), this group of compounds from microbial sources should receive more attention.

If mycorrhizae are sites of action for allelochemicals, this is an important indirect aspect of allelopathic interaction among plants. Inhibition of mycorrhizal formation or a reduction in the efficiency of mycorrhizal association would reduce the nutrient level of the mycorrhizal plant and subsequently its competitiveness, stress tolerance or nodulation. Although allelochemicals have been implicated in the reduction of nodulation and nodule size, possible mycorrhizal involvement has not been examined. This is a difficult area of research but one that will provide better understanding of this complex situation.

Nitrification in the Rhizosphere

Conflicting reports on the question of nitrification in the rhizosphere exist. Goring and Clark (149) found that nitrogen was immobilized in the rhizosphere giving an apparent inhibition of nitrification, but there was a rapid accumulation of nitrate after removal of the roots from the rhizosphere. Other workers found that exudates did not inhibit nitrification by pure cultures of nitrifying bacteria and that the number of Nitrosomonas and Nitrobacter increased in the rhizosphere (150). This is contrary to Rice's results (Table 2) in which root extracts of several plants inhibited nitrification in pure cultures and in soil. Moore and Waid (151) eliminated immobilization and denitrification as possible causes for low content of nitrate and confirmed an earlier report (152) that inhibition of nitrification by grass roots was responsible for low levels of nitrate in permanent grassland soils. Munro (107, 108) found that root extracts from Hyparrhenia filipendula and several other species inhibited the growth of nitrate- and ammonia-oxidizing bacteria. Contrary to this, Purchase (153, 154), using root washings of H. filipendula, found no evidence of toxicity to Nitrobacter and Nitrosomonas. Since Nitrobacter is more sensitive to phosphorus deficiency than Nitrosomonas, and because phosphorus deficiency is sufficiently severe in some soils to restrict growth, its ability to compete for nitrogen is diminished. Inhibition was found in the root extract where the concentration of the allelochemical(s) may have been greater than in the root wash. This later controversy points out a major problem in much of the allelopathy literature, i.e. the lack of identification and quantification of toxic components.

Allelochemical Concentration in Soil

Allelochemical complexes, pool size, and turnover in soils. Many chemicals occur in soils that have been released from living and decaying plant tissues and soil microorganisms. Generally these consist of nearly all of the common amino acids, common sugars, aliphatic acids, nucleotides, some enzymes, and various benzoic and phenolic acids and their derivatives (8). Little information is available on concentration levels of many of these components. Reports (155, 156) indicate that free amino acid levels seldom exceed 2 ug/g soil, but these levels can be 7-fold higher in rhizosphere soils. This is a concentration range of about 10^{-6} to 10^{-5} M. Most of the phytotoxic compounds isolated from plants and soils are the phenolic acids and their derivatives (157, 158). Aromatic amino acids are precursors of phenolic and benzoic acids in plants, and some microorganisms and soil microbes readily metabolize these latter compounds to various products (159, 160). Free phenolic acids can occur in the soil solution but only ferulic, p-hydroxybenzoic, p-coumaric, and vanillic acids are commonly found, and at amounts less than 0.01% of the total soil organic matter (158, 161, 162). These four phenolic acids are readily utilized by soil microorganisms (163-165). There is little direct evidence (or research) relating the effects of soil microbial metabolism of phenolics to phytotoxicity to higher plants (166). Recent studies do show that amelioration of phenolic phytotoxicity could be achieved with some microbes in solution culture (167). There is sufficient evidence that some plants stimulate the growth of phenolic acid-degrading organisms (165), and that many soil fungi can use phenolic acids as sole carbon sources (168). This indicates that some soil microbes can exert their maximal effect on phenolic acid metabolism during plant growth.

Maximal levels for p-coumaric and ferulic acids of 30.0 and 6.5 μmol/100 g of soil have been reported (158) and concentrations of 4×10^{-5} M and 3×10^{-5} M, respectively, for these two acids in other soils (161). Other studies indicate a similar concentration range of 2.3×10^{-7} to 10^{-6} M for p-hydroxybenzoic, vanillic and p-coumaric acids (169). These levels may be too low to have direct measurable allelopathic effects on plants in greenhouse or growth chamber studies (non-rhizosphere soils, low microbial population). However, in field rhizosphere soils (high microbial population) these levels could be sufficient to influence microbial growth (positively or negatively) resulting in an indirect effect via alteration of mycorrhizal-root associations.

Complexing of various compounds commonly occurs in soil, transforming low molecular weight materials into high molecular weight polymers. Phenolic acids are intermediates in the formation of lignins and humic substances (170) and are also important in stabilizing nitrogen in organic forms in soils (171). Associated higher molecular weight compounds (i.e. aggregates, polymers of phenols) can alter the concentration of low molecular weight phenolics in the rhizosphere (160, 172). One proposed scheme of such polymerization consists of condensation of amino acids with phenolic compounds and sugars to yield complex polymers (Fig. 2) (173).

Phenolic Binding and Availability in Soils. Low molecular weight phenolic and polymeric phenolic complexes occur in soils and can be bound to soil particulates. Wang et al. (174) have shown that clay minerals can act catalytically to influence phenolic polymerization. These polymers can then form clay-organic complexes resulting in reduced availability of these materials to plant roots and thus decreased phytotoxic or allelopathic potential. Soil mineral content (i.e. Al^{+3} or $Fe^{+2,+3}$) also has a great influence on absorption of phenolics and is implicated in reducing phytotoxicity, increasing biotic degradation, and in catalytically transforming these compounds into humic materials (175). The common low molecular weight phenolics have varying affinities for soils depending on soil type and phenolic structure (175, 176). Phenolics (especially polyphenolics) exuded from plants or produced during decomposition can be rapidly leached from the soil surface, become bound, and contribute to humus formation (174). Phenolic acid toxicity is also dependent on soil nutrient status, especially with regard to nitrogen and phosphorus levels (177).

Since lignins are polymers of phenolics and are major plant constituents with resistance to microbial decomposition, they are the primary source of phenolic units for humic acid synthesis (178, 179). Once transformed, these humic acids become further resistant to microbial attack and can become bound to soils (180); form interactions with other high molecular weight phenolic compounds (ex. lignins, fulvic acids) and with clays (181); and influence the biodegradation of other organic substrates in soils (182, 183). Some compounds, i.e. benzoic and cinnamic acids are not protected against biodegradation to a high degree by linkage and/or absorption on soil constituents such as clay or humus (184), hence they may have a rapid turnover rate in soils.

Incorporation of some xenobiotics (herbicides) into soil humus-complexes occurs via pathways analogous to those for incorporation of naturally occurring phenolic and benzoic acids and do indeed involve phenolic and humus-like constituents

(Figure 3) (173). Although these complexes are resistant to degradation and re-entry into the rhizosphere, some microorganisms and other microfauna can cause degradation of these complexes with potential phytotoxicity (185, 186). This concept is pertinent to this discussion in that xenobiotics and/or bound naturally occurring phenolics could be potential pools of toxic compounds or allelochemics. If these compounds are released from bound complexes, they may not only affect plants directly, but also affect rhizosphere organisms and microbial-root interactions. The nature and toxicity of these humus-bound residues is unresolved as is whether these polymeric substances are potentially beneficial or harmful. Although there is considerably more information on degradation and transformation of potential allelochemicals than presented here, this brief presentation points to the importance of further research in this area.

Biochemical Sites and Modes of Action of Allelochemicals

There are numerous reported effects of various allelochemicals (phenolics) on plants, but relatively little work on effects of these compounds on microorganisms. Even though some reports deal with physiological and biochemical effects on these systems, most are concerned only with toxicity studies (i.e. growth inhibition) and don't consider possible sites of action, or mode and mechanisms of action, especially at the molecular level. Allelochemical mode of action has been reviewed (1, 2, 60, 187), and although mode of action studies have increased since 1966, there is a general void in this allelochemical research area. Possible sites of action for allelochemicals should be quite similar, if not identical to those of herbicides. For plants these include effects on: cell walls, membranes, major organelles (mitochondrion, chloroplast, nucleus, nucleolus, etc.), major processes (cell division, photosynthesis, respiration, protein synthesis, lipid synthesis, etc.) and on key enzymes. For mode of action studies of allelochemicals on microorganisms most of these also apply. When the intimate associations of microorganisms with roots is considered, other important possible sites of action of allelochemicals (and herbicides) are apparent. These are: biochemical and physiological action on mycorrhizal binding, infection processes of nodulating bacteria, nodulation development, and important alterations of key enzyme activities associated with these processes. Allelochemical research in these areas is woefully lacking.

Summary and Conclusions

To further clarify the actual role, impact, and expression of allelopathic phenomena in natural environments, much more research

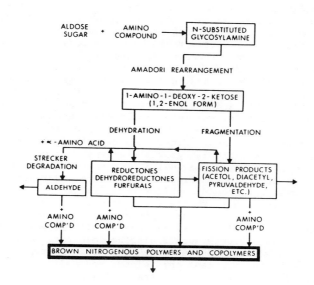

Figure 2. Scheme for formation of nitrogenous polymers in soils by condensation of amino acids with polyphenols and sugars. (From ACS Symposium Series No. 29, 1976) (<u>173</u>).

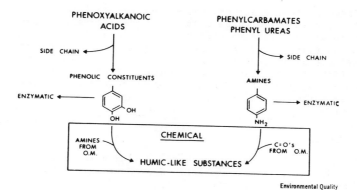

Environmental Quality

Figure 3. Proposed chemical reactions leading to stabilization of phenylamide, phenylcarbamate and phenylurea pesticides in soils. (From ACS Symposium Series No. 29, 1976) (<u>173</u>).

and information is needed on several parameters. Highly effective methods for isolation, identification, purification, and quantitation of allelochemicals (both water soluble and insoluble) in soils are available, but are not used in most allelochemical studies. Use of these techniques is a prerequisite for needed molecular modes of action research. Determination of pool size, turnover rate and degradation pathways (fate) of allelochemics in soils would also provide much insight. Effects of allelochemicals on rhizosphere microflora, and in particular on mycorrhizae are needed. Knowledge of possible modification of root exudates by rhizosphere microflora and the biochemistry and physiology of mycorrhizal-root associations (especially on infection processes, nodulation, the impact of hyphal bridge formation, etc.) could provide major breakthroughs. Since turnover of allelochemical pools, and plant and microbial growth are dynamic processes, the dynamic nature of soil and the environment should be considered for possible allelopathic interactions among plants. Further allelopathic research should consider the effects of multiple allelochemicals because there are interactions of allelochemicals on both the target plant and rhizosphere organisms (188, 189). Interactions between allelochemicals and fertility, moisture stress and shading on plant growth should also be investigated. To accomplish these goals, the expertise of several disciplines is needed; i.e., chemistry, plant physiology, and soil microbiology.

The information presented here shows that rhizosphere microorganisms play a major role in the production and metabolism of secondary plant compounds, especially phenolics. The rhizosphere contains numerous organisms that have direct and indirect associations with major crop and weed species. Some of these associations play major roles in plant nitrogen metabolism, nutrient uptake, and water relations. Microorganisms that form associations with plant roots are sensitive and are affected by various factors (water stress, water logging, soil aeration, temperature, pH, agricultural chemicals, and naturally occurring plant materials). Studying effects of allelochemicals in soil only on plant roots, seed germination, etc., may not give an accurate appraisal of toxicity, site of action, or mechanisms of action because of the intimate and complex nature of the rhizosphere. This complexity is further demonstrated by considering a series of 31 transfer pathways that interconnect plants, animals, soil organic matter, and soil mineral pools (190).

Acknowledgments

The authors thank Ruth Jones and Doris McKenzie for their excellent assistance in literature-searching. Judy Fava is thanked for her expert and rapid typing of the manuscript.

Literature Cited

1. Rice, E. L. "Allelopathy"; Academic Press: New York, 1974.
2. Rice, E. L. Bot. Review 1979, 45, 15–109.
3. Odum, E. P. "Fundamentals of Ecology"; W. B. Saunders: Philadelphia, 1959, p. 257.
4. Stevenson, F. J. In "Nitrogen in Agricultural Soils"; Stevenson, F J., Ed., Am. Soc. Agronomy: Madison, Wisc., 1982; chap. 1 & 3.
5. Sprent, J. I. In "Advanced Plant Physiology"; Wilkins, M. B., Ed., Pitman Publ. Ltd: London, 1984; chap. 12.
6. Papavizas, G. C.; Davey, C. B. Plant Soil 1961, 14, 215–236.
7. Bowen, G. D.; Rovira, A. D. Annual Rev. Phytopathol. 1976, 14, 121–144.
8. Rovira, A. D.; Davey, C. B. In "The Plant Root and Its Environment"; Carson, E.W., Ed., Univ. of Virginia Press: Charlottesville, 1974; chap. 7.
9. Bowen, G. D.; Rovira, A. D. In "Modern Methods in the Study of Microbial Ecology"; Rosswall, T., Ed.; Swedish Natural Science Research Council: Stockholm, 1973; pp. 443–450.
10. Hale, M. G.; Moore, L. D.; Griffin, G. J. In "Interactions Between Non-pathogenic Soil Microorganisms and Plants"; Dommergues, Y. R.; Kenpa, S. V., Eds., Elsevier: Amsterdam; 1978; pp. 163–203.
11. Bowen, G. D. In "Soil-Borne Plant Pathogens"; Schippers, B.; Gams, W., Eds.; Academic Press: New York, 1979; pp. 209–227.
12. Parkinson, D. In "Soil Biology"; Burges, A.; Raw, R., Eds.; Academic Press: New York, 1967; Chap. 15.
13. Mosse, B.; Stribleny, D.P.; LeTacon, F. In "Advances in Microbial Ecology"; Alexander, M., Ed.: Plenum Press: New York, 1981; Chap. 4.
14. Moser, M.; Haselwandter, K. In "Physiological Plant Geology III": Lange, O. L.; Nobel, P. S.; Osmond, C. B.; Ziegler, H., Eds., Springer-Verlag; New York; 1983; Chap. 9.
15. Marks, G. C.; Kozlowski, T. T., Eds. "Ectomycorrhizae"; Academic Press: New York, 1973.
16. Harley, J. L. "The Biology of Mycorrhiza"; Lenard Hill: London, 1969 (334 pages).
17. Harley, J. L.; Smith, S. E. "Mycorrhizal Symbiosis"; Academic Press: New York, 1983.
18. Sanders, F. E.; Mosse, B.; Tinker, P. B., Eds. "Endomycorrhizas"; Academic Press: New York, 1975.
19. Rouatt, J. W. Can. J. Microbiol. 1959, 5, 67–71.
20. Starkey, R. L. Soil Sci. 1929, 27, 319–334.
21. Starkey, R. L. Soil Sci. 1929, 27, 355–378.
22. Starkey, R. L. Soil Sci. 1929, 27, 433–444.
23. Gyllenberg, H. Can. J. Microbiol. 1957, 3, 131–134.

24. Lochhead, A. G.; Rouatt, J. W. Proc. Soil Sci. Soc. Am. 1955, 19 48-49.
25. Gerdemann, J. W. In "The Plant Root and Its Environment"; Carson, E. W., Ed.; Univ. of Virginia Press: Charlottesville, 1974; Chap. 8.
26. Gerdemann, J. W. In "The Development and Function of Roots"; Torrey, J. G.; Clarkson, D. T., Eds.; Academic Press: London, 1975, pp. 575-591.
27. Mosse, B. Ann. Rev. Phytopathol. 1973, 11, 171-196.
28. Hayman, D. S. Can. J. Bot. 1983, 61, 944-963.
29. Gerdemann, J. W. Ann. Rev. Phytopathol. 1968, 6, 397-418.
30. Khan, A. G. New Phytol. 1972, 71, 613-619.
31. Powell, C. L.; Daniel, J. N. Z. J. Agric. Res. 1978, 21, 675.
32. Yost, R. S.; Fox, R. L. Agron. J. 1982, 74, 475-481.
33. Yost, R. S.; Fox, R. L. Agron. J. 1979, 71, 903-908.
34. Mosse, B. New Phytol. 1973, 72, 127-136.
35. Mosse, B. New Phytol. 1977, 78, 277-288.
36. Ross, J. P.; Gilliam, J. W. Soil Sci. Soc. Am. J. 1973, 37, 237-239.
37. Guttay, A. J. R. J. Amer. Soc. Hort. Sci. 1983, 108, 222-224.
38. Gray, L. E.; Gerdemann, J. W. Plant Soil 1973, 39, 687.
39. Rhodes, L. H.; Gerdemann, J. W. Soil Biol. Biochem. 1978, 10, 361-364.
40. McIlveen, W. D.; Spotts, R. A.; Davis, D. D. Phytopath. 1975, 65, 645.
41. Powell, D. L. in "Endomycorrihizas"; Sanders, F. E.; Mosse, B.; Tinker, P. B., Eds.; Academic Press: New York, 1975, pp. 461-468.
42. Ross, J. P. Phytopath. 1971, 61, 1400-1403.
43. Hiriel, M. C.; Gerdemann, J. W. Soil Sci. Soc. Am. J. 1980, 44, 654-655.
44. Safir, G. R.; Boyer, J. S.; Gerdemann, J. W. Science 1971, 172, 581-583.
45. Safir, G. R.; Boyer, J. S.; Gerdemann, J. W. Plant Physiol. 1972, 49, 700-703.
46. Nelsen, C. E.; Safir, G. R. Planta 1982, 154, 407-413.
47. Green, N. E.; Smith, M. D.; Beavis, W. D.; Aldon, E. F. J. Range Management 1983, 36, 576-578.
48. Rovira, A. D. Plant Soil 1956, 7, 178-194.
49. Rouatt, J. W.; Katznelson, H. Nature 1960, 186, 659-660.
50. Peterson, E. A. Can. J. Microbiol. 1961, 7, 2-6.
51. Hale, M. G.; Foy, C. L.; Shay, F. J. In "Advances in Agronomy"; Brady, N. C. Ed.; Academic Press: New York, 1971; Vol. 23, pp. 89-109.
52. Hale, M. G.; Moore, L. D. In "Advances in Agronomy", Brady, N. C. Ed., Academic Press: New York, 1979; Vol. 31, pp. 93-124.

53. Muller, C. H. In "Biochemical Interactions Among Plants"; National Academy of Sciences: Washington, D.C., 1971; pp. 64–71.

54. Hoffman, G. R.; Hazlett, D. L. J. Range Management 1977, 30, 134–137.

55. Weaver, T. W.; Klavich, D. Am. Midland. Nat. 1977, 97, 508–512.

56. Lill, R. E.; Waid, J. S. N. Z. J. For. Sci. 1975, 5, 165–170.

57. Martin, J. P. Aust. J. Biol. Sci. 1971, 24, 1131–1142.

58. Whittaker, R. H.; Feeny, P. P. Science 1971, 171, 757–770.

59. Swain, T. Ann. Rev. Plant Physiol. 1977, 28, 479–501.

60. Moreland, D. E.; Egley, G. H.; Worsham, A. D.; Monaco, T. J. Adv. in Chemistry 1966, 53, 112–141.

61. Moje, W. In "Diagnostic Criteria for Plants & Soils"; Chapman, H. D., Ed.; Univ. Calif. Press: Berkeley, 1966, pp. 533–569.

62. Hoagland, R. E.; Duke, S. O. In "Biochemical Responses Induced by Herbicides"; Moreland, D. E.; St. John, J. B.; Hess, F. D., Eds., Amer. Chem. Soc.: Washington, D.C., 1982; pp. 175–205.

63. Drost, D. C.; Doll, J. D. Weed Sci. 1980, 28, 229–233.

64. Jangaard, N. O.; Sckerl, M. M.; Schieferstein, R. H. Weed Sci. 1971, 19, 17–20.

65. Friedman, T.; Horowitz, M. Weeds 1971, 19, 398–401.

66. Colton, C. E.; Einhellig, F. A. Amer. J. Bot. 1980, 67, 1407–1413.

67. Elmore, C. D. Weed Sci. 1980, 28, 658–660.

68. Bokhari, U. G. Ann. Bot. 1978, 42, 127–136.

69. Jackson, J. R.; Willemsen, R. W. Amer. J. Bot. 1976, 63, 1015–1023.

70. Schumacher, W. J.; Thill D. C.; Lee, G. A. J. Chem. Ecol. 1983, 9, 1235–1245.

71. Tinnin, R. O.; Muller, C. H. Bull. Torrey Bot. Club 1972, 99 287–292.

72. Bonasera, J.; Lynch, J.; Leck, M. A. Bull. Torrey Bot. Club 1979, 106, 217–222.

73. Campbell, G.; Lambert, J. D. H.; Arnason, T.; Towers, G. H. N. J. Chem. Ecol. 1982, 8, 961–972.

74. Chu, C.-H.; Young, C.-C. J. Chem. Ecol. 1975, 1, 183–193.

75. Alsaadawi, I. S.; Rice, E. L. J. Chem. Ecol. 1982, 8, 1011–1023.

76. Parenti, R. L.; Rice, E. L. Bull. Torrey Bot. Club 1969, 96, 70–78.

77. Schreiber, M. M.; Williams, J. L., Jr. Weeds 1967, 15, 80–81.

78. Ballester, A.; Vieitez, A. M.; Vieitez, E. Bot. Gaz. 1979, 140, 433–436.

79. Irons, S. M.; Burnside, O. C. Weed Sci. 1982, 30, 372–377.

80. Schon, M. K.; Einhellig, F. A. Bot. Gaz. 1982, 143, 505–510.
81. Wilson, R. E.; Rice, E. L. Bull. Torrey Bot. Club 1969, 95,
    432–448.
82. Lodhi, M. A. K. Can. J. Bot. 1979, 57, 1083–1088.
83. Datta, S. C.; Chatterjee, A. K. Flora 1980, 169, 456–465.
84. Einhellig, F. A.; Rasmussen, J. A. Amer. Midl. Nat. 1973, 90,
    79–86.
85. Abdul–Wahab, A. S.; Rice, E. L. Bull. Torrey Bot. Club 1967,
    94, 486–497.
86. Lolas, P.C.; Coble, H. D. Weed Sci. 1982, 30, 589–593.
87. Caussanel, J.-P.; Kunesch, G. Z. Pflanzenphysiol. 1979, 93,
    229–243.
88. Tang, C.-S.; Young, C.-S. Plant Physiol. 1982, 69, 155–160.
89. Alsaadawi, I.S.; Rice, E. L.; Karns, T. K. B. J. Chem.
    Ecol. 1983, 9, 761–775.
90. Lovett, J. V.; Levitt, J.; Duffield, A. M.; Smith, N. G.
    Weed Res. 1981, 21, 165–170.
91. Gliessman, S. R. Bot. J. Linn. Soc. 1976, 73, 95–104.
92. Gliessman, S. R.; Muller, C. H. J. Chem. Ecol. 1978, 4,
    337–362.
93. Stewart, R. E. J. Chem. Ecol. 1975, 1, 161–169.
94. Lodhi, M. A. K. J. Chem. Ecol. 1979, 5, 429–437.
95. Friedman, J.; Rushkin, E.; Waller, G. R. J. Chem. Ecol.
    1982, 8, 55–65.
96. Gressel, J. B.; Holm, L. G. Weed Res. 1964, 4, 44–53.
97. Gabor, W. E.; Veatch, C. Weed Sci. 1981, 29, 155–159.
98. Kommedahl, T.; Old, K. M.; Ohman, J. H.; Ryan, E. W. Weed
    Sci. 1970, 18, 29–32.
99. Oswald, H. J. Ecol. 1948, 36, 192–193.
100. Penn, D. J.; Lynch, J. M. J. Appl. Ecol. 1981, 18, 669–674.
101. Toai, T. V.; Linscott, D. L. Weed Sci. 1979, 27, 595–598.
102. Hoffman, G.R.; Hazlett, D. L. J. Range Management 1977, 30,
     134–137.
103. Muller, W. H.; Hauge, R. Bull. Torrey Bot. Club 1967, 94,
     182–191.
104. Lovett, J. V.; Sagar, G. R. New Phytol. 1978, 81, 617–625.
105. Ras, V. R.; Ras, N. S. S.; Mukerji, K. G. Plant & Soil 1973,
     39, 449–452.
106. Husaina, A.; Mallik, M. A. B. J. Sci. 1972, 1, 139–145.
107. Munro, P. E. J. Appl. Ecol. 1966, 3, 227–229.
108. Munro, P. E. J. Appl. Ecol. 1966, 3, 231–238.
109. Jobidon, R.; Thibault, J.-R. Amer. J. Bot. 1982, 69,
     1213–1223.
110. Thibault, J.-R.; Fortin, J.-A.; Smirnoff, W. A. Amer. J. Bot.
     1982, 69, 676–679.
111. Lodhi, M. A. K.; Killingbeck, K. T. Amer. J. Bot. 1980, 67,
     1423–1429.
112. Murthy, M. S.; Nagodra, T. J. Appl. Ecol. 1977, 14,
     279–282.

113. Murthy, M. S.; Ravindra, R. Oecologia 1974, 16, 257–258.
114. Murthy, M. S.; Ravindra, R. Oecologia 1975, 18, 243–249.
115. Rychert, R. C.; Skujins, J. Soil Sci. Soc. Amer. Proc. 1974, 38, 768–771.
116. Lewis, J. A.; Papauizas, G. C. Soil Sci. 1974, 118, 156–163.
117. Chan, E. C. S.; Basavanand, P.; Liivak, T. Can. J. Microbiol. 1970, 16, 9–16.
118. Stiven, G. Nature 1952, 170, 712–713.
119. Chou, C.-H.; Lin, T.-J.; Kao, C.-I. Bot. Bull. Acad. Sinica 1977, 18, 45–60.
120. Tan, K. H.; Tantiwiramanond, D. Soil Sci. Soc. Amer. J. 1983, 47, 1121–1124.
121. Angle, J. S.; Pugashetti, B. K.; Wagner, G. H. Agron. J. 1981, 73, 301–306.
122. Robinson, R. K. J. Ecol. 1972, 60, 219–224.
123. Rice, E. L. Ecology 1964, 45, 824–837.
124. Rice, E. L. Physiol. Plant. 1965, 18, 255–268.
125. Rice, E. L.; Parenti, R. L. Southwest. Nat. 1967, 12, 97–103.
126. Floyd, G. L.; Rice, E. L. Bull. Torrey Bot. Club 1967, 94, 125–129.
127. Rice, E. L. Bull. Torrey Bot. Club 1968, 95, 346–358.
128. Rice, E. L. Amer. J. Bot. 1971, 58, 368–371.
129. Rice, E. L.; Pancholy, S. K. Amer. J. Bot. 1973, 60, 691–702.
130. Rice, E. L.; Pancholy, S. K. Amer. J. Bot. 1974, 61, 1095–1103.
131. Rice, E. L.; Lin, C.-Y.; Huang, C.-Y. Bot. Bull. Acad. Sinica 1980, 21, 111–117.
132. Olsen, R. A.; Odham, G.; Lindberg, G. Physiol. Plant. 1971, 25, 122–129.
133. Hattingh, M. J.; Louw, H. A. Can. J. Microbiol. 1969, 15, 361–364.
134. Damergi, S. M.; Johnson, H. W. Agron. J. 1966, 58, 223–224.
135. Elkan, G. H. Can. J. Microbiol. 1961, 7, 851–856.
136. Barea, J. M.; Brown, M. E. J. Appl. Bact. 1974, 37, 583–593.
137. Schenck, S.; Stotzky, G. Can. J. Microbiol. 1975, 21, 1622–1634.
138. Menzies, J. D.; Gilbert, R. G. Soil Sci. Soc. Amer. Proc. 1967, 31, 495–496.
139. Sullia, S. B. Geobios 1974, 1, 175–177.
140. Rhodes, L. H.; Gerdemann, J. W. New Phytol. 1975, 75, 555–561.
141. Crush, J. R. New Phytol. 1974, 73, 743–749.
142. Fitter, A. H. New Phytol. 1977, 79, 119–125.
143. Hatling, M. S.; Gray, L. E.; Gerdemann, J. W. Soil Sci. 1973, 116, 383–387.

144. Glass, A. D. M. Plant Physiol. 1973, 51, 1037-1041.
145. Glass, A. D. M. Phytochem. 1975, 14, 2127-2130.
146. Glass, A. D. M. J. Exp. Bot. 1974, 25, 1104-1113.
147. Hutchinson, S. A. Trans. Brit. Mycol. Soc. 1971, 57,
     185-200.
148. Stotzky, G.; Schenck, S. Can. J. Microbiol. 1976, 21,
     1622-1634.
149. Goring, C. A. I.; Clark, F. E. Soil Sci. Soc. Amer. Proc.
     1948, 13, 261-266.
150. Molina, J. A. E.; Rovira, A. D. Can. J. Microbiol. 1964,
     10, 249-257.
151. Moore, D. R. E.; Waid, J. S. Soil Biol. Biochem. 1971, 3,
     69-83.
152. Theron, J. J. J. Agr. Sci. 1951, 41, 289-296.
153. Purchase, B. S. Plant & Soil 1974, 41, 527-539.
154. Purchase, B. S. Plant & Soil 1974, 41, 541-547.
155. Sowden, F. J. Soil Sci. 1969, 107, 364-371.
156. Sowden, F. J. Can. J. Soil Sci. 1970, 50, 227-232.
157. Borner, H. Bot. Rev. 1960, 26, 393-424.
158. Wang, T. S. C.; Yeh, K. L.; Cheng, S. Y.; Yang, T. K. In
     "Biochemical Interactions Among Plants"; Nat'l. Acad.
     Sciences: Washington, 1971; pp. 113-120.
159. Dagley, S. In "Soil Biochemistry"; McLaren, A. D.;
     Peterson, G. H., Eds.; Marcel Dekker: New York, 1967; pp.
     289-317.
160. Turner, J. A.; Rice, E. L. J. Chem. Ecol. 1975, 1, 41-58.
161. Whitehead, D. C. Nature, London 1964, 202, 417-418.
162. Vaughan, D.; Ord, B. G. Soil Biol. Biochem. 1980, 12,
     449-450.
163. Knosel, D. Zeit. Pfanzenernahrung Bodenkunde 1959, 85,
     58-66.
164. Batistic, L.; Mayaudon, J. Anns. l'Itst. Pasteur 1970, 118,
     190-206.
165. Sparling, G. P.; Ord, B. G.; Vaughan, D. Soil Biol.
     Biochem. 1981, 13, 455-460.
166. Sparling, G. P.; Vaughan, D. J. Sci. Food Agric. 1981, 32,
     625-626.
167. Vaughan, D.; Sparling, G. P.; Ord, B. G. Soil Biol.
     Biochem. 1983, 15, 613-614.
168. Henderson, M. E. K.; Farmer, V. C. J. Gen. Microbiol. 1955,
     12, 37-46.
169. Guenzi, W. D.; McCalla, T. M. Proc. Soil Sci. Soc. Amer.
     1966, 30, 214-216.
170. Haider, K.; Martin, J. P.; Filip, Z. In "Soil Biochemistry";
     Paul, E. A.; McLaren, A. D., Eds.; Marcel Dekker: New York,
     1975; pp. 195-224.
171. Parsons, J. W.; Tinsley, J. In "Soil Components";
     Gieseking, J. E., Ed.; Spring-Verlag: New York, 1975; pp.
     263-304.

172. Henderson, M. E. K. Pure Appl. Chem. 1963, 7, 589–602.
173. Stevenson, F. J. In "Bound & Conjugated Pesticide
     Residues", Kaufman, D. D.; Still, G. G.; Paulson, G. D.;
     Bandal, S. K., Eds., Amer. Chem. Soc.: Washington, D. C.;
     1976; pp. 180–207.
174. Wang, T. S. C.; Li, S. W.; Ferng, Y. L. Soil Sci. 1978,
     126, 15–21.
175. Hauang, P. M.; Wang, T. S. C.; Wang, M. K.; Wu, M. H.; Hsu,
     N. W. Soil Sci. 1977, 123, 213–219.
176. Shindo, H.; Kuwatsuka, S. Soil Sci. Plant Nutr. 1976, 22,
     23–33.
177. Stowe, L. G.; Osborn, A. Can. J. Bot. 1980, 58, 1149–1153.
178. Oglesby, R. T.; Christman, R. F.; Driver, C. H. Adv. Appl.
     Microbiol. 1967, 9, 111–184.
179. Hurst, H. M.; Burges, N. A. In "Soil Biochemistry";
     McLaren, A. D.; Peterson, G. H., Eds.; Marcel Dekker: New
     York, 1967; pp. 260–286.
180. Martin, J. P.; Richards, S. J.; Haider, K. Soil Sci. Soc.
     Amer. Proc. 1967, 31, 657–662.
181. Greenland, D. J. Soil Sci. 1971, 111, 34–41.
182. Martin, J. P.; Parsa, A. A.; Haider, K. J. Biol. Biochem.
     1978, 40, 483–486.
183. Martin, J. P.; Haider, K. Soil Sci. Soc. Amer. J. 1976, 40,
     377–380.
184. Haider, K.; Martin, J. P. Proc. Soil Sci. Soc. Amer. 1975,
     39, 657–662.
185. Bollag, J.-M.; Loll, M. J. Experimentia 1983, 39,
     1221–1231.
186. Kaufman, D. D.; Still, G. G.; Paulson, G. D.; Bandal, S. K.,
     Eds. "Bound & Conjugated Pesticide Residues"; Amer. Chem.
     Soc.: Washington, D. C., 1976.
187. Horsley, S. B. In "Proc. Fourth N. Amer. Forest Biology
     Workshop"; Wilcox, H. E.; Hamer, A. F., Eds.; State Univ.
     NY: Syracuse, 1977, pp. 93–136.
188. Einhellig, F. A.; Rasmussen, J. A. J. Chem. Ecol. 1979, 5,
     815–824.
189. Williams, R. D.; Hoagland, R. E. Weed Sci. 1982, 30,
     206–212.
190. Frissel, M. J., Ed. "Cycling of mineral nutrients in
     agricultural ecosystem". Elsevier Scientific Publ. Co: New
     York, 1978.

RECEIVED August 10, 1984

Antimicrobial Agents from Plants: A Model for Studies of Allelopathic Agents?

ALICE M. CLARK, CHARLES D. HUFFORD, FAROUK S. EL-FERALY, and JAMES D. McCHESNEY

The Department of Pharmacognosy, School of Pharmacy, The University of Mississippi, University, MS 38677

A general scheme for the bioassay directed isolation and characterization of antimicrobial agents from plants is presented and discussed. The utility of the procedure is demonstrated by the characterization of two antimicrobial alkaloids from the tulip tree, Liriodendron tulipifera L. The generalization of the approach to the study of allelopathic agents in plants is suggested.

Traditionally, the search for new antimicrobial substances has emphasized the need to discover agents for use in human medicine. Even with the imposing array of substances now available which are useful to treat infectious diseases in man, it is not difficult to document the continued need for new agents. Infectious diseases rank only behind cancer and cardiovascular disease as causes of death and considering that mortality represents the smaller portion of those injured by infection, the magnitude of the need for additional agents is evident. Further, consideration of the increasing occurrance of strains resistant to currently used antibiotics emphasizes even more the need to find new, useful antimicrobial agents for medical use. Such considerations have by-and-large caused researchers to overlook the probable reasons that antimicrobial substances are present in plants but rather to view plants only as a source of biologically active substances to be exploited. Indeed we, ourselves, have only recently begun to take a broader view and to consider the possible role of these biologically active substances in the ecology of the plants producing them. For example, it is interesting to note that many of these plant-derived substances are of moderate antibiotic potency when compared to microbial-derived antibiotics but that they also often possess additional activities as insect antifeedants, allelopathic substances, etc.

A number of summaries of early work as well as recent successes of the examination of higher plants as sources of antimicrobial agents have appeared (1-4). Many of the earlier investigators in the field used strains of microorganisms as bioassay organisms

0097–6156/85/0268–0327$06.00/0

which were not generally available or were not representative of
their genera and therefore, it is difficult or impossible to repro-
duce their observations of antimicrobial activity in specific plant
extracts. This has led to the recognition that carefully selected
and reproducible bioassay procedures are necessary to successful programs of discovery.

In our laboratories the following organisms have been selected
as representative and qualitative antimicrobial assays are carried
out using the agar well diffusion assay procedure (5):
Bacillus subtilis (ATCC 6633), Gram positive bacterium; Staphylococcus aureus (ATCC 6538), Gram positive bacterium; Escherichia coli,
(ATCC 10536), Gram negative bacterium; Pseudomonas aeruginosa (ATCC
15442), Gram negative bacterium; Mycobacterium smegmatis (ATCC 607),
Acid-fast bacterium; Candida albicans (ATCC 10231), Yeast-like fungus; Saccharomyces cerevisiae (ATCC 9763), Yeast-like fungus; Aspergillus niger (ATCC 16888), Filamentous fungus; Trichophyton mentagro-
phytes (ATCC 9972), Dermatophyte; Polyporus sanguineus (ATCC 14622),
Plant pathogen; Helminthosporium species (ATCC 4671), Plant pathogen.
Crude extracts and chromatographic fractions are routinely tested at
a concentration of 20 mg/ml in ethanolic or aqueous ethanolic solution. Results of the qualitative screen are reported as the average
radius of the zone of inhibition surrounding the well containing the
test solution. For compounds which show significant activity in the
qualitative screen, minimum inhibitory concentrations (MIC) are determined using the two-fold serial dilution technique previously described (5). All compounds are tested initially at concentrations
of 100 μg/ml in the first tube. Streptomycin sulfate is used as a
positive control for bacteria and amphotericin B is used as a positive control for fungi.

Plant materials are extracted to exhaustion by percolation with
ethanol. Some very water soluble products may possibly be missed by
this procedure but the convenience of the process outweighs this
possibility. Many times highly water soluble substances such as
carbohydrates crystallize from the extracts during concentration so
clearly some quantity of even very water soluble substances are extracted by the ethanol percolation.

Extracts which demonstrate sufficient activity in the bioassay
are next fractionated by a simple partitioning procedure (Figure 1)
which separates the components into gross chemical classes. The
various fractions are re-assayed and those showing activity are then
further fractionated and purified by appropriate chromatographic procedures.

Of particular utility in our experience is adsorption column
chromatography on silica gel, alumina or florisil depending upon the
nature of the active partition fraction. The column chromatography
is monitored by a combination of thin layer chromatography and bioassay of fractions. For specific cases other chromatographic procedures are very useful: gas chromatography for volatile and, or heat
stable substances, partition chromatography or the recently developed
droplet counter current chromatography for polar, water soluble substances and high performance liquid chromatography which is applicable to a broad range of substances but still largely limited to
analytical scale separations.

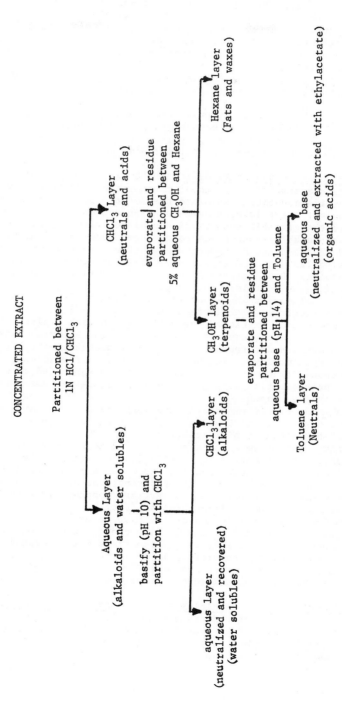

Figure 1. Partitioning procedure.

As active substances are separated and purified they are characterized by a combination of spectroscopic analyses and chemical correlations. Particularly useful spectroscopic analysis techniques are
nuclear magnetic resonance (proton and carbon), mass spectrometry and
infra-red and ultraviolet spectrophotometry.

The characterization of two antimicrobial alkaloids from the
heartwood of the tulip tree, Liriodendron tulipifera L. (Magnolia-
ceae), is illustrative of the general procedure (5).

Plant Material - The yellow heartwood was separated from the sapwood
of L. tulipifera and air dried before grinding. Other plant parts
(leaf, stem bark, root bark, root, fruit, and flower) were collected,
but alcoholic extracts showed no antimicrobial activity.

Extraction and Separation of Alkaloids - The air-dried ground heartwood (2.2 kg) was extracted by percolation at room temperature with
alcohol USP until a negative alkaloid test of the percolate was observed. Removal of the solvent at reduced pressure and at 40° left
71 g of residue that exhibited antimicrobial activity. A 35 g sample
of the alcohol-soluble residue was partitioned between 125 ml each
of ether and 2% citric acid in water. The ether layer was extracted
twice more with 125 ml of 2% citric acid, filtered to remove some
interfacial solids (5.8 g alkaloid negative, no antimicrobial activity), dried (sodium sulfate), and evaporated to dryness, giving
8.6 g of ether solubles that had no antimicrobial activity.

The aqueous citric acid layers were combined, adjusted to pH
9-10 with ammonia, and extracted three times each with 1 liter of
chloroform. The aqueous layer was neutralized, and a portion was
evaporated to dryness and found to have no antimicrobial activity.
The remaining aqueous layer was then acidified with acetic acid; a
saturated aqueous solution of ammonium reineckate was added to a portion, but no precipitation occurred. No precipitate was noted with
Valser's or Mayer's reagent.

The combined chloroform layers were dried (sodium sulfate) and
evaporated to dryness to give 6.2 g of chloroform solubles that
showed all of the antimicrobial activity.

The active chloroform-soluble residue (6.2 g) was separated into
tertiary phenolic and nonphenolic fractions by dissolving the residue
in 250 ml of chloroform and extracting three times each with 250 ml
of 5% sodium hydroxide solution. After drying, the chloroform solution was evaporated to leave 4.7 g of tertiary nonphenolic alkaloids
that possessed all of the antimicrobial activity.

The combined aqueous solution of the base layers was treated
with an excess of ammonium chloride until a cloudy suspension was
noted. This suspension was extracted three times with an equal
volume of chloroform. The chloroform layer, after washing with water
and drying (sodium sulfate), was evaporated to give 1.4 g of tertiary
phenolic bases that had no antimicrobial activity.

Isolation of Antimicrobial Alkaloids from Tertiary Non-phenolic Base
Fraction - A 2 g portion of the crude nonphenolic base fraction was
dissolved in chloroform and chromatographed over 200 g of aluminum
oxide (Woelm, neutral, grade III). The solvents used were 300 ml of
chloroform, 500 ml of 1% methanol in chloroform, 300 ml of 2% metha-

nol in chloroform, and 400 ml of 16% methanol in chloroform; finally the column was washed with 50% methanol in chloroform. The fractions (20 ml each) were evaporated in tared flasks, combined according to their weights and to their similarity on TLC, and then assayed (Table I).

(+)-Glaucine and Dehydroglaucine from Fractions 12-19 - TLC analyses of these fractions indicated the presence of two alkaloidal constituents. These were obtained pure by chromatography (2.4-g sample compiled from several columns as described previously) over 200 g of silica gel G (Silica gel G for TLC according to Stahl was slurried with water first, dried at $110°$ for 12 hr, and sieved through an 80-mesh sieve before use.) using ether as the eluent.

The first 125 ml of eluent contained no alkaloids, but the next 150 ml yielded a crude alkaloid (115 mg). This alkaloid was crystallized from alcohol to yield 59 mg of slightly colored plates, mp $113-115°$. Subsequent recrystallizations raised the melting point to $121-122°$ (pale-yellow plates). The mass spectrum exhibited a parent ion at m/z 353. The UV spectrum showed maxima at 260 and 332 nm, while NMR indicated a 1H singlet at δ 9.60 (ArH) and a 3H singlet at δ 3.01 (NCH_3). These data are characteristic of dehydroaporphine alkaloids (6), and the physical data agree with those reported for dehydroglaucine (7). Direct comparison (melting point, TLC, UV, and IR) of a sample of dehydroglaucine prepared from glaucine by potassium permamganate oxidation (7) with that obtained from the separation confirmed the identity.

The next 1 liter of eluent yielded no alkaloids, but the following 1 liter yielded the second alkaloid (1.78 g), which was crystallized from alcohol-hexane to yield 0.855 g of needles, mp $119-120°$. This was identified as (+)-glaucine by direct comparison (melting point, mixture melting point, TLC, IR, and circular dichroism) with an authentic reference sample of (+)-glaucine. (+)-Glaucine was reported previously to be the major alkaloidal constituent of the heartwood (8).

After the glaucine had been eluted, the column was washed with 50% methanol in ether. All fractions were assayed, and only the fraction containing dehydrogalucine was active. Dehydroglaucine was subsequently assayed and was shown to be the antimicrobial agent present in fractions 12-19 (Table II).

Liriodenine from Fractions 33-40 - Crystallization of the residue of these fractions (118 mg) from chloroform yielded 85 mg of yellow needles, mp $280-281°$. The melting point, IR, and UV data were consistent with that reported for the yellow alkaloid, liriodenine, previously reported from the heartwood (9). Direct comparison (melting point, mixture melting point, IR, and UV) with an authentic sample of liriodenine confirmed the identity. Antimicrobial assay showed liriodenine to be the active component present in these fractions (Table II).

Liriodenine oxime (9) and methiodide (10) were prepared as described in the references and were tested for antimicrobial activity (Table II).

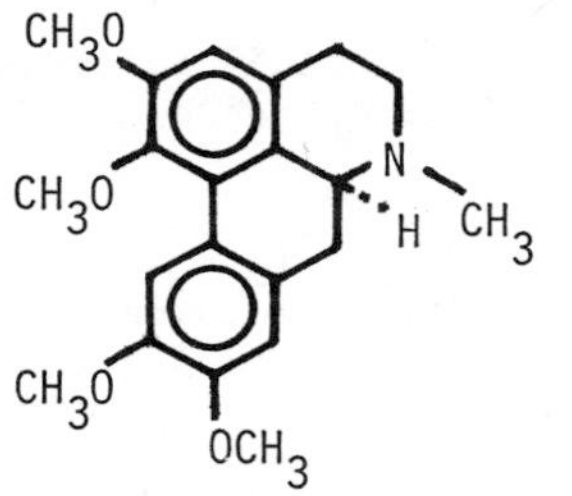

(+)-Glaucine

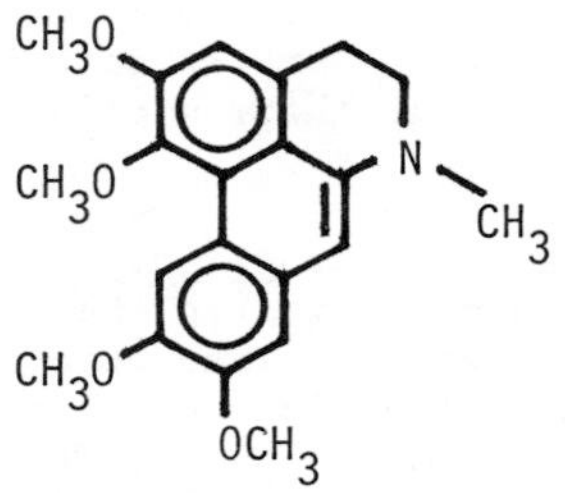

Dehydroglaucine

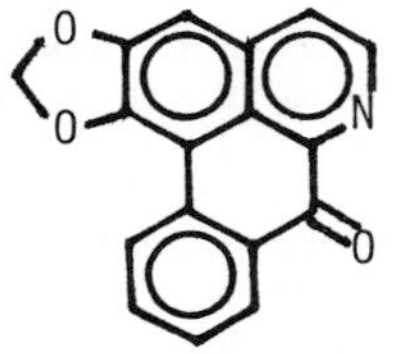

Liriodenine

The values reported in Table III represent readings taken after incubation times of 24 hr for all organisms except M. smegmatis, which was read at 72 hr. The concentration of the tube of highest dilution that was free from growth was recorded as the minimum inhibitory concentration (micrograms per milliliter).

That compounds of the potency of liriodenine are present in plant tissues underscores the potential of plants to affect dramatically other organisms in their environment. Release of liriodenine or similarly potent agents from decomposing plant materials may significantly modify the microbial flora of the root zone. This in turn may have direct and indirect effect the plants which will germinate and grow in the affected soil, the classical allelopathic effect (11).

We wish to emphasize that our success for characterization of the substances responsible for the antimicrobial activity found in the tulip tree was due to utilization of a plan of procedure which incorporates the following elements:
1. A bioassay which is generally accepted in the discipline as representative and sensitive for the biological effect.
2. Preparation of extracts by a standardized solvent extraction.
3. Preliminary fractionation of the extract to separate broad classes of compounds.
4. Bioassay directed purification of the active fractions employing appropriate chromatographic procedures.
5. Spectroscopic and chemical characterization of pure active substances.

This approach has allowed us to characterize other antimicrobial substances as well (12-14). The application of a similarly designed procedure will facilitate identification of substances in plants responsible for allelopathy.

Table I — Chromatographic Separation of Tertiary Nonphenolic Base Fraction

Fraction Number	Eluent	Weight of Residue, mg	Remarks
1–11	CHCl₃	65	Nonalkaloidal, inactive[a]
12–19	CHCl₃	400	Crystalline residue, glaucine, dehydroglaucine, active
20–32	1% CH₃OH in CHCl₃	728	Amorphous residue, inactive[a]
33–40	1% CH₃OH in CHCl₃	118	Yellow solid, liriodenine, active
41–51	2% CH₃OH in CHCl₃	206	Crystalline residue, michelalbine, inactive[a]
52–65	2% CH₃OH in CHCl₃	84	Amorphous residue, inactive[a]
66–90	16% CH₃OH in CHCl₃	30	Amorphous residue, inactive[a]
Wash	50% CH₃OH in CHCl₃	300	Amorphous residue, inactive[a]

[a] No activity was observed against any of the test organisms.

Table II - Antimicrobial Activity of Extracts, Fractions, and Compounds

Sample[a]	Zone Diameter, mm					
	Staph. aureus	B. subtilis	M. smegmatis	C. albicans	S. cerevisiae	A. niger
Alcohol extract	4	N.T.[b]	10	2	N.T.	5
Tertiary nonphenolic fraction	8	N.T.	12	5	N.T.	9
Fractions 12-19	3	N.T.	2	1	N.T.	-
Fractions 33-40	5	N.T.	10	3	N.T.	10
Liriodenine	5	8	11	3	5	11
Liriodenine methiodide	6	6	14	17	15	4
Dehydroglaucine	4	7	8	6	6	-
Oxoglaucine methiodide	9	9	10	10	11	-

[a]All samples were also tested against E. coli and P. aeruginosa and were found to be inactive. All other fractions also were assayed, but none showed any activity against any of the microorganisms listed in this table. Galucine, glaucine methiodide, liriodenine oxime, michelalbine, and oxoglaucine were tested, but none showed any activity against any of the microorganisms.
[b]N.T. = not tested

Table III – Minimum Inhibitory Concentration (Micrograms per Milliliter) of Active Compounds

Compound	Staph. aureus	B. subtilis	M. smegmatis	C. albicans	S. cerevisiae
Liriodenine	3.1	0.39	1.56	6.2	6.2
Liriodenine methiodide	6.2	3.1	3.1	0.78	3.1
Oxoglaucine methiodide	25	25	25	1.56	25
Dehydroglaucine	25	25	25	25	50
Streptomycin sulfate	3.1	1.56	0.78	—	—
Amphotericin B	—	—	—	0.78	0.78

Literature Cited

1. Mitscher, L.A.; Leu, R.P.; Bathala, M.S.; Wu, W.N.; Beal, J.L.;
    White, R.; Lloyida, 1972, 35, 157-166.
2. Mitscher, L.A.; Al-Shamma, A.; In "Annual Reports in Medicinal
    Chemistry"; Hess, H.J., Ed.; Academic:New York, 1980; Vol. 15,
    pp 255-266.
3. Mitscher, L.A., In "Recent Advances in Phytochemistry";
    Runeckles, V.C., Ed.; Plenum:New York, 1975; Vol. 9, pp 243-282.
4. de Souza, N.J.; Gangnli, B.N.; Reden, J., In "Annual Reports in
    Medicinal Chemistry"; Hess, H.J., Ed.; Academic:New York, 1982;
    Vol 17, pp. 301-310.
5. Hufford, C.D.; Funderburk, M.J.; Morgan, J.M.; Robertson. J.
    Pharm. Sci. 1975, 64, 789-792.
6. Sharma, M. "The Isoquinoline Alkaloids"; Academic:New York,
    1972; p 224.
7. Kiryakov, H.G. Chem. Ind. (London) 1968, 1807.
8. Cohen, J.; Von Langenthal, W.; Taylor, W.I., J. Org. Chem. 1961,
    26, 4143-.
9. Buchanan, M.A.; Dickey, E.E.; J. Org. Chem. 1960, 25, 1389-.
10. Yang, T.H., J. Pharm. Soc. Japan. 1962, 82, 804-.
11. Rice, E.L., "Allelopathy"; Academic:New York, 1974.
12. Hufford, C.D.; Lasswell, W.L.; Lloydia, 1978, 41, 156-160.
13. Clark, A.M.; El-Feraly, F.S.; Li, W.; J. Pharm. Sci., 1981, 70,
    951-952.
14. McChesney, J.D.; Silveira, E.R.; "Antimicrobial Diterpene
    Constituents of Croton sonderianus", presented at the 24th
    Annual Meeting of the American Society of Pharmacognosy,
    Oxford, MS, July 23-28, 1983.

RECEIVED July 17, 1984

A Survey of Soil Microorganisms for Herbicidal Activity

R. M. HEISEY, J. DeFRANK, and A. R. PUTNAM

Pesticide Research Center, Michigan State University, East Lansing, MI 48824

Soil microorganisms, particularly actinomycetes, produce a diversity of unusual metabolites. We report here initial investigations to discover microbially-produced compounds having potential for development as herbicides. Microorganisms, primarily actinomycetes, were isolated from soils collected in Michigan, Pennsylvania, and California and tested for phytotoxin production on solid and in liquid medium. Of 347 isolates screened on solid medium, 10-12% severely inhibited growth of indicator seedlings. Cycloheximide, a phytotoxic antibiotic with little herbicidal value, was produced by many of the most inhibitory isolates in broth culture. Several isolates, however, produced highly toxic broth that did not contain detectable amounts of cycloheximide. We therefore believe certain soil microorganisms have potential for production of unique natural product herbicides.

Soil microorganisms produce many compounds that are potentially toxic to higher plants. Examples include members of the following: antibiotics (1-6), fatty and phenolic acids (7-12), amino compounds (13-15), and trichothecenes (16, 17). "Soil sickness" and "replant problems" have been reported where certain crops or their residues interfere with establishment of a subsequent crop (18, 19). Toxins resulting from microbial activity sometimes are involved, but it is often unclear whether these are synthesized de novo in microbial metabolism or are breakdown products of the litter itself (20).

Microorganisms associated with the roots of certain plants may produce or facilitate release of phytotoxins. For example, microbes in the rhizosphere of chamise (Adenostoma fasciculatum H. & A.) appear to contribute to suppression of herbs near these shrubs (21), a phenomenon previously attributed to toxins washed from the chamise foliage (22, 23). Similarly, hydrogen cyanide, a potent phytotoxin,

0097–6156/85/0268–0337$06.00/0
© 1985 American Chemical Society

was released in greater amounts when dead peach root tissue was incubated with soil from peach and cherry orchards than when root tissue was incubated with soil from other sites (24).

The recent growth of biotechnology has stimulated interest in microorganisms as a potential source of natural product herbicides (25-27). The actinomycetes, unicellular microorganisms taxonomically lying between true bacteria and true fungi, are especially interesting in this regard because they produce a wide range of antibiotics. Actinomycetes are abundant in many soils and composts. Several herbicidal compounds produced by these organisms have already been discovered: herbicidin A and B (28, 29), herbimycin A and B (30, 31), anisomycin and toyocamycin (32, 33). Two amino acid derivatives produced by actinomycetes, N-{4-[Hydroxy(methyl)phosphinoyl]homoalanyl}-alanylalanine (Bialaphos, Meiji Seika) and DL-Homoalanin-4-yl(methyl) phosphinic acid (Glufosinate, Hoechst AG) are currently undergoing development as commercial herbicides (27).

This paper describes the initial phase of research directed toward the discovery of compounds produced by soil microorganisms, primarily actinomycetes, having potential for development as herbicides. Our interests include both new compounds, as well as previously known ones having heretofore unrecognized herbicidal properties.

Materials and Methods

Rationale. Microorganisms were isolated from soil and screened for toxin production according to the scheme in Figure 1. Some of the organisms causing strong inhibition on solid medium were tested for toxin production in liquid medium. Liquid culture will be required to obtain large amounts of material for commercial production of herbicides, however, the ability to produce toxins on solid medium does not necessarily imply toxin production in broth (34). Cycloheximide, a phytotoxic but relatively nonspecific antibiotic with little value as a herbicide, is produced by many actinomycetes. Liquid cultures were tested for cycloheximide to determine whether it caused the observed toxicity.

Isolation of Actinomycetes from Soil. Soil samples were collected from depths of 0-7 cm from a variety of sites in Michigan, Pennsylvania, and California. Selective isolation of actinomycetes was achieved by drying and alkalifying the soil and inoculation of soil suspensions onto an arginine-glycerol-salt (AGS) medium (35). In this procedure, 1 g of pulverized air-dried soil was mixed with 1 g of calcium carbonate. The mixture was incubated 7-9 d at 28 C in sterile parafilm-sealed petri dishes. A damp atmosphere was maintained by suspending several sterile water-saturated sheets of filter paper over the soil. Serial dilutions of the soil/calcium carbonate mixture, prepared with sterile distilled water, were spread on AGS plates after agitation, and the plates were incubated approximately 10 d at 28 C. Inoculations of 1, 10, and 100 µg soil per 1.5 X 10 cm petri plate were usually optimal. As colonies developed, at least one representative of all actinomycetous forms (having powdery, crustose, or leathery texture) that appeared different for each particular soil sample was transferred onto new AGS plates. These isolates were streaked and grown on AGS or A-9 medium until pure cultures were obtained.

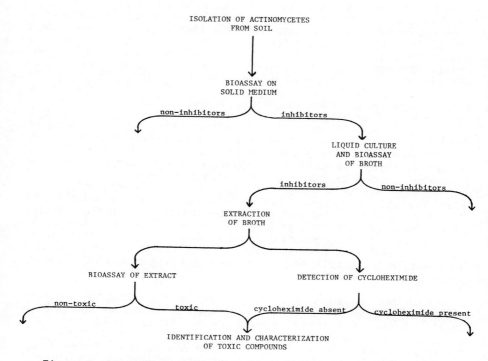

Figure 1. Flow diagram showing protocol followed to screen soil microorganisms for production of phytotoxic compounds.

Bioassay on Solid Medium. A-9, a medium previously shown to be favorable for antibiotic production by actinomycetes in shake flasks (36), was modified for bioassays on solid medium. We halved the concentration of components in A-9 and adjusted the pH to 6.9-7.1 with KOH to reduce the possibility of osmotic or toxic effects of medium components themselves on seed germination and seedling growth. The medium was amended with 15 g agar per liter and poured into 10 x 10 x 1.5 cm square plastic petri dishes, about 60 ml per plate.

Square plates containing solid A-9 medium were inoculated in a 2-cm-wide band along one edge with the various microbial isolates. The plates were incubated for 14-15 d at 28 C, during which time metabolites produced by the isolates diffused into the medium. Fifteen surface-sterilized seeds of cucumber (Cucumus sativus L. "Pikmaster"), and 20 each of barnyardgrass (Echinochloa crusgalli Beauv.) and garden cress (Lepidium sativum L. "Curlycress") were placed on the non-inoculated portion of the medium after incubation (Figure 2). Seeds were placed equidistant in rows of five perpendicular to the microbial band. Control plates, identical to treatments except not inoculated with a microorganism, were also prepared. Surface sterilization of cress seeds was achieved with 10 hr of propylene oxide fumigation. Barnyardgrass and cucumber seeds were fumigated 2 and 6 hr, respectively, with propylene oxide, soaked 15 min in 0.1% mercuric chloride solution, and rinsed thrice with sterile distilled water. The planted plates were placed back into the incubator, and after 3-4 d seed germination and seedling growth were evaluated. Seedling growth was rated: 1) growth minimal or germination nil, 2) pronounced inhibition, 3) slight inhibition, 4) growth comparable to that of controls, 5) stimulation over controls. Germination in inoculated plates was expressed as a percentage of germination in controls.

Liquid Culture and Bioassay of Broth. Some of the microbial isolates causing strong inhibition or stimulation on solid medium were subsequently tested for toxin production in liquid medium. Liquid cultures were established in 2-liter baffle-bottomed erlenmeyer flasks containing 500 ml of half-strength A-9 medium (pH 6.9-7.1) by addition of 20-25 ml of inoculum from seed cultures. The plugged flasks were placed on an orbital shaker at 110-125 rpm and incubated at 28 C. Culture broth was tested for inhibition or stimulation after 8-12 d of incubation. Ten surface-sterilized seeds each of cress and barnyardgrass were placed on five sheets of sterile filter paper in sterile 10 x 1.5 cm glass petri dishes. Seeds and paper were moistened with 7 ml distilled water and 2 ml of culture broth (containing microbial cells). The dishes were incubated in darkness at 28 C, and after 3-4 d radicle length of the seeds was measured.

Extraction of Broth. Cultures screened for biological activity in liquid medium were tested for cycloheximide production. A 400-ml vol of the various broth cultures was partitioned with two 200-ml vol of dichloromethane. The dichloromethane extract, filtered through Whatman #50 paper, was concentrated under vacuum on a rotary evaporator and taken to dryness with N_2. Temperatures during concentration were 20-40 C. Non-inoculated broth for controls was similarly extracted.

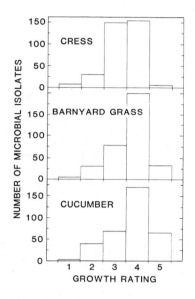

Figure 2. Bioassay of several actinomycetous isolates on solid medium. Note striking inhibition of cress (top of plates), barnyard grass (middle), and cucumber (bottom) caused by isolates E5 and E8, compared to non-inoculated control (C). Cucumber was stimulated by D22 and F6.

<u>Detection of Cycloheximide</u>. The presence of cycloheximide in the dichloromethane extracts was determined with one- and two-dimensional thin layer chromatography (TLC). One dimensional TLC was on LK6 (Whatman) silica gel plates with acetone:ethyl acetate (1:1) as the developing solvent. Two dimensional TLC was on CS5 (Whatman) plates, which have a C_{18} reversed phase strip in the first dimension and silica gel in the second. The CS5 plates were developed with methanol:water (2:1) in the C_{18} direction, dried 30-60 min at 110 C, and developed with acetone:ethyl acetate:dichloromethane (1:1:1) on the silica gel. Cycloheximide was made visible by spraying the plates with 5% vanillin in sulfuric acid and turned a distinctive green, reaching maximum intensity after about 4 hr. Spots believed to be cycloheximide were evaluated by comparing their reaction time, color, and R_f with co-chromatographed samples of authentic cycloheximide.

<u>Bioassay of Extracts</u>. Extracts tested for the presence of cycloheximide were also bioassayed for phytotoxicity. The extracts were redissolved in acetone, and 0.2 mg in 2 µl was applied to 6-cm-dia disks of filter paper. The extract was distributed on the paper with 0.2 ml of methanol. The disks were dried with warm air, placed in 1.5 x 6 cm petri dishes, and moistened with 1.5 ml distilled water. Ten cress seeds were placed on the paper, and after incubation for 3 d at 28 C radicle length of the seedlings was measured.

Results

More actinomycetous isolates were obtained from soils with a high organic matter content than from sandy soils with low organic matter (Table I). Sample I, collected under black locust trees, and sample P, collected under fescue sod in a young apple orchard, yielded the most isolates. Fewest isolates were obtained from sample A, which contained little organic matter due to having previously been burned, and samples C, H, and U, all sandy soils with low organic matter content.

The response of cress, barnyardgrass, and cucumber to compounds exuded into solid medium by the microbial isolates ranged from striking inhibition to stimulation (Figure 2, 3). Of a total of 347 isolates tested, 10-12% reduced seedling growth to $\leq$ 2, our criterion of severe inhibition. The majority of isolates had little or no effect on seedling growth, 69-87% causing a growth rating of 3 or 4 for the three indicators. The number of stimulatory isolates (growth rating of 5) ranged from 19% for cucumber to 10% for barnyard grass to only 2% for cress.

There was a strong correlation between toxin production on solid medium and toxin production in liquid culture (Table II). Almost all isolates that when grown on solid medium strongly inhibited both barnyardgrass and cress also severely reduced radicle growth of these indicators when grown in liquid culture. Barnyardgrass and cress typically were similar in growth rating response in bioassays on solid medium. However, in germination on solid medium and radicle growth in bioassays of broth, cress was often more sensitive than barnyardgrass.

Isolates B9wh, C10, I10, J20, K11, and L16 strongly stimulated growth, germination, or both of one or more of the indicators when grown on solid medium. When grown in liquid culture, C10, J7, and

Table I. Comparison of soil samples with regard to total number of microorganisms isolated and the percentage of total causing severe toxicity (growth rating ≤ 2 or germination ≤ 75% of controls) to cucumber (Cucu), barnyard grass (Bygr), and cress

Soil sample	Soil sample collection site	Total	Microorganisms isolated					
			% causing growth rating ≤ 2			% reducing germination ≤ 75%		
			Cucu	Bygr	Cress	Cucu	Bygr	Cress
A	CA, El Dorado Co.; previously burned ponderosa pine site.	4	0	0	0	0	0	25
B	MI, Ingham Co.; Miami loam; alfalfa-orchard grass field.	15	13	27	13	0	27	27
C	MI, Ingham Co.; Spinks loamy sand; bare field.	6	0	0	0	0	0	0
D	MI, Ingham Co.; near manure pile in cattle feeding area.	22	14	23	9	0	14	9
E	MI, Ingham Co.; Edwards muck; natural vegetation.	13	15	15	15	0	15	23
F	MI, Ingham Co.; Tawas muck; bare field.	13	0	8	8	0	8	8
G	MI, Ingham Co.; Carlisle muck; swampy edge of pond.	11	0	0	0	0	18	0
H	MI, Ingham Co.; Boyer loamy sand; top of anthill.	7	0	0	0	0	14	8
I	MI, Clinton Co.; under black locust saplings.	27	19	7	7	0	15	26
J	MI, Clinton Co.; under elm saplings, periodic flooding.	18	6	11	11	0	6	28
K	MI, Clinton Co.; in quaking aspen grove.	12	0	0	8	0	0	17
L	MI, Clinton Co.; base of large black walnut tree.	22	0	0	5	0	18	23
M	MI, Clinton Co.; under thick quackgrass sod.	21	5	5	10	0	29	29
N	MI, Ionia Co.; under dead rye, young apple orchard.	17	6	6	0	0	6	18
P	MI, Ionia Co.; under fescue sod, young apple orchard.	27	4	0	11	0	11	30
Q	MI, Lake Co.; fibrous peat; swamp along shallow pond.	13	31	23	23	0	23	39
R	MI, Wexford Co.; sandy soil under jack pine.	10	30	20	20	0	20	20
S	MI, Wexford Co.; under thick litter, red pine stand.	20	25	15	25	0	15	25
T	MI, Benzie Co.; muck, white cedar-hemlock swamp.	14	14	14	7	0	29	14
U	MI, Wexford Co.; sandy soil, quaking aspen-bracken fern.	8	25	25	13	0	38	25
V	PA, Lancaster Co.; base of hackberry tree.	11	18	18	18	0	36	36
W	PA, Lancaster Co.; water-saturated mud, farm pond drain.	22	36	18	18	0	9	23
X	PA, Lancaster Co.; mud mortar, 200-yr-old barn wall.	14	7	0	14	0	7	43

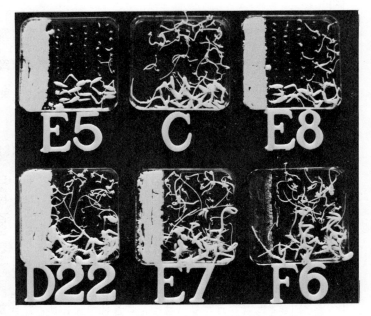

Figure 3. Histograms showing growth response of cress, barnyard grass, and cucumber to 347 microbial isolates bioassayed on solid medium. Growth response: 1) seedling growth minimal or germination nil, 2) pronounced inhibition, 3) slight inhibition, 4) growth comparable to controls, 5) stimulation over controls.

Table II. Effect of microorganisms on solid and in liquid medium on growth and germination of cucumber (Cucu), barnyard grass (Bygr), and garden cress. Data are for all microbial isolates tested in liquid medium

| Microbial isolate | Bioassay of solid medium | | | | | | Bioassay of liquid medium | |
| | Growth rating* | | | Germination# | | | Radicle length# | |
	Cucu	Bygr	Cress	Cucu	Bygr	Cress	Bygr	Cress
A6	4	4	4	100	87	16	35	30
B9wh	3	4	3	105	132	38	88	27
B12	1	1	1	100	63	48	0	0
B13	1	1	1	100	6	11	2	1
C10	5	5	5	103	100	92	111	64
D6	2	2	2	93	87	70	34	15
E5	2	1	1	100	38	27	0	0
E8	2	2	2	100	69	65	57	39
F5	3	2	2	100	100	104	56	47
I9	2	4	3	103	116	89	74	45
I10	4	4	2	103	123	21	101	34
I17	2	2	3	102	49	42	12	8
I18	1	1	2	97	0	21	2	3
I27	2	3	3	103	108	73	2	2
J7	4	4	3	105	110	32	174	54
J8	2	2	2	105	91	47	36	8
J16	4	2	2	105	80	65	63	42
J20	4	5	5	103	154	104	118	78
K9	3	3	2	105	86	6	55	58
K11	5	5	4	105	97	100	81	69
L8	5	4	4	102	91	89	97	51
L16	5	5	5	103	95	95	94	33
M18	3	3	3	97	74	100	41	4
N6	2	2	3	105	83	81	79	42
R1	2	3	3	95	77	89	28	10
R6	2	2	2	88	28	37	0	0
U3	2	2	3	102	42	37	1	2
V7	2	2	2	105	17	17	0	0
V9	2	3	2	98	58	41	2	2
W1	4	5	3	105	108	44	99	37
W4	2	3	2	105	102	111	56	10
W8	2	3	3	98	102	94	43	24
W17	2	3	2	102	98	21	1	2
X7	2	3	2	88	56	21	85	16

*Growth rating: 1) seedling growth minimal or germination nil, 2) pronounced inhibition, 3) slight inhibition, 4) growth comparable to controls, 5) stimulation over controls.

#Germination and radicle length expressed as percentage of germination and radicle length of controls.

J20 stimulated radicle growth of barnyard grass. Cress radicle growth was not stimulated by culture broth of any of the isolates tested.

Dichloromethane extracts of most broth cultures caused some degree of phytotoxicity. Table III shows the results for all isolates tested in liquid culture whose broth reduced cress radicle growth to < 50% of controls. Cycloheximide was present in many of the most phytotoxic extracts. Only one isolate, V9, out of the 12 most toxic (those whose dichloromethane extract reduced cress radicle growth to < 5%) lacked detectable amounts of cycloheximide. Conversely, all extracts containing detectable amounts of cycloheximide were extremely toxic and reduced cress radicle growth to < 5%. Authentic cycloheximide for comparison, applied 5 µg/petri dish, inhibited cress radicle growth to 6%. Three isolates, M18, W4, and W17, produced broth that inhibited cress radicle growth to < 10%, but dichloromethane extracts of their broth were not strongly toxic. This suggests water soluble phytotoxins.

Discussion

Actinomycetes are typically most abundant in well drained, circumneutral to alkaline soils having abundant organic matter. Water-logging and low pH may reduce populations (37, 38). The numbers of actinomycetous organisms isolated from the various soil samples in our study follow this pattern. No clear trend emerged as to a particular edaphic or biotic factor causing an increase in the proportion of inhibitory isolates in a soil sample.

Because of the arbitrary isolation procedure, and because many microorganisms are macroscopically indistinguishable, it is likely that we tested some of the same species more than once. Some also may have been microorganisms other than actinomycetes. This uncertainty does not negate our approach, however, since it is likely that other microorganisms in addition to actinomycetes also produce herbicidal compounds and that toxin production may differ even among strains or races within a species (39). Therefore, testing microorganisms other than actinomycetes and screening isolates of the same species, especially those isolated from different soil samples, enhance the possibility of success. Definitive taxonomy, which can be very time consuming, may better be left until an organism is known to produce useful compounds.

Dichloromethane extraction of culture broth, thin layer chromatography of the extract, and visualization with 5% vanillin/sulfuric acid spray is effective for detecting cycloheximide in culture broth. Cycloheximide applied to TLC plates in amounts as low as 1 µg/spot will produce visible color with the vanillin spray.

The soil microorganisms apparently produced more than one phytotoxin. Cycloheximide was present in the culture broth of a high proportion, but not all, of our most toxic isolates. It is therefore imperative that microorganisms be tested for cycloheximide production in the early stages of a herbicide discovery program. Cycloheximide production need not eliminate particular isolates from further study, since other toxins may also be produced. It may be prudent, however, to devote most initial effort to organisms that produce potent inhibitors other than cycloheximide. We are pursuing this approach and are in the process of isolating and identifying phytotoxins produced by V9, our most phytotoxic isolate that does not produce detectable amounts of cycloheximide.

Table III. Effect of culture broth (2 ml/petri dish) and dichloromethane extract of culture broth (0.2 mg/petri dish) on cress radicle growth, as related to presence of cyclohex- imide. Data are for all isolates tested in liquid culture whose broth inhibited cress radicle growth to $\leq$ 50% of controls

Microbial isolate	Cress radicle length*		Cycloheximide in extract#
	Broth	Extract	
A6	30	47	-
B9wh	27	71	-
B12	0	2	+
B13	1	1	+
D6	15	13	-
E5	0	1	+
E8	39	2	+
F5	47	88	-
I9	45	89	-
I10	34	28	-
I17	8	0	+
I18	3	1	+
I27	2	4	+
J8	8	38	-
J16	42	39	-
L16	33	103	-
M18	4	110	-
N6	42	77	-
R1	10	1	+
R6	0	1	+
U3	2	2	+
V7	0	1	+
V9	2	5	-
W1	37	63	-
W4	10	97	-
W8	24	81	-
W17	2	93	-
X7	16	18	-

*Expressed as percentage of control radicle length.

#Presence (+) or absence (-) as determined in thin layer chromatography.

Acknowledgments

We thank Jean Kentner and Jeff Mohr for their excellent assistance.

Journal article no. 11301 of the Michigan Agricultural Experiment Station.

Literature Cited

1. Wright, J. M. Ann. Bot. 1951, 15, 493-499.
2. Nickell, L. G.; Finlay, A. C. J. Agric. Food Chem. 1954, 2, 178-182.
3. Brian, P. W. Ann. Rev. Plant Physiol. 1957, 8, 413-426.
4. Norstadt, F. A.; McCalla, T. M. Science 1963, 140, 410-411.
5. Evans, G.; Cartwright, J. B.; White, N. H. Plant Soil 1967, 26, 253-260.
6. McCalla, T. M.; Norstadt, F. A. Agric. Environm. 1974, 1, 153-174.
7. Guenzi, W. D.; McCalla, T. M. Agron. J. 1966, 58, 303-304.
8. Whittaker, R. H. In "Chemical Ecology;" Sondheimer, E.; Simeone, J. B., Eds.; Academic: New York, 1970; p.43-70.
9. Patrick, Z. A. Soil Sci. 1971, 111, 13-18.
10. Chou, C.-H.; Lin, H.-J. J. Chem. Ecol. 1976, 2, 353-367.
11. Chou, C.-H.; Chiou, S.-J. J. Chem. Ecol. 1979, 5, 839-859.
12. Harper, S. H. T.; Lynch, J. M. Plant Soil 1982, 65, 11-17.
13. Owens, L. D.; Wright, D. A. Plant Physiol. 1965, 40, 927-930.
14. Owens, L. D.; Guggenheim, S.; Hilton, J. L. Biochim. Biophys. Acta 1968, 158, 219-225.
15. Seto, H; Toru, S.; Imai, S.; Tsuruoka, T.; Ogawa, H.; Satoh, A.; Inouye, S.; Niida, T.; Otake, N. J. Antibiot. 1983, 36, 96-98.
16. Cutler, H. G.; LeFiles, J. H. Plant Cell Physiol. 1978, 19, 177-182.
17. Jarvis, B. B.; Midiwo, J. O.; Tuthill, D.; Bean, G. A. Science 1981, 214, 460-462.
18. Rice, E. L. "Allelopathy;" Academic: New York, 1974.
19. Rice, E. L. Bot. Rev. 1979, 45, 15-109.
20. Patrick, Z. A.; Toussoun, T. A.; Snyder, W. C. Phytopathology 1963, 53, 152-161.
21. Kaminsky, R. Ecol. Monogr. 1981, 51, 365-382.
22. McPherson, J. K.; Muller, C. H. Ecol. Monogr. 1969, 39, 177-198.
23. Christensen, N. L.; Muller, C. H. Ecol. Monogr. 1975, 45, 29-55.
24. Patrick, Z. A. Can. J. Bot. 1955, 33, 461-486.
25 Misato, T. J. Pesticide Sci. 1982, 7, 301-305.
26. Demain, A. L. Science 1983, 219, 709-714.
27. Fischer, H.-P.; Bellus, D. Pestic. Sci. 1983, 14, 334-346.
28. Arai, M.; Haneishi, T.; Kitahara, N.; Enokita, R.; Kawakubo, K.; Kondo, Y. J. Antibiot. 1976, 29, 863-869.
29. Haneishi, T; Terahara, A.; Kayamori, H.; Yabe, J.; Arai, M. J. Antibiot. 1976, 29, 870-875.
30. Omura, S.; Iwai, Y.; Takahashi, Y.; Sadakane, N.; Nakagawa, A.; Oiwa, H.; Hasegawa, Y.; Ikai, T. J. Antibiot. 1979, 32, 255-261.
31. Iwai, Y.; Nakagawa, A.; Sadakane, N.; Omura, S.; Oiwa, H.; Matsumoto, S.; Takahashi, M.; Ikai, T.; Ochiai, Y. J. Antibiot. 1980, 33, 1114-1119.
32. Yamada, O.; Kaise, Y.; Futatsuya, F.; Ishida, S.; Ito, K.;

Yamamoto, H.; Munakata, K. Agr. Biol. Chem. 1972, 36, 2013-2015.

33. Yamada, O.; Ishida, S.; Futatsuya, F.; Ito, K.; Yamamoto, H.; Munakata, K. Agr. Biol. Chem. 1974, 38, 1235-1240.

34. Shomura, T.; Yoshida, J.; Amano, S.; Kojima, M.; Inouye, S.; Niida, T. J. Antibiot. 1979, 32, 427-435.

35. El-Nakeeb, M. A.; Lechevalier, H. A. Appl. Microbiol. 1963, 11, 75-77.

36. Warren, H. B., Jr.; Prokop, J. F.; Grundy, W. E. Antibiot. Chemother. 1955, 5, 6-12.

37. Alexander, M. "Introduction to Soil Microbiology;" John Wiley and Sons: New York, 1977.

38. Goodfellow, M.; Williams, S. T. Ann. Rev. Microbiol. 1983, 37, 189-216.

39. Scheffer, R. P.; Livingston, R. S. Science 1984, 223, 17-21.

RECEIVED July 11, 1984

Use of Bioassays for Allelochemicals in Aquatic Plants

LARS W. J. ANDERSON

Aquatic Weed Control Research Laboratory, Botany Department, Agricultural Research Service, U.S. Department of Agriculture, University of California, Davis, CA 95616

Lack of appropriate bioassays has hampered progress in detecting inhibitory or growth regulatory compounds in leachates or extracts of aquatic macrophytes. Assays relying on seed germination or callus growth in terrestrial plants are not necessarily relevant to the "target" aquatic plants. In these studies, sensitive bioassays were developed using explants of Hydrilla verticillata and vegetative propagules of Potamogeton nodosus and P. pectinatus. These systems respond to many growth regulators and herbicides much as do whole, intact aquatic plants. Responses to some potential allelochemicals have been demonstrated in small-volume exposures. Typical quantifiable responses include: frequency of new shoot production, elongation of new shoots, chlorosis, root production, necrosis of "parent" explant, and inhibition of sprouting. Explant systems should permit inexpensive large-scale evaluation of phytoactive compounds whether from natural sources or from synthetic, agrichemical production.

The proliferation of various aquatic weeds in the United States causes losses and damages exceeding $300 million annually (1). In spite of this, there are very few herbicides available for use in aquatic sites. There are two main reasons for this dearth of effective products: potential sensitivity of the aquatic environment, and lack of sufficient research and development by agrichemical companies for what is perceived to be a "minor use" and "high risk" venture. Although there is some truth in this perception, it has also resulted in the omission of economically important aquatic weeds in the primary chemical screening and identification programs of most companies. The general exception to this is use of duckweed (Lemna sp.) since it is easy and inexpensive to culture. Unfortunately, duckweed is also one of the least important aquatic weeds and it is not even in the same family as any of the ten most noxious aquatic weeds: hydrilla (Hydrilla verticillata),waterhyacinth (Eichhornia crassipes), Eurasian watermilfoil (Myriophyllum spicatum), pondweed species (Potamogeton spp. -there are at least five important species), elodea (Elodea canadensis), and coontail (Ceratophyllum demersum) (2). However, culture and whole-plant testing using these more important weeds requires much more space and time and higher costs compared to duckweed.

These circumstances have provided impetus to investigate the potential for using beneficial aquatic plants which might out-compete more noxious weeds (3). One such plant, dwarf spike-rush (Eleocharis coloradoensis) produces chemical(s) which exhibit inhibitory effects on other aquatic plants (4). The further characterization and identification of these allelochemicals requires sensitive bioassays which are also relevant to the "target" aquatic weed species. Although many convenient bioassays have been developed for plant growth regulators and allelopathic compounds, they often bear no resemblance to the target plant.

The objective of the research reported here was to develop sensitive bioassays which utilize near-whole plant systems of appropriate target aquatic weeds and which require little space and low volumes of incubation medium. Such bioassays could be used to help identify active fractions of chromatographically partitioned allelochemicals and could also be used in primary screening procedures for newly synthesized agrichemicals.

Methods and Materials

1. Plant sources. Two general types of plant bioassay systems were used: 1) vegetative propagules (winterbuds) of American pondweed (Potamogeton nodosus) and sago pondweed (P. pectinatus; 2) explants of hydrilla (Hydrilla verticillata) (Figure 1). American pondweed "winterbuds" were collected in the fall in de-watered irrigation canals in Richvale, CA. and were maintained at ca. 6°C in plastic bags. This treatment breaks fall dormancy in 6 to 8 weeks. Sago pondweeds were obtained from a commercial supplier in Oshkosh, WI., and were kept at ca. 6°C to prevent sprouting. Vigorously-growing hydrilla was collected from the Imperial Irrigation District, CA., and shipped to Davis within 24 hours after harvesting. Hydrilla was kept in 51 cm square x 38 cm deep plastic tubs and kept thoroughly flushed with tap water. Intact shoots of the collected material were cut and planted into a standard soil "U.C. Mix" (.5 ft^3 Delta peat, .5 ft^3 coarse sand, 45 g KNO$_3$, 30g K$_2$SO$_4$, 713 g dolomite, 180 g gypsum, 320 g super phosphate). Potted cuttings were allowed to root 3 to 5 weeks in a large fiberglass tank in a greenhouse under 14-h light (natural light supplemented with cool-white fluorescent lamps = 250 μE m^{-2} sec^{-1}). Only those plants with healthy-appearing, vigorous new growth were used in subsequent bioassays.
2. Preparation of vegetative propagules. Winterbuds of American pondweed were surface-sterilized by rinsing them free of soil and debris and soaking in 1% hypochlorite (ca. 1:4 v/v dilution of

Figure 1:   Left: two-node explants of hydrilla showing new shoots (arrows) 2 weeks after excision (explants 2 cm long); middle: ungerminated "winterbuds" of American pondweed (5 cm long); right: ungerminated tubers of sago pondweed (5 cm long).

Chlorox) for 20 min. Winterbuds were then rinsed 3 times with
sterile distilled water to remove the hypochlorite. Sago pondweed
"tubers" were rinsed in distilled water, but not surface sterilized.
3. Preparation of hydrilla explants. A stainless steel razor was
used to remove the distal 4 to 5 cm ("apical explants") from
either rooted hydrilla, or hydrilla freshly received from the
field. Subtending sections containing two adjacent intact nodes
("2-node explants") with whole whorls of leaves were removed, but
1 to 2 intervening nodes were left above the next cut (Figure 2).
No 2-node explants were taken below (proximal to) a subtending
lateral branch. All explants were kept in tap water for 24 hours
in a growth chamber before use in bioassays (25°C, 14-h day, ca.
200 µE m^{-2} sec^{-1}).
4. Exposure to growth regulators, herbicides and allelochemicals.
All test compounds were dissolved in sterile 1% Hoagland's medium
(5) or in an organic solvent (methanol, acetone or ethanol) and
subsequently diluted to test concentration with sterile 1%
Hoaglands medium. All experiments included 1% Hoagland's medium
controls, and controls containing appropriate concentrations of
organic solvent. Generally, 5 winterbuds or tubers or 10 explants
were placed in 500-ml foam-stoppered Erlenmyer flasks with 250 ml
of medium. Treatments were run in triplicate in a growth chamber
at 25°C, 14-h day, 150-200 µE m^{-2} sec^{-1} cool-white fluorescent
light. Only explants from a common batch of collected or transplanted hydrilla were used in an experiment. In some sets of
experiments, light levels were varied with neutral density filters
or temperatures were varied. Plant growth regulators used were:
GA$_3$ (gibberellic acid), ABA (abscisic acid) BA (benzyladenine), KT
(Kinetin) (All from Calbiochem, Inc.). Allelochemicals were DAD
(dihydroactinodiolide (6,7)) and solstitialin. Herbicides used
were fluridone, chlorsulfuron, and sulfometuron.
5. Observation of effects. Seven to 14 days after exposure to
test compounds, length, number of leaves and presence of stomata
in sprouted winterbuds were determined. For sago pondweed tubers,
length, number of new daughter plants and presence of roots were
determined. Hydrilla explants were evaluated for number of and
length of new shoots. In some experiments the chlorophyll-a
content of the apical 2 cm on apical explants was determined by 3
successive extractions in 90% acetone using a power-driven Teflon
pestle. Absorbance of Millipore-filtered (.45µ) acetone extracts
was determined at 630, 645, 665 nm on a spectrophotometer.
Chlorophyll-a was calculated by equations of Strickland and
Parsons (8).

Results

American pondweed responded to GA$_3$ and the
herbicide sulfometuron within 7 days after exposure (Table I). As
with terrestial plants, GA$_3$ caused elongation of shoots (interno-
des) compared to controls, whereas sulfometuron almost completely
blocked elongation since no significant lengthening occurred
between day 7 and 14. The presence of GA$_3$ did not counteract the
inhibition caused by sulfometuron. This is reasonable since

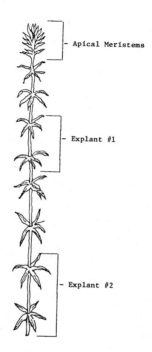

Figure 2:   Diagram of <u>Hydrilla</u> <u>verticillata</u> showing location from which apical meristems ("apical explants" and "two-node" explants were excised.

Table I

Shoot length of American Pondweed 1 and 2 weeks after continuous
exposure to sulfometuron alone and in combination with different
plant growth regulators (PGR)

| Treatment[a] | Shoot Length (cm) | |
	1 week after treatment	2 weeks after treatment
Control[b]	10.3 ± 0.8[c]	16.6 ± 0.6
Sulfometuron	7.2 ± 0.6	7.6 ± 0.3
N-6 Benzyl Adenine (BA)	11.6 ± 1.1	15.7 ± 0.8
Gibberellic Acid (GA)	16.3 ± 0.4	28.4 ± 1.5
Kinetin (KT)	11.1 ± 0.7	17.2 ± 1.5
Zeatin (ZT)	12.1 ± 1.1	15.8 ± 1.4
Sulfometuron + BA	6.3 ± 0.8	7.4 ± 1.0
Sulfometuron + GA	6.5 ± 0.4	7.1 ± 0.3
Sulfometuron + KT	6.4 ± 0.5	7.9 ± 0.7
Sulfometuron + GT	6.9 ± 0.7	7.4 ± 0.4

[a] Concentration of sulfometuron was 1.0 ppb (2.7×10^{-9}M);
     PGRs was 10^{-5}M
[b] 1% Hoaglands medium
[c] Value represents $\overline{X} \pm$ SD; n=20

sulfometuron interferes with cell division via blockage of amino acid synthesis (9) while GA₃ affects cell expansion. In fact, none of the growth regulators counteracted sulfometuron.

Another type of response in American pondweed is shown in Table II. Abscisic acid (ABA) causes a change in leaf development and morphology whereby exposed plants produce leaves having stomata on their upper surfaces. These "floating" type leaves are more complex than the simple 2-cell thick "submersed" leaves that are normally produced upon sprouting of winterbuds (10). Solstitialin, a compound isolated from Centaurea solstitialis, had no apparent effect on leaf length, number of leaves or ABA-induced float leaf formation in American pondweed.

In another bioassay (Table III), an active compound (DAD) isolated from whole dwarf spikerush (Eleocharis coloradoensis) (6,7) appeared to induce floating leaf formation in American pondweed. The response to DAD was somewhat erratic and much less pronounced than that observed for ABA at even higher concentrations than ABA. (Previous work has shown that high concentrations of ABA, e.g. 5-13 ppm, inhibit sprouting of winterbuds).
The response to DAD is particularly interesting since submersed "floating-type" leaves were observed on American pondweed when it was grown with dwarf spikerush (3).

GA₃ caused elongation in sago pondweed, even in the presence of solstitialin (Table IV). At 1 to 50 ppm, solstitialin reduced the number of leaves per plant even in the presence of GA₃. At 50 ppm solstitialin caused a reduction in the number of daughter plants, and seemed to have slightly inhibited root production as well.

The above results were obtained using volumes of 100-250 ml in Erlenmyer flasks. However, much smaller volumes can be used (2-5 ml) if winterbuds are individually exposed in glass tubes. Table V shows the response of American pondweed winterbuds exposed to ABA for 72 h in 8 mm (i.d) x 14.5 cm tubes and then transferred to fresh 1% Hoaglands for 4 days. The effect on leaf development is the same as with larger volume exposures, but this system would allow use of much smaller quantities of allelochemicals.

Hydrilla explant bioassays. The general effects of temperature on production and elongation of new shoots on 2-node hydrilla explants is shown in Figure 3. No shoots were produced at 15 C, but at 36 C ca. 50% of the explants had new shoots averaging 7 cm in length within 7 days. In most cases, only one new shoot was produced per explant, but occasionally, two shoots were produced (either one at each node or two at one node).

Most plants exhibit "apical dominance" which means that the presence of a terminal (distal) meristem tends to suppress lateral shoot initiation (11). Since lateral shoot production is an important characteristic to assess in hydrilla, the frequency of shoot production was determined in sequentially cut (distal to proximal) explants (Table VI). Even though the 4 cm apical meristem contained several nodes, almost none of these produced new shoots. However, nearly half the 2-node explants subtending the cut apical meristem produced new shoots. There was no apparent difference in percent of new shoots produced once the apical meristem was removed.

Table II   Effect of Solstitialin and Abscisic Acid on American Pondweed (Potamogeton nodosus)[a]

Treatment	Length(cm)		Leaves/plant		Float leaves/plant	
	11-day	25-day[b]	11-Day	25-Day	11-Day	25-Day
Control	16.5±4.3	18.1±	5.9±0.2	--	0	0
ABA 10^-6M	6.3±0.5	10.1±1.8	3.6±0.3	--	1.1±0.2	2.3±0.4
Solstitialin						
50 (ppm) (1.8X10^-4M)	14.5±1.4	19.5±1.9	5.5±0.5	--	.01[c]	0
20 (ppm) (7.1X10^-5M)	14.1±0.5	17.0±0.9	6.1±0.2	--	0	0
1 (ppm) (3.6X10^-6M)	14.3±0.9	21.9±1.4	4.3±0.4	--	0	0
ABA 10^-6M + Solstitialin						
50 (ppm)	6.2±0.5	10.2±0.5	3.3±0.4	--	0.9±0.2	3.9±1.3
20 (ppm)	7.2±2.6	11.7±0.6	3.6±0.2	--	1.0±0.3	3.1±0.3
1 (ppm)	6.5±0.5	10.9±0.4	3.1±0.3	--	0.6±0.1	2.2±0.4

a   Winter buds were exposed to compounds in 1% Hoagland for 7 days. Values are mean ± SE; 5 plants/replicate; 3 replicates/treatment.
b   11 Days posttreatment, plants were washed and transferred to fresh 1% Hoagland not containing test compounds. Data are from plants kept 25 days after transfers.
c   One plant had one leaf with stomata.

TABLE III Effect of DAD (Dihydroactinodiolide) on Leaf Development in Potamogeton nodosus

Treatment (Exposure Period) Experiment	Leaves with Stomata/Total Leaves[a]					% Floating Leaves
	(5 days) No.1	(72 H) No.2	(72 H) No.3	(72 H) No.4	Mean Exp.2,3,4	
Control	0/36	7/30	2/29	0/29	3/29	10.3
$10^{-5}M$ ABA (2.63 ppm)	-	18/28	16/28	-	17/28	60.7
$10^{-6}M$ ABA (.263 ppm)	12/34	8/25	14/23	19/31	14/26	53.8
$5 \times 10^{-7}M$ ABA (.131 ppm)	16/37	9/16	21/33	7/26	12/25	48.0
10ppm DAD ($3.5 \times 10^{-5}M$)	4/40	16/22	1/29	1/13	6/21	28.5
25 ppm DAD ($0.9 \times 10^{-4}M$)	-	12/23	0/30	6/29	6/27	22.2
50 ppm DAD ($1.8 \times 10^{-4}M$)	-	14/27	5/28	3/31	7/29	24.1

a   Each treatment consisted of two replicates, five plants per replicate. Plants were scored 7 days after start of exposure to ABA or DAD.

Table IV

Effect of Solstitialin and Gibberellic Acid on
Sago Pondweed (Potamogeton pectinatus)[a]

Treatment	Length (cm)	No. Leaves/plant	% with daughter plant	% with roots
Control	14.9±1.3	19.2±1.8	86.7±7	100±0
Solstitialin				
50 ppmw (1.8X10^{-4}M)	14.8±1.0	14.2±1.5	53.3±17	73±7
20 ppmw (7.1X10^{-5}M)	14.5±1.0	13.8±1.5	60.0±11	73±7
1 ppmw (3.6X10^{-6}M)	15.1±1.1	13.7±1.9	66.7±7	66±7
Gibberellic Acid				
10^{-4}M (35 ppmw)	27.0±2.2	17.7±1.9	100±0	93±7
10^{-5}M (3.5 ppmw)	26.3±1.1	18.9±1.9	93±7	100±0
10^{-6}M (.35 ppmw)	22.3±1.3	18.9±1.8	80±0	93±7
Solstitialin 20 ppm +Gibberellic Acid				
10^{-4}M (35 ppm)	24.5±2.1	14.2±2.1	66.7±13	80 ±12
10^{-5}M (3.5ppm)	27.7±1.5	14.8±1.9	66.7±13	100 ±0
10^{-6}M (.35ppm)	19.2±1.6	15.1±2.0	66.7±18	80 ±20

[a] Tubers were exposed to compounds for 7 days in 1% Hoagland nutrient
Values are means ±SE.; 5 plants/replicates, 3 replicates per treatment

TABLE V

Response of American Pondweed to 72 h exposure to 2 ml ABA in 8 mm x 14.5 cm Bioassay Tubes[a]

Treatment:	Plant Length (cm)	% Leaves with Stomata
Control	10.2±.5	0
5×10⁻⁷ M ABA (.11 ppm)	8.2±.4	29.8±8
10⁻⁶ M ABA (.26 ppm)	7.4±.5	39.8±7

a   For each treatment 10 winterbuds were individually sprouted in tubes for 72 h, then rinsed and transferred to a 500 ml Erlenmyer flask containing 200 ml fresh 1% Hoagland's medium for 3 more days. Values are mean ± S.E.

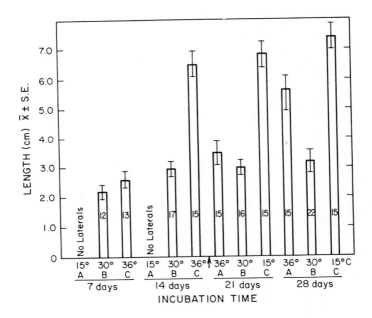

Figure 3: Effect of temperature on production and elongation of new shoots on hydrilla two-node explants. Arrow indicates when explants originally maintained at 15 or 36 C were exxchanged. Values are means ± S.E.

Table VI

Relationship between Hydrilla explant position and new shoot production

Explant No.	% New Lateral Shoots on Two-Node Explants[a]	
	Consecutive Explants	5-Nodes Between Explants
Apical Meristem (4 cm)	0	5
1 (distal end)	20	40
2	35	40
3	40	45
4	30	
5	22	
6 (proximal end)	50	

a Explants maintained in 1% Hoagland Medium in 500 ml Erlenmeyer flasks.

Results presented in Table VII show that sufficient light is needed for new shoot formation on explants and that the herbicide fluridone causes chlorosis in new growth just as in whole plants (12,13). These data also confirm that the apical explant, which contains the terminal meristem, is a poor system for assaying inhibitors of new shoot production.

Tables VIII and IX show responses of hydrilla explants to various plant growth regulators and to solstitialin. GA$_3$, BA and solstitialin enhanced elongation at 10^{-4} or 10^{-5} M. Zeatin, on the other hand, dramatically increased new shoot initiation in apical explants but had little effect on 2-node explants.

DAD at 50 ppm caused a significant reduction in new shoot length on hydrilla explants, but did not affect initiation of new shoots (Table X). ABA was more effective than DAD in inhibiting shoot elongation, but also did not block shoot initiation.

The hydrilla 2-node explants were also sensitive to the herbicides chlorsulfuron. One part per billion chlorsulfuron (ca. 3 X 10^{-9}M reduced growth of new shoots by almost 80% but had no effect on new shoot initiation (Table XI). When the herbicide was removed after 14 days, new shoots began to elongate.

Discussion

A wide variety of plant growth bioassays are available for detecting growth regulatory and phytotoxic activity of chemicals in terrestrial plants (14,15). These assays were developed primarily for the purpose of characterizing hormonal responses and for quantification of endogenous hormones. However, when they are utilized to detect allelochemicals the experimenter assumes a physiological and biochemical similarity in mode of entry and mechanism of action which may or may not exist. This is particularly true for economically important aquatic weeds, since their growth and development has been sparsely investigated. Given the range of secondary products in aquatic plants (16) which might be allelopathic, it is important to tailor bioassays to fit a given "target" plant and to use an assortment of quantifiable observations.

The results presented here demonstrate that these approaches can be applied to Potamogeton sp. and Hydrilla verticillata. In both assay systems, typical responses to known plant growth regulators were observed in most cases. The unique response of P.nodosus to ABA however, clearly shows the importance of not relying solely on "classical" bioassays.

Since aquatic plants in general (and particularly those which have "weedy" growth) are well adapted to conditions very different from terrestrial sites, one might expect different growth responses. Thus, the nature of allelopathic interactions needs to be determined. In the case of dwarf spikerush for example, leachate experiments showed that production of daughter plants was inhibited, not germination of tubers (4). This implies that using a seed germination (or tuber germination) assay alone would not be fruitful.

Hydrilla explants responded to herbicides essentially the same as whole plants: fluridone caused chlorosis in new shoots,

Table VII

Effect of Fluridone on 2-node Explants and Apical tips of Hydrilla verticillata exposed to different light levels[a]

Treatment	Two-Node Explants (New Shoot (%))		4-cm Apical Tips					Chlorophyll[b] (mg/g. fr. wt.)
			New Shoot (%)		New Shoot Length (cm)			
Treatment μE m⁻² sec⁻¹	Day 7	Day 11	Day 7	Day 11	Day 7	Day 11		Day 11
Control								
No light	2.5±1.9	3.0±2.4	0	3.3	5.8±.7	4.6±.2		.382 .06
10	7.5±2.5	15.0±2.9	6.7	3.3	5.3±3.3	5.5±.3		.867±.05
60	35.02±2.8	37.5±2.5	10.0	16.7	5.9±.3	5.8±.3		1.392±.38
135	25.0±6.4	27.8±8.4	16.1	19.4	5.2±.2	5.4±.2		.778±.07
Fluridone (0.25 ppmw)								
No light	0	5.0±2.8	0	0	4.9±.2	5.2±.2		.386±.03
10	7.5±4.8	17.5±2.5	3.3	10.0	6.4±.4	6.6±.5		.542±.05
60	10.0±7.1	22.5±8.5	16.7	30.0	5.7±.3	5.6±.3		.409±.06
135	30.0±14.7	35.0±1.8	10.0	10.0	5.6±.2	5.7±.2		.337±.04

a Various light levels were achieved using open mesh shade cloth. Explants were exposed to fluridone in 1% Hoaglands medium for 11 days. Treatments consisted of 10 4-cm Apices/replicates, 3 replicates per treatment and 10 2-node explants/replicates, 4 replicates per treatment. Values are mean ± SE.

b After 11 days, the terminal 2-cm were removed from 5 apices in each replicate. Values are mean + SE (n=15 for each treatment).

TABLE VIII

Effect of ABA and Zeatin on New Shoot Production from Apical
Shoots and Two-node Sections of <u>Hydrilla</u> <u>verticillata</u>[a]

| | Number of new shoots per explant | |
Source:	Apical Explants	Two Node Explants
Control	$.33 \pm .2$	$.40 \pm .1$
10^{-6} ABA (.26 ppm)	$1.0 \pm .2$	$.46 \pm .1$
5×10^{-7} ABA	$0.5 \pm .31$	$.60 \pm .1$
5×10^{-8} ABA	$0.10 \pm .1$	$.27 \pm .2$
10^{-6}M Zeatin (.26ppm)	$1.7 \pm .2$	$.27 \pm .07$
10^{-6}M Zeatin + 5×10^{-7}M ABA	$1.5 \pm .2$	$.47 \pm .0$

a   Values are means $\pm$ S.E. from three replicates of five
    explants per replicate (15 explants).
    Light length = 24 hour day at 28°C

TABLE IX

Effect of Plant Growth Regulator on Hydrilla
Two-node Explants[a]

Treatment	New Shoot Length (cm)	% New Shoots [b]	
Control	2.32± .8	36.7	a
Solstitialin			
50 ppmw (1.8×10^{-4}M)	4.20± .9	40.0	a
20 ppmw (7.1×10^{-5}M)	5.67± .7	50.0	a
1 ppmw (3.6×10^{-6}M)	0.94± .1	30.0	a
ABA			
10^{-4} M (26.0 ppm)	1.33± .2	30.0	a
10^{-5} M (2.6 ppm)	2.0 ± .7	16.7	b
10^{-6} M (.26 ppm)	1.17± .1	20.0	b
Ga$_3$			
10^{-4} M (35.0 ppm)	7.87±1.6	50.0	a
10^{-5} M (3.5 ppm)	2.7 ±1.0	43.3	a
10^{-6} M (.35 ppm)	1.56± .5	26.7	b
BA			
10^{-4} M (23.0 ppm)	4.7 ± .9	50.0	a
10^{-5} M (2.3 ppm)	5.10±2.3	33.3	a
10^{-6} M (.23 ppm)	2.30±1.3	23.3	b

a Values are means of 3 replicates, 10 explants per replicate
   7 days posttreatment.
b Values with the same letter are not significantly different at the 5%
   level using Duncan's Multiple Range Test.

TABLE X

Response of Hydrilla Two-node Explants to Dihydroactinodiolide (DAD)[a]

Treatment:	Length of New Shoots (cm)		% New Shoots	
	Experiment 1	Experiment 2	Experiment 1	Experiment 2
Control	1.09 A	2.42 A	60 A	30 A
DAD				
50 ppm (1.8×10^{-4}M)	0.42 C	1.2 B	53 A	35 A
25 ppm (0.9×10^{-4}M)	0.69 B	1.5 B	65 A	35 A
10 ppm (3.5×10^{-5}M)	0.72 B	1.83 AB	55 A	25 A
ABA 10^{-5}M (2.6ppm)	–	0.16C	–	25 A

a Values are for 7 days posttreatment, means of 20 explants per
  treatment.
  Values with same letter are not significantly different at 5%
  level using Duncan's Multiple Range Test.

TABLE XI

Effect of Chlorosulfuron on Growth of Hydrilla verticillata Two-Node Explants[a]

| | Days Posttreatment | | | | | |
| | Length of New Shoots | | | % New Shoots | | |
Treatment	10	14	24[b]	10	14	24[b]
Control	0.91±0.03	1.18±0.15	1.65±0.02	30.0±10	30.0±10	30.0±10
1 ppb	0.21±0.06	0.24±0.07	0.98±0.14	46.6±15	43.3±5	43.4±12
5 ppb	0.24±0.09	0.21±0.09	0.75±0.11	36.7±6	33.3±3	34.0±17
10 ppb	0.19±1.06	0.19±0.08	0.72±0.10	36.7±9	36.6±9	43.3±18

a Values are mean ± S.E. for thirty 2-node explants, ten per replicate. All apices were 4 cm long at start of treatment.

b Between Days 14 and 24 explants were transferred to fresh medium not containing Chlorosulfuron.

sulfometuron stopped shoot elongation. With the exception of subterranean turion ("tuber") production, one could view the explants as wholeplant "analogues" which are easily manipulated and which require much less space, time and resources to use. It may even be possible, with the correct plant growth regulator ratios, to induce tuber production on explants. Furthermore, with sufficient care in sterilizing hydrilla tubers as a source, one could maintain axenic explants. However, it is important to be aware of variability which can affect the responses of explants. These include external variables such as culture condition (temperature, light, day length, nutrient medium) and endogenous variables such as plant vigor, plant age, position on shoot (i.e. where the explant was removed relative to the apical meristem) and internode length. By using "standard" (consistent) sets of external and endogenous conditions, reproducible responses are obtained.

Ultimately, it is important to develop explant bioassays which can help quantify allelochemicals. This will require greater availability of purified active products. However, at this time, the explant systems can be used to show qualitative effects whether in crude extracts, leachates or from HPLC fractionation (17). Once target species effects are characterized, other bioassays which may be more sensitive could be used as well.

Acknowledgments

I am indebted to Dr. Kenneth Stevens and Ms. Gloria Merril, USDA/ARS, Western Regional Research Center, for providing DAD and solstitialin. The technical assistance of Ms. Doreen Gee and Mr. Nathan Dechoretz is greatly appreciated.

Literature Cited

1. Anderson, L.W.J. et al. USDA/ARS Aquatic Weed Research Planning Conference Report. 1982. 30pp.
2. Sculthorpe, C.D. The Biology of Aquatic Vescular Plants 1967, Edward Arnold, Ltd., London, p.16-20
3. Yeo, R.R. J. Aquat Plant Manage. 1984. 22,
4. Frank, P.A., Dechoretz. N. Weed Sci. 1980 28, 499
5. Hoagland, D.R., Arnon, D.I. Calif. Agric. Exp. Stn. Circ. 347 1950
6. Stevens K.L.,Merrill, G.B. J. Food Chem.1980, 28, 644
7. Stevens K.L., Merrill, G.B. Experimentia 1981 37, 1133
8. Strickland, J.D.H. Parsons, T.R. Handbook of Seawater Analysis, Fish. Res.Board Can. Bull.No.167 1968, 311 pp
9. Ray, T.B. Weed Sci. Soc. Amer. 1984 Abstract No. 223.
10. Anderson, L.W.J. Science 1978 201,1135
11. Phillips, I.D., Ann. Rev. Plant Physiol. 1975 26, 341
12. Anderson, L.W.J. Weed Sci: 1981 29 723
13. Bartels, P.G,. Watson, C.W. Weed Sci. 1979 26, 198
14. Evans, M.L. Ann. Rev. Plant Physiol 1975 26, 241
15. Mitchell, J.W., Livingston, G.A., Agriculture Handbk,- No.336U.S.D.A. 1968, 140 pp.
16. McClure, J.W. "Secondary Constituents of Aquatic Angio- sperms"in Photochemical Phylogeny, J.B. Harborne (ed.) 1970 Academic Press, N.Y. 335 pp.
17. Martin, D.F.Misc.Report No.A-883-2 U.S. Army Corps of Engineers 1983, 37 pp.

RECEIVED July 31, 1984

Allelopathic Substances from a Marine Alga (*Nannochloris* sp.)

RALPH E. MOON and DEAN F. MARTIN

Chemical and Environmental Management Services (CHEMS) Center, Department of Chemistry, University of South Florida, Tampa, FL 33620

Substances elaborated by a marine alga (Nannochloris sp.) adversely affect a red tide organism, Ptychodiscus brevis, in two ways. One component causes cytolysis and a second causes P. brevis to shrink and enter a non-motile resting stage. The latter behavior is probably responsible for the mitigation of ichthyotoxic activity noted when P. brevis and the marine alga were mixed. Various studies indicate the substances elaborated by Nannochloris sp. are not toxic to fish and other marine organisms tested, though the tests, of course, have not been exhaustive. The cytolytic material has significant ecological implications inasmuch as a serious red tide along the west coast of Florida can be responsible for massive mortalities of marine animals and costs associated with a red tide outbreak have been estimated to exceed $17 million. Limited studies are available that indicate the distribution of cytolytic agents inversely matches the locations of so-called "seed beds" of P. brevis.

The release of allelopathic substances in the marine environment is of particular interest to those who are affected by red tides, i.e. periodic blooms of micro-organisms that give rise to discolored water and that may result in intoxications. The background of the problem has been provided elsewhere (1-3). Red tides are found in a variety of marine environments, but several red tides, notably those in Japan in the Seto Inland Sea, and in U.S. coastal waters (cf. California, New England, and the Gulf of Mexico) have attracted particular attention, owing to the impact of red tide organisms on shellfish (paralytic shellfish poisoning) or mass mortality of marine animals associated with outbreaks of Ptychodiscus brevis (4).

This organism, an unarmored dinoflagellate, has been observed in sporadic outbreaks in west coast Florida waters and in other near-shore waters of the Gulf of Mexico (1). Outbreaks have a considerable social, public health and economic impact because of fish kills. For example, the outbreak that lasted for three summer months during

0097–6156/85/0268–0371$06.00/0

1970 was estimated to cost about 17 million dollars (5) because of
lost tourist revenues, cleanup costs, and other economic impacts.
While this may be considerably smaller than the costs associated with
an outbreak of Chattonella subsalsa in the Seto Inland Sea (6) that
affects food fish, it nevertheless represents a serious impact.

The possibility of managing a bloom of Ptychodiscus brevis has
been considered for many years, and various limitations have been
noted to the use of chemical control (3) or the use of control by
some organisms (7). Chiefly, the problems are two-fold: the volume
to be treated and the limitations of the control agent. For example,
a red tide may cover hundreds of square miles, it will be present in
patches, and may be unevenly distributed through the water column.
One response may be that it is possible to manage the red tide at the
source of the bloom (8) or that there is no need to manage a red tide
over vast areas, merely in localized ones that are of special interest. The point is moot until a suitable control agent is
available. A considerable number of chemicals have been reviewed for
possible red tide control substances, (9), and in retrospect this
type of research is subject to ultimate failure because the first
criterion of a successful control agent (chemical or biological) has
not been considered.

A successful control agent probably should be sought in the
marine environment for three reasons. First, this is where the
problem is and if a natural control agent has developed, this is
where it should be found. Second, the most potent chemical agents
seem to be found in the ocean or in the atmosphere; potency appears
to be required because of the tremendous dilutions. Third, this is
where we found what we believe to be a potentially useful control
agent.

An alga, isolated from a red tide outbreak area, produced
chloroform-soluble material capable of causing lysis of P. brevis.
The original observation reported by Kutt and Martin (10) was subjected to a more thorough study by McCoy and Martin (11). The
organism initially was identified as Gomphosphaeria aponina, but based
upon SEM studies was reidentified as a Nannochloris sp. (12). It has
been demonstrated (11) that concentrated organisms when added to
P. brevis culture led to the cytolysis of the red tide organism; that
concentrated suspensions of destroyed cells (destroyed by successive
freeze-thawing) led to cytolysis, and that chloroform extracts of the
cell-free medium produced a residue that caused cytolysis of P.
brevis (11). The term aponin (apparent oceanic natural cytolin) has
been used to describe the substance(s) responsible for cytolysis.

Some effort has been directed toward understanding the growth
requirements of the alga of interest, allelopathic alga (13), the
optimum salinity (14), the distribution of allelopathic agents (15),
and the temperature optimum (16). This paper reviews the approaches
taken to separate the allelopathic agents from the other materials
and the methods used to characterize the biological activity of
aponin from Nannochloris sp.

Materials and Methods

Isolation of the Crude Extract. An isolation of Nannochloris sp.
(previously referred to as Gomphosphaeria aponina) was cultured from

samples of seawater from the west coast of Florida. Unialgal stock cultures were maintained in an enriched seawater media (17) in large scale semi-continuous cultures, grown under optimized culture conditions (18).

Nannochloris sp. cultures were harvested in 30-35 liter volumes shortly after the culture attained a stationary cell count (5000 - 7500 cells/ml) after 2 or 3 weeks. Cells were separated from the seawater media using a continuous centrifugation technique. The clear, cell-free centrifugate was extracted for 24 hours (1 hour of mixing) with redistilled chloroform (50 ml/1). Following a 24-hour settling period, the aqueous phase was decanted and the remaining chloroform extract was poured into a cylindrical separatory funnel. This chloroform-organic phase was carefully separated for residual aqueous portions, collected and reduced in volume (to 5 to 10 mls) using a rotary evaporator (Buchi-rotovapor) and held at 4°C. This crude preparation was termed the "crude extract" and was used extensively for cytolytic and inhibitory studies.

Organismal Studies. The allelopathic interaction between Nannochloris sp. and several species of algae, bacteria and fungi were performed using the bioassay technique and paper disc method (cf. 19). The bioassay technique demonstrated a cytolytic effect (up to 100% kill) when test species of Ptychodiscus brevis and Chattonella subsalsa (6) were exposed to various concentrations of the crude extract. Test results of Prymnesium parvum showed little or no effect from the crude extract (28). The inhibitory activity of the crude extract using the disc method was shown among all species of fungi tested including: Saccharomyces cerevisiae, Rhizopus stolonifera, Candida albicans, Allischeria boydii, and Aspergillus fumigatus. No inhibitory effects were noted for bacterial species (Escherichia coli, Pseudomonas aeruginosa, Staphylococcus aureus, and Streptococcus aureus) tested in the same manner.

Evaluation of HPLC Components from the Crude Extract. Individual peaks collected from HPLC injection of the crude extract (Figure 1) showed that only one component (fraction 7) exhibited an inhibitory effect upon Saccharomyces cerevisiae when tested using the paper disc method. A similar study using the germination test method (19) showed quite different responses. Observations of dark green Boston lettuce seeds were viewed from three perspectives including: 1) number of seeds germinated, 2) developmental stage achieved, and 3) final morphological orientation.

During the six-day study period, most compounds showed little or no significant inhibition of seed germination when compared to control plants. Several fractions (Nos. 1-5, 9, 21, and 29, see Figure 1) were notably inhibitory with 60% (6 out of 10) not forming a hypocotyledon after the fourth day. All preparations of the crude extract series and the initial injection fraction showed total inhibition of germination during the four day period. Following this observation, 100 µl of distilled water was added to each test dish to see if inhibition of seed germination would remain permanent. On the fifth day, previously dormant seeds impregnated with compounds from fractions 105, 9, 21, and 26 and 10 µl of crude extract, showed development of a hypocotyledon. On the sixth day, almost all test

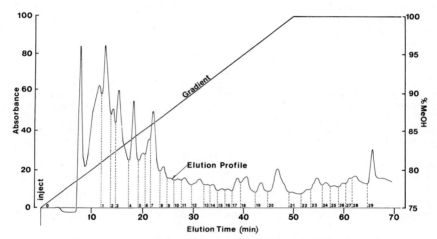

Figure 1.   HPLC profile of materials elaborated from cultures
of Nannochloris sp.   Fraction 7 is cytolytic toward P. brevis.
After Moon and Martin (19). Copyright 1981, Microbios Lett.

systems showed development of both hypocotyledon and cotyledons in 70% of the seeds tested. Total inhibition was demonstrated by fractions 0, 29, and crude extract series 25 μl through 100 μl.

Results and Discussion

Lettuce Seed Studies. Variations in final morphological development were noted when germinated test plants appeared different from control plants. Four orientation patterns were observed and include: a flattened appearance with dicotyledons upright but flat against the bottom of the weighing boat (Type 1), complete disorientation of the seedling where orientation appeared to be either horizontal, vertical, and/or inverted (Type 2), oriented in an inverted manner i.e., hypocotyledon up and dicotyledon down (Type 3) and the normal orientation with dicotyledons raised upwards at least 1 cm by the hypocotyledons from the filter paper (Type 4). These orientations are summarized in Table I, together with examples of fractions involved.

Table I. Behavior of Dark Green Boston Lettuce Seeds in the Presence of Algal Extracts ([19])

Behavior Type	Appearance	Example[a]
I	Dicotyledons upright, but flattened	5,6,8,13–19,21,24, 29
II	Seedling completely disoriented; orientation of dicotyledon could be horzontal, vertical and/or inverted. Considerable randomness noted.	1,4,7,9,10,12,25
III	Inverted. Hypocotyledon oriented upward, dicotyledon oriented downward.	2,3,11,20,26–28
IV	Normal orientation.	control

[a]Fraction numbers, Figure 1

A Comparative Evaluation of the Crude Extract and an Antibiotic. Inhibition and cytolytic studies of the polyene antibiotic Filipin (UpJohn Company of Kalamazoo, Michigan, U.S.A) assessed the cytolytic activity of an antibiotic with antifungal characteristics. Filipin was chosen as a test substance since it lacked attached sugar moieties (as did the crude extract) and exhibited fungicidal activity by binding to cell membrane sterols, destabilizing the cell membrane resulting in a lysed cell ([20]). The crude extract was observed microscopically to lyse cells also. Fungicidal testing demonstrated the inhibitory effect of both Filipin and the crude extract toward plated preparations of S. cerevisiae ([21]). Cytolytic bioassay preparations of the crude extract and Filipin confirmed cytolytic activity in both preparations. The cytolytic effect of Filipin was

initiated at 100–300 ppb with near total cytolysis at concentrations
of 1 to 10 ppm. The cytolytic activity of the crude extract was dem-
strated between 800 ppb and 1 ppm.

The inhibitory effects of in vitro sterol addition (21) showed
large percentage kills (83% and 91% respectively). The in vitro
addition of ergosterol (10.1 mM) to P. brevis cell cultures with
Filipin (1.5 mM) showed complete inhibition of the cytolytic effect.
Ergosterol (10.1 mM) added to P. brevis cell cultures with cell extract showed a 10% reduction in cell mortality. The addition of
ergosterol alone (control) showed no cytolytic effect at the experimental concentration (10.1 mM).

Mechanism of Cytolysis. These results are particularly significant
for understanding the mechanism of cytolysis. It is known that
aponin and ergosterol form an associated species (22) and that P.
brevis contains ergosterol (23). Interaction of aponin and
ergosterol (or a sterol closely related to ergosterol) should disrupt
the integrity of a P. brevis cell membrane and adversely affect the
membrane osmoregulatory capabilities.

The hypothesis is supported by two additional observations (24).
First, mean cell volume for P. brevis in the absence of aponin remained constant for 8 hours, but, in the presence of aponin, a
notable increase was observed within an hour and continued for eight
hours. Second, Trypan blue (CI 23850) tests indicated increased cell
permeability in the presence of aponin: viable, motile cells were
only slightly stained; swollen cells and cell debris were highly
stained.

Other Mechanism(s) of Red Tide Limitation. Cytolysis is not the only
mechanism of allelopathic activity of Nannochloris sp., and, indeed,
it may not be the best activity to focus attention on. The release
of toxins from P. brevis in confined volumes by aponin would hardly
engender an enthusiastic response from all ecologists. On the other
hand, we suspected that some fraction of the crude extract was
responsible for mitigating the ichthyotoxic activity of P. brevis in
the presence of aponin (11, 24). For example, Nannochloris sp. cells
mixed with P. brevis produced two effects, depending upon the relative concentrations (24). Relatively high concentrations of Nannochloris sp. resulted in cytolysis and cell debris, but relatively
low concentrations resulted in sessile or "resting-stage" P. brevis
cells that were small, thick-walled, and non-motile.

The latter behavior is significant because the P. brevis would
be rendered ineffective and ichthyotoxins would presumably not be
released from sessile cells. All fractions (Fig. 1) were investigated by Pabon and Martin (25). After initial screening of fractions
(Figure 1) for activity, it appeared that sessile-formation activity
was associated with first ten fractions (minus cytolytic fraction 7),
and further investigation indicated that fraction 4 was responsible
for induction of cytolysis (at 500 ppb). Clearly, additional studies
are in order, but the evidence of a mechanism that causes an additional disfunction of P. brevis cells is encouraging.

Field Evaluation Studies for Cytolytic/Fungicidal Substances (26).
The cytolytic effect of the crude extract on P. brevis prompted an

investigation to isolate and detect cytolytic compounds in the marine environment. Marine sediments were selected since this would be most likely to contain the greatest concentration of lipophilic substances in contrast to the evaluation of seawater. The study (26) involved the comparison of marine sediment extracts from fifteen stations along the West Coast of Florida. Comparative tests using the bio-assay technique on P. brevis showed that all sediment extracts were cytolytic toward P. brevis. Sediment extracts from shallow stations showed the highest amount of cytolytic activity when compared to deeper station locations. In a general way, cytolytic activity showed an inverse relationship with location for red tide "seed beds". Those areas that are thought to be the points of origin of red tide (27) were the locations that showed the least cytolytic activity in the sediments. The implications of this have been re-viewed elsewhere (8).

Fungicidal activity was determined by the disc method and zones of inhibition were recorded by measuring the diameter (mm) of the inhibition zone. Yeast cultures (S. cerevisiae) showed growth in-hibition (clear area surrounding disc) by sediment extracts from all stations when compared to control discs. HPLC analysis of sediment extracts showed more than 20 components in the migration profile of each station. Of these components, a fraction demonstrated to possess cytolytic activity in the crude extract (V_e = 24 ml) was present in all stations when compared to the migration profile of the active fraction (Figure 2).

Other Organisms Tested with Aponin. It is probably impossible to satisfy all persons interested in the effect of aponin or other organisms because the range of potentially affected organisms is so great. Nevertheless a number of organisms have been tested. We tested fish which were adapted to salt water (Poecilia sphenops) (11), and other fish have been tested at Mote Marine Laboratory (cf. 8); none were adversely affected by aponin or Nannochloris sp. As noted earlier, two marine algae have been tested: a Chrysomonad, Prymnesium parvum was unaffected by aponin (28), though a Japanese red tide organism Chattonella subsalsa (6) was cytolyzed by aponin. Brine shrimp, Artemia salina, were not adversely affected by Nannochloris sp. nor aponin (29). In addition, it has been reported clams and some other organisms were not adversely affected in studies done at Mote Marine Laboratory, Sarasota, FL (cf. 8). In summary, organisms representing a number of trophic levels have been studied and though there is always room for additional research in this area, results to date are promising.

Comparative Evaluation of Related Compounds. Structural studies are incomplete, but NMR and gas chromatographic — mass spectroscopic analysis of a component (fraction 7) isolated by HPLC indicated bi-phenyl characteristics (30). In an attempt to evaluate the cytolytic effect of compounds with similar structure, 4-hydroxybiphenol and 2-hydroxybiphenol were tested for lytic effects on P. brevis. Bio-assays of 2- and 4-hydroxybiphenol indicated that both compounds were cytolytic to P. brevis (31).

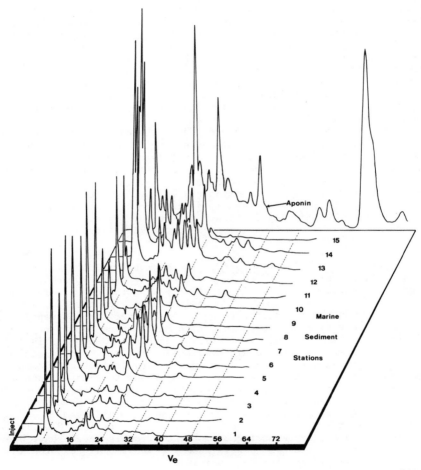

Figure 2. HPLC chromatograms (isocratic mode, 60% methanol, 40% water) of sediment extracts from 15 study sites in west Florida coastal waters. Migration profile are compared among sediment extracts and crude extract of <u>Nannochloris</u> sp. cell-free culture [See Moon and co-workers (<u>26</u>) for specific sites].

Summary

The results presented here indicate the existence of environmentally significant allelopathic substance or substances that affect a red tide organism in laboratory studies. It would be difficult to imagine that such substances do not have some impact in the natural environment, particularly in view of the observed (26) distribution of aponin vis-a-vis the presence of "seed beds" of red tide (8). The cytolytic agent termed aponin has been characterized in terms of biological activity, and the results to date indicate a material that does not adversely affect the organisms tested. Material isolated from cultures of Nannochloris sp. also has some phytotoxic activity as evidenced by the assays with lettuce seeds, and some antifungal activity. The materials elaborated by Nannochloris sp. thus have environmental significance, but it must be admitted that the full significance of these materials, like others, remains to be fully appreciated.

Acknowledgments

We gratefully acknowledge the support of the National Institute of Environmental Health Sciences (through grant number ES02810-03).

Literature Cited

1. Rounsefell, G. A.; Nelson, W. R. U.S. Fish Wildl. Serv. Spec. Sci. Rpt. Fisheries 1966, No. 535.
2. Martin, D. F.; Martin, B. B. J. Chem. Educ. 1976, 53, 614-617.
3. Steidinger, K. A. Prog. Physiol. Res. 1983, 2, 435-442.
4. Steidinger, K. A. Proc. 2nd Internat. Conf. Toxic Dino-flagellate Blooms 1979, p. 435-442.
5. Habas, E. J.; Gilbert, C. K. Environ. Letters 1974, 6, 134-147.
6. Halvorson, M. J.; Martin, D. F. J. Environ. Sci. Health 1981, A16, 373-379.
7. Steidinger, K. A.; Joyce, Jr., E. A. Fla. Dept. Nat. Resources, Educ. Ser. 1973, No. 17.
8. Martin, D. F. J. Environ. Sci. Health 1983, A18, 685-700.
9. Marvin, K. T.; Proctor, Jr., R. R. U.S. Fish Wildl. Serv. 1964, Data Rpt. 2.
10. Kutt, E. C.; Martin, D. F. Environ. Letters 1975, 9, 195-208.
11. McCoy, Jr., L. F.; Martin, D. F. Chemico-Biol. Interactions 1977, 17, 17-24.
12. Sakamoto, Y.; Krzanowski, J. J.; Lockey, R. F.; Martin, D. F. J. Environ. Sci. Health 1983, A18, 721-728.
13. Eng-Wilmot, D. L.; Martin, D. F. Microbios 1978, 19, 167-179.
14. Martin, D. F.; Gonzalez, M. G. Water Res. 1978, 12, 951-955.
15. McCoy, Jr., L. F.; Martin, D. F. J. Environ. Sci. Health 1978, A13, 517-525.
16. Eng-Wilmot, D. L.; Hitchcock, W. S.; Martin, D. F. Mar. Biol. 1977, 41, 71-77.
17. McCoy, Jr., L. F.; Eng-Wilmot, D. L.; Martin, D. F. J. Agri. Food Chem. 1979, 27, 69-74.
18. Eng-Wilmot, D. L.; Martin, D. F. Fla. Sci. 1977, 40, 193-197.
19. Moon, R. E.; Martin, D. F. Microbios Letters 1981, 18, 103-110.
20. Gottlieb, D.; Carter, H. E.; Lung-Chi, W.; Sloneker, J. H. Phytopathol. 1960, 50, 594-603.

21. Moon, R. E.; Martin, D. F.   Microbios Letters   1980, 10, 115–119.
22. Barltrop, J.; Martin, D. F.   Microbios Letters   in press.
23. Hamdy, A.; Jahoda, S. W.; Martin, D. F.   Microbios Letters   in press.
24. Eng-Wilmot, D. L.; Martin, D. F.   Microbios Letters   1981, 17, 109–116.
25. Pabon´de Majid, L.; Martin, D. F.   Microbios Letters   1983, 22, 59–65.
26. Moon, R. E.; Krumrei, T. N.; Martin, D. F.   Microbios Letters   1980, 14, 7–15.
27. Steidinger, K. A.   Crit. Rev. Microbiol.   1973, 3, 4
28. Moon, R. E.; Martin, D. F.   Bot. Mar.   1981, XXIV, 591–593.
29. Eng-Wilmot, D. L.; Martin, D. F.   J. Pharma. Sci.   1979, 963–966.
30. Moon, R. E.   Doctoral Dissertation, University of South Florida, Tampa, 1980.
31. Moon, R. E.; Martin, D. F.   J. Environ. Sci. Health   1981, A16, 373–379.

RECEIVED July 12, 1984

Naturally Occurring Substances That Inhibit the Growth of *Hydrilla verticillata*

PATRICIA M. DOORIS and DEAN F. MARTIN

Chemical and Environmental Management Services (CHEMS) Center, Department of Chemistry, University of South Florida, Tampa, FL 33620

Hydrilla verticillata (Royle), a rooted, submersed perennial exotic aquatic plant was introduced into Florida about 1959. Subsequently this plant has been spread to a number of areas of the United States. It can easily be a serious problem in a waterway or lake, owing to its ability to form dense mats and occupy the top meter of water thus shading out benthic native palnts. A search for natural control chemicals has revealed that in some lakes hydrilla, although present, does not thrive, owing to the presence of natural products that may be derived from cypress (Taxodium distichum) or other plants. Material that inhibits the growth of hydrilla has been obtained from extraction of cypress-derived lake sediment, and HPLC has been used to obtain a single component that is bio-active. The fraction inhibits the photosynthesis of hydrilla leaves in a Warburg apparatus, and the rate of respiration is accelerated. The inhibiting material has been characterized in terms of HPLC chromatograms, and it appears possible to "fingerprint" a body of water for hydrilla growth potential. Information available on structural features is reviewed.

Hydrilla verticillata (family Hydrocharitaceae) was introduced to the United States from Africa by way of South America (1). Since the 1960's, the plant has become a major nuisance in Florida, a state in which hydrilla does not suffer a winter die-back each year. For example, it is estimated that approximately 40,000 hectares of the State's waterways are directly affected by heavy hydrilla growth (2) necessitating the expenditure of millions of dollars annually for control and research. Costs are not only economic but environmental as well because hydrilla is able to compete successfully with native submerged plants and to displace them in some situations (3).

For many years, the primary means of hydrilla control has been the application of herbicides to affected waterbodies. Various

chemical agents have been routinely employed for over ten years by
agencies that have responsibility for weed control in order to reduce
hydrilla's impact on major lakes and streams. The long-term use of
herbicides has been recognized as producing material changes in
aquatic systems, some of which are particularly undesirable in shallow waterbodies which are located in temperate and sub-tropical
climates. This recognition has led to the search for more acceptable
control agents within the aquatic environment itself in the hope that
such agents would prove to be specifically inhibiting to the exotic
hydrilla while being innocuous to native plants and wildlife.   Efforts in this area have taken two primary directions.

The first involves capitalizing upon the weed-eating habits of
certain species of fish, eg. Ctenopharyngodon idella and Tilapia sp.,
both of which are, themselves, exotic species. Field studies and
laboratory experiments have been carried out to determine the value
of these and other fish species in hydrilla control (4). Some considerable success has been achieved in Florida with C. idella; however, some costs to the aquatic environment have recently become
apparent in the form of algal blooms and a suspected decline in the
sports fishery in lakes stocked with C. idella (5). Additional
studies have used hybrid carp (resulting from the cross of a female
C. idella and a male bighead carp, Hypophthalmichthys nobilis Rich.)
(6) The use of this fish limits worries about uncontrolled reproduction, but the effectiveness of the hybrid is controversial at present.

A second direction taken in the attempt to identiy a naturally-
occurring hydrilla control agent involves the observation that not
all water bodies to which hydrilla is introduced succumb to infestation by the plant. In particular, some lakes in Florida fail to support the plant's usually heavy growth (7,8). Years of observation in
Florida have revealed several lakes which fall into the category of
being "hydrilla non-supportive", and investigations of the physical,
biological, and chemical components of these lakes have led to the
conclusion that the major hydrilla-inhibitory mechanism is chemical
but one probably assisted by the processes of biological degradation
(8). Further work has identified a material present in lake waters
(derived from lake sediments in the laboratory) which substantially
reduces hydrilla growth in the laboratory.

Materials and Methods

Preparation of the Crude Inhibitory Solution. For laboratory experiments, the hydrilla-inhibiting solution is generated by extracting
highly organic lake sediment particles in distilled, deionized water
at a temperature of 121°C and a pressure of one atmosphere for 20-25
minutes. This process was chosen because it is believed to simulate
the natural, prolonged leaching of lake sediments while preventing
microbial action in the solution (9). Following extraction, the now
highly colored solution is separated from the sediments by filtration
through an 8μ filter. The filtrate is used as the base material in
growth studies, chemical characterizations, and further separations.

Demonstration of Hydrilla Inhibition. Initial laboratory studies of
approximately eight lakes confirmed the field observation that some
lakes do not promote heavy hydrilla growth (7,10). Expanded to

examine 34 lakes, these studies showed that the hydrilla-inhibitory tendency was uncommon but geographically widely distributed in Florida (11).

In this study, hydrilla, collected from area ponds and streams was grown under controlled conditions in closed systems in water collected from each of the lakes. Growth success was measured either by a change in plant wet weight and/or in the dissolved oxygen concentration of the lake water. Of all of the lakes studied, 14 showed a relatively low capacity to support hydrilla. Studies testing the inhibitory action of the sediment extract were done in the same manner, using the same experimental apparatus (12). In this work, however, the growth medium consisted of modified Hoagland's solution (13) with the experimental flasks being supplemented by addition of the crude sediment extract to achieve a final extract concentration of 0.02%. Plants grown in the presence of the sediment extract exhibited significant decreases in wet weight over a seven-day period, while plants grown in Hoagland's solution alone responded with vigorous growth. This response to the presence of the inhibitor appeared not to be common to all the Hydrocharitaceae, as similar studies with other species (eg. *Egeria*) failed to produce any evidence of growth reduction.

The basis for the inhibitory-action of the sediment extract has been sought in later studies using separated components of the crude extract (14,15).

Chemical Analyses of Aqueous Extracts. Subsequent to the demonstration of hydrilla growth inhibition by the crude sediment extract, basic information concerning select characteristics was obtained. The extracts, as prepared, were concentrated with respect to $C_{organic}$ and were highly colored (Table I). Passage of the extract over a chelating cation-exchange resin (Bio-Rad Chelex 100) did not change the character of the extract with respect to inorganic and organic carbon, dissolved iron, or color; however, similar treatment with an anion-exchange resin (Bio-Rad Cellex-A) removed over one-half of the $C_{organic}$ as well as considerable color. Treatment with Chelex resulted in no alteration in the hydrilla inhibitory action of the extract when tested in the usual manner, whereas treatment with Cellex produced a material lower in organic carbon and color and which possessed no inhibitory characteristics (16). In these studies, total and inorganic carbon was measured using a Beckman Model 915 total organic carbon analyzer; a Technicon AutoAnalyzer II was used for dissolved iron, inorganic phosphate, nitrate-nitrite nitrogen analyses. Protein was determined using a Folin-Lowry method, and color was determined using a Beckman DB-GT spectrophotometer.

Chromatographic Separations. Analyses were made using a Beckman (ALTEX) Model 110A liquid chromatograph equipped with a solvent programmer (Model 1601) and a multi-wavelength detector set at 254 nm. The DuPont preparative-scale Zorbax column (21.2 x 350 mm) used may have degraded or have been conditioned by repeated runs during the course of a year. Aqueous extracts (or natural water samples) were passed over a C-18 Bond-Elut cartridge (Analytichem International) before being injected. In all runs, a linear gradient was used during a 20 minute run starting with 60% water – 40% methanol and

ending with 100% water. Two fractions were obtained, usually one at
50% of the run (fraction 1) and a second very sharp peak at 83 ± 4%
of the programmed run (fraction 2).

Table I. Data for Selected Chemical and Physical Parameters for
          Crude Sediment Extracts Prepared from Two Sediments (18)[a]

Parameter	Extract I	Extract II
Fe(mg/l)	0.93	0.417
$C_{organic}$ (mg/l)	480.97	457.51
$C_{organic}$ (mg/l)	14.03	10.58
PO_4-P (μg/l)	320.8	465.81
NO_2-NO_3-N (μg/l)	23.72	28.79
Protein (mg/ml)	0.81	0.65
Color (absorbance at 400 nm)	0.22	N.A.

[a]Ext. I from Lake Starvation sediment, Hillsborough County, Florida;
Ext. II from Lake White Trout sediment, Hillsborough County, Florida.

Results and Discussion

Fractionation of the Sediment Extract. Ultrafiltration methods (17)
have produced some useful data about the nominal molecular size of the
active component. Crude sediment extract was filtered (Amicon, 50
psi) and separated into three fractions having the following nominal
molecular fractions: <10,000, <2,000, <500. The fractions were tested
for hydrilla-inhibition, and the two lower M.W. fractions proved to
be inactive against the plant. The fraction having a M.W. greater
than 2,000 but less than 10,000 showed inhibitory action somewhat enhanced over that of the crude extract (12).

Infrared Spectra. Limited effort has been made to identify the major
chemical group components of the extract. Infrared spectra (Perkin-
Elmer 913) of the evaporated extracts from two lakes were similar in
having peaks at the following frequencies: 3,000-3,100, 3,200-3,600,
1,500-1,650, and 1,690-1,760 cm^{-1}. These preliminary data would suggest the presence of polyphenolic compounds in the extract which may
be related to humic materials occurring in nature (16).

Chromatographic Separations and Implications. Fractionation of
hydrilla-inhibiting extracts has perhaps produced the most interesting results concerning the basis for hydrilla inhibition. Two
fractions were generated, and best results were obtained with an
older, preparative-scale Zorbax column. The effect of both fractions
on hydrilla was tested using Warburg apparatus: the first fraction
did not affect either plant respiration or photosynthesis. The

second fraction, however, reduced the photosynthetic activity of hydrilla leaves and increased respiratory activity relative to controls
(15). Work is currently underway investigating an organnelle-level
basis for the effect on photosynthesis. It is pertinent to note that
the same increased respiration effect noted with the second fraction
has been verified for a separately prepared crude extract and in
another laboratory (17). Moreover, the crude extract contains a
singlet-oxygen sensitizer that has an adverse effect on germination
of lettuce seeds (14). Thus, photodynamic action may be involved in
the adverse effect on hydrilla, but this point remains to be
verified.

Obtaining consistent chromatograms of aqueous extracts provides
a means of characterizing the presence of a hydrilla-growth inhibitor
in natural waters (19). Some examples are provided in Table II,
where the location of the second fraction is indicated as percentage
of the programmed run. Three different sediment samples were compared for Lake Starvation with consistent results. The fraction obtained for the extract of Lake White Trout is similar to the fraction
for Lake Starvation. Hydrilla disappeared from Lake Kerr (located in
Marion County, Florida, and the sediment obtained at the time yielded
an aqueous extract that appears to contain the hydrilla inhibitor.

Table II.   Chromatographic Characterization of Extracts for Hydrilla-
            Inhibitor Fraction

Sample	Peak Location, Percent of Programmed Run[a]
Lake Starvation Sediment Extract	
Sample I	84.4 ± 2.7
Sample II	82.1 ± 0.7
Sample III	82.3 ± 4.7
White Trout Lake Sediment Extract	87.7 ± 4.7
Lake Kerr Sediment Extract	86.3 ± 2.3
Hillsborough River water	75.5 ± 1.4

[a]Mean ± standard deviation

All aqueous extracts are brown colored, and so is the Hillsborough
River where samples of hydrilla are found. Examination of Table II,
however, suggests that the location of the second peak in this water
sample is significantly different from the other five samples. Presently, chromatograms of natural waters and the distribution of
hydrilla are being obtained and compared with a view to being able to
understand those factors that may limit the growth and spread of this
noxious aquatic plant.

It is clear that, at this point, the definitive chemical characterization of the inhibitory material has yet to be done. Much is
known concerning the distribution and the source of the material, and
work on the biological basis if the observed inhibition is promising.
However, the potential for the use of the material as an hydrilla-
control agent will depend upon the identification and purification of
the bio-active components detected in HPLC chromatograms.

Acknowledgments

We gratefully acknowledge the past support of our research by the
Southwest Florida Water Management District, The Florida Department
of Natural Resources (Bureau of Aquatic Plant Research and Control),
and the Aquatic Plant Control Research Program at the U.S. Engineer
Waterways Experiment Station.

Literature Cited

1. Andress, L. Proc. Fla. Dept. Nat. Resources Aquatic Plant Re-
    search Rev. Conf. 1977, Orlando, FL.
2. Gangstad, E. O. In "Weed Control Methods for Recreational
    Facilities Management"; Gangstad, E. O., Ed.; CRC Press, Boca
    Raton, FL 1982, pp. 77–94.
3. Haller, W. T.; Sutton, D. L. Hyacinth Contr. J. 1975, 13,
    48–50.
4. Osborne, J. A.; Sassic, N. M. J. Aquat. Plant Manage. 1979, 17,
    45–48.
5. Miller, W. J.; VanDyke, J.; Riley, D. Proc. Grass Carp. Conf.
    1979, Gainesville, FL, pp. 159–176.
6. Cassani, J. R.; Caton, W. E.; Hassen, Jr., T. H. J. Aquat.
    Plant Manage. 1982, 20, 30–32.
7. Martin, D. F.; Doig, III, M. T.; Millard, D. K. Hyacinth Contr.
    J. 1971, 9, 36–39.
8. Dooris, P. M.; Martin, D. F. In "Weed Control Methods for
    Recreational Facilities Management"; Ganstad, E. O., Ed.; CRC
    Press, Boca Raton, FL, 1982, pp. 61–75.
9. Martin, D. F.; Dooris, P. M. J. Environ. Sci. Health 1983,
    A18, 519–525.
10. Martin, D. F.; Victor, D. M.; Dooris, P. M. Water Res. 1976,
    10, 65–69.
11. Dooris, P. M.; Dooris, G. M.; Martin, D. F. Fla. Sci. 1982,
    45 (Suppl.), 30.
12. Dooris, P. M.; Martin, D. F. Water Res. Bull. 1980, 16, 112–
    117.
13. Stewart, K. K.; Elliston, R. A. Fla. Sci. 1973, 228–233.
14. Barltrop, J.; Martin, D. F. J. Environ. Sci. Health 1983, A18,
    29–36.
15. Barltrop, J.; Martin, B. B.; Martin, D. F. J. Aquat. Plant
    Manage. in press.
16. Dooris, P. M. Doctoral Dissertation, University of South
    Florida, Tampa, 1978.
17. Gjessing, E. Environ. Sci. Technol. 1970, 4, 437–438.
18. Dooris, G. M. personal communication 1984.
19. Martin, D. F.; Dooris, P. M.; Dooris, G. M.; Bova, Jr., R. J.
    Annual Meeting, Florida Section, ACS, Lakeland, FL, May 11, 1984.

RECEIVED June 12, 1984

Oxygenated Fatty Acids: A Class of Allelochemicals from Aquatic Plants

ROBERT T. VAN ALLER, GEORGE F. PESSONEY, VAN A. ROGERS,
EDWARD J. WATKINS, and HAROLD G. LEGGETT

Departments of Chemistry and Biology, University of Southern Mississippi, Hattiesburg, MS 39401

Allelochemic effects of aquatic macrophytes on algae
are discussed. Bioassays of chromatographic fractions
from Eleocharis microcarpa Torr. indicate that oxygenated fatty acids are the causative agents. Methods
of isolation of these materials from aquatic macrophytes and from natural waters are described. Purification and structure determinations show that prom-
inent components of the fraction are C_{20} tri-
hydroxycyclopentyl and C_{18} hydroxycyclopentenone
fatty acids. Similar components were extracted from
other aquatic plants. In addition, these components
were extracted from pond waters. Implications to algal
autoinhibition, algal succession in eutropic waters,
and control of algal diversity are discussed.

Ecologists for many years have been fascinated by the possible causes
and controls of phytoplankton diversity and seasonal succession.
Even from the early days, investigators have recognized that numerous
factors play a part and that pH, major nutrient ions, and physical
factors are important causes. See review articles by Hutchinson [1],
Dugelate [2] and Tilman, et al. [3]. Allelochemic effects between
algal species in eutropic systems have also been recognized as important factors [4-6]. Keating [6] studied algal succession in a
eutropic pond and concluded that dominant blue-green algae secreted
substances that inhibited predecessor species and stimulated succes-
sor species. Extracts of cultures were studied but no attempt was
made to identify chemicals. Earlier, Proctor [7] grew two member
algal cultures in each of several combinations from which he established a dominance pattern. He then steam-distilled a yellowish-
white substance from one of the cultures and concluded the substance
was a mixture of fatty acids. Spoehr [8] isolated "chlorellin" from
Chlorella sp. and concluded that the substance was a mixture of
photooxidized unsaturated fatty acids.
For several years, we have been interested in the effect that

0097–6156/85/0268–0387$06.00/0

higher aquatic plants exhibit in controlling plankton diversity in
eutropic systems. In one of the few studies of allelochemic interactions between higher plants and algae, Hasler and Jones in 1949 (9)
studied the effects Potamogeton foliosus and Anacharis canadensis on
phyto- and zooplankton and found blue-green algae to be inhibited by
these plants. In a study by Fleming in these laboratories (10), an
inverse correlation was made between the amount of coverage of Chara
vulgaris and numbers and diversity of blue-green algae in fish
hatchery ponds.

Our investigations of water quality in many commercial aquacul-
ture ponds (11) have shown that such systems are highly eutropic and
that most experience a dense bloom of blue-green algae during the
summer months. These blooms range from 2×10^5 TO 5×10^7
cells/l of usually one or sometimes two species of the genera Micro-
cystis, Oscillatoria, Anabaena, and Lyngbya. Very few rep-
resentatives from the other divisions of algae are evident during
these blooms. A few ponds are seen to have dense growths of the
higher plants Najas sp. and Eleocharis sp. These ponds are also
eutropic but characteristically contain fewer than 5×10^4 cells/l
that consist mostly of a normal diversity of green algae and diatoms
with blue-green algae present only in small numbers. Macronutrients
exist in both sediment and water column in sufficient quantities to
promote algal blooms. The nutrient concentrations in these ponds are
comparable to adjacent ponds with heavy algal blooms. Zooplankton
are generally present in these ponds in normal amounts.

Blue-green algae are notorious for causing oxygen depletion,
toxic fish kills, and odors in eutropic systems. They are also implicated in the widespread occurrence of Legionella pneumophila, the
causative organism of Legionaires' Disease (12). We began our investigations of the allelochemic effects of aquatic higher plants on
blue-green algae in the hope of identifying selective blue-green
algal inhibitors.

Preliminary Investigations

Agar-paper disk bioassay was used to test fractions from various
chromatographic separations. Twenty-four species of algae were
screened in order to select at least one that: would show good inhibition by a crude extract of E. microcarpa, could be maintained in
culture, and would produce results whithin a reasonable period of
time. Table I lists organisms used in preliminary bioassays.
Cultures were obtained from the Texas Collection of Algae at the
University of Texas, or isolated from local habitats. Cultures were
grown under standard laboratory conditions on algal growth media sol-
idified with agar. Both axenic and unialgal cultures were used.
Inhibition zones were noted from 7 to 14 days after inoculation. It
can be seen that the extract was effective in inhibiting blue-green
algae to a greater extent than other taxa. After initial screening,
Anabaena flos-aquae was selected to be used exclusively to assay for
biological activity of chromatographic fractions.

Separation and Structural Studies

A crude extract of E. microcarpa was prepared by boiling the fresh

Table I. Inhibition of Selected Algae

Cyanochloronta		Chlorophycophyta	
Anabaena catenula	++	Chlamydomonas eugametos	–
A. cylindrica	++	C. reinhardtii	–
A. flos-aquae	++	Chlorella sp.	–
Anabaena sp.	++	Chlorococcum hypnosporum	+
Anacystis nidulans	+	Scenedesmus quadricauda	–
Cylindrospermum sp.	++	Scenedesmus sp.	–
Lyngbya sp.	++	Stichococcus sp.	+
Nostoc muscorum	++	Ulothrix fimbriata	–
Oscillatoria tenuis	++	Oedogonium sp.	+
Phormidium sp	++		
Plectonema notatum	++		
		Chrysophycophyta	
Euglenophycophyta		Xanthophyceae	
		Tribonema sp.	+
Euglena gracilis	––	Bacillariophyceae	
		Nivicula sp.	+

(++) good inhibition, (+) detectable zone, (–) no inhibition

plant in water for one hour and filtering. After acidifying, active materials were extracted into chloroform and the excess chloroform removed by evaporation. The resulting crude extract was used in column and TLC separations. Figure 1 shows the results of column chromatography on silica gel followed by two consecutive preparative TLC systems. The chloroform fraction from column chromatography gave a slightly positive bioassay and contained simple saturated and unsaturated fatty acids by comparison of RRTs with authentic samples by GLC. The chloroform/acetone fraction contained most of the extracted material and activity. More polar fractions did not show activity and were therefore not investigated further. The oily chloroform/acetone fraction gave typical broad carboxylic IR absorption between 2850 cm^{-1} and 2350 cm^{-1} and at 1710 cm^{-1}. It was also soluble in dilute NaOH and $NaHCO_3$. Esterification eliminated the broad carboxylic absorption and produced a normal carbonyl ester absorption at 1730 cm^{-1}. A positive 2,4-DNP test and a strong IR absorption at 3400 cm^{-1}, which was unchanged by esterification, confirmed that this mixture contained keto and hydroxy fatty acids.

Preparative TLC (System I, 85:15:2 chloroform/methanol /water followed by System IV, 50:50:2 hexane/ethyl ether/formic acid) of the chloroform/acetone fraction produced 43 separate bands, 33 of which had definite activity. IR spectra of these active bands were strikingly similar, exibiting the same major features as noted for the unseparated chloroform/acetone fraction. UV spectra were also similar: absorption at 275 and 220 nm, with the latter being strongest.

Autoxidation of unsaturated fatty acids is well known. Modifications of the extraction and separation procedures were made to investigate the possibility that these oxygenated fatty acids (OFAs) arise as artefacts. When exposure to light and air were minimized, no changes were noted in TLC and HPLC.

The relatively large number and small amounts of these OFAs were surprising and made further structural studies difficult. Since fraction 4:6 had good activity and was well separated from other bands, it was used for further studies. Ozonolysis, low and high resolution MS, ^{1}H-NMR and ^{13}C-NMR data obtained on 4:6 allowed structure (I) (13-15) to be assigned.

Structure of I

This compound has not been previously reported in the literature. The compound is unusual in that the substitution pattern of alkyl and hydroxyl groups on the ring is analogous to the prostaglandin F series.

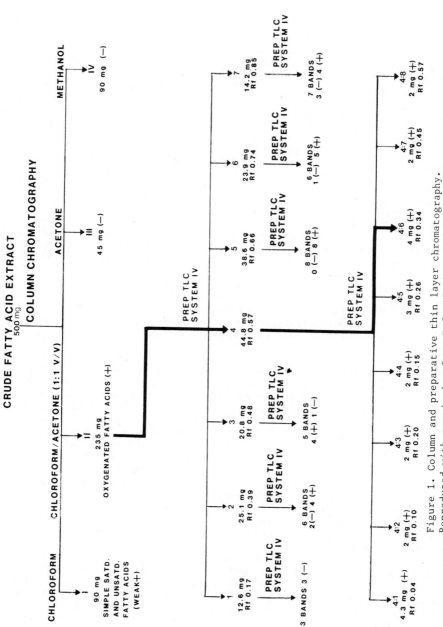

Figure 1. Column and preparative thin layer chromatography. Reproduced with permission from Ref. 15. Copyright 1983, The Hormel Institute.

Other fractions from preparative TLC were in smaller amounts and many were not distinctly separated from adjacent bands. It was obvious that preparative TLC would not produce sufficient purity and amounts for the identification of other components. Thus, a method of extracting prostaglandins from biological fluids with ODS silica was modified and reverse phase HPLC used to separate components (16). The aqueous extract from E. microcarpa reproduced essentially the chloroform/acetone column fraction by absorption onto ODS silica, followed by elution with methyl formate. This method gave substantially higher yields, i.e., 125-150 mg of OFAs from 2 kg of plant. Analytical RP-HPLC on C_{18} silica is shown in Figure 2. RRTs of PGE_1 and PGA_1 are shown by vertical lines. (I) elutes at about 28.5 min using a linear gradient of water/methanol starting with 2:1 v/v and ending with 100% methanol.

Elution of OFAs begins at 24 min and ends at about 36 min. Many more minor peaks can be detected by slowing the gradient. The peak at 31.5 min, component II, was chosen for further study because its polarity was similar to the prostaglandins and was a major component of the mixture. It corresponded to band 6:5 from preparative TLC.

Preparative HPLC of component II was difficult. Component II was collected with a resolution comparable to that shown in Figure 2. Higher resolution showed the presence of at least 5 minor peaks. Slowing the water/methanol gradient finally produced a single peak, which when collected, totaled 14 mg of slightly yellow oil from 2.1 gms OFAs extracted with ODS silica (16). Structural work on II has not yet appeared in the literature, so more detail is included.

The IR spectra of II and the methyl ester of II were very similar to the unseparated OFA fraction and the corresponding methyl esters as expected. The polymeric carboxylic acid absorption between 2800 cm^{-1} and 2350 cm^{-1} was eliminated by esterification. Strong absorption of II between 1715 cm^{-1} and 1690 cm^{-1} suggested the presence of a conjugated carbonyl in addition to the acid carbonyl. The methyl ester showed a strong doublet at 1735 cm^{-1} (ester carbonyl) and at 1705 cm^{-1} (conjugated five-membered ring carbonyl) (17). Absorption in the $3500 \text{ cm}^{-1} - 3200 \text{ cm}^{-1}$ region was changed by esterification but remained strong at 3400 cm^{-1}, indicating the presence of an alcohol group. In addition, there were olefinic absorptions at 3007 cm^{-1}, 1650 cm^{-1} (cis) and 1600 cm^{-1} (carbonyl conjugated). The UV spectrum showed a strong absorption at 222 nm, indicative of a carbonyl group conjugated with a double bond.

Electron and chemical ionization MS (isobutane) on II is shown in Figure 3. Comparison of these spectra shows two M^+ ions from EI at 306 and 308 and the corresponding 307 and 309 (M+1)$^+$ from CI. Relative peak heights gave an estimated 60-40 mixture of two very similar components. High resolution MS produced empirical formulas for ions above m/e 76. The ion at m/e 306.1820 (IIa) had a calculated formula of $C_{18}H_{26}O_4$ with 6 R+D, and the ion at m/e 308.1963 (IIb) had a calculated formula of $C_{18}H_{28}O_4$ with 5 R+D. The 2 proton difference was assumed to be the result of one less double bond in IIb. GC/MS of the methylated mixture produced peaks having M^+ ions of m/e 320 and m/e 322 corresponding to the esters of IIa and IIb. GC/MS of the hydrogenated methyl esters produced one M^+ of m/e 326 showing IIa and IIb were converted to the same hydrogenated ester. The assumption of a double bond difference in IIa

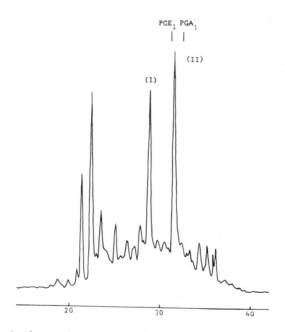

Figure 2. Analytical reversed phase HPLC of methyl formate fraction.
Reproduced with permission from Ref. 15. Coyright 1983, The Hormel
Institute.

a. Electron ionization

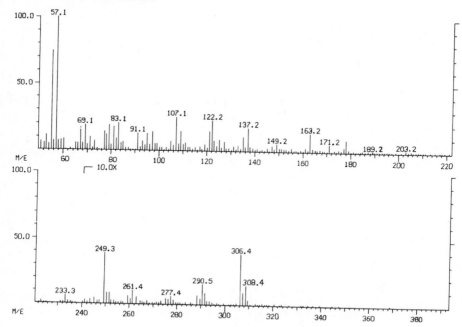

b. Chemical ionization

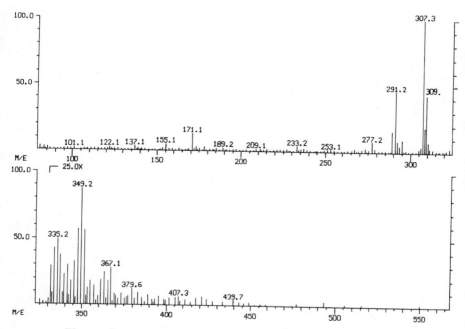

Figure 3. Low Resolution Mass Spectra of Component II

and IIb was reasonable, considering the difficulty in separation; thus MS fragmentation patterns of the free acids, the methyl esters, and the hydrogenated methyl ester can then be interpreted. Figure 4 shows proposed structures and fragmentation patterns (* from methyl esters, ** from the hydrogenated methyl ester).

The ion m/e 171, $C_9H_{15}O_3$ (2 R+D) and those corresponding to the methyl esters and hydrogenated ester derivatives and the unbroken chain of dioxy ions to $C_4H_3O_2$ establish the positions of the hydroxyl group and the double bond. The other half of the compounds, corresponding to m/e 135 and m/e 137, had emperical formulas of $C_9H_{11}O$ and $C_9H_{13}O$ with 4 and 3 R+D respectively, suggesting the possibility of a cyclic structure. The ion m/e 81, C_5H_5O, 3 R+D is consistent with a cyclopentenone structure. An m/e 81 was also seen in the MS of the methyl esters and an m/e 83 from the hydrogenated methyl ester. Ions at m/e 97, C_6H_9O, 2 R+D and m/e 99, $C_6H_{11}O$ indicate the C_4 ring substituents in IIa and IIb are located alpha to the carbonyl group.

Carbon thirteen NMR of the mixture of IIa and IIb was complex but correlated with results from IR, UV and MS. The attached proton test was useful in making the assignments shown in Table II. Assignments confirm that the oxygen on the cyclopentenyl ring is a ketone. Also, the absorption at 13.7 ppm (methyl, to a double bond) confirms the location of the double bond in the C_4 group attached to the ring to be between carbons 15 and 16. The additional two absorptions at 29.3 and 29.7 ppm (ATP up) when compared with MS data indicate one of the molecules to be saturated in the C_4 group. It is of interest that a <u>cis</u> conformation with respect to the ring substituents can be assigned to both compounds on the basis of one strong absorbance at 44.2 ppm (ATP down) for carbon 10; if <u>trans</u>, the chemical shift for carbon 10 would occur downfield around 47 ppm. Chemical shifts for carbon 14 were hidden by solvent (CH_3OH-d) in the 50-52 ppm region. The <u>cis</u> assignment agrees with that of Vick and Zimmerman (<u>18</u>), who assigned the <u>cis</u> conformation to the cyclopentenone fatty acids isolated by incubating trienoic acids with a flaxseed enzyme extract. IIa and IIb have much structural similarity to these enzymatically produced fatty acids, but differ from the A series prostaglandins which have the <u>trans</u> conformation of substituents. Bioassay showed the mixture of IIa and IIb to be active against <u>A. flos-aquae</u>.

Occurance of OFAs in Other Aquatic Plants and Natural Waters

HPLC of other plant genera by ODS silica extraction produced complex chromatograms similar to that from <u>E. microcarpa</u>. Pond waters from various sources were also screened for OFAs. These chromatograms were less complex than those from the plants and, notably, showed mostly compound I, and component II. Table III shows the occurrence of these compounds in plants and pond waters.

OFAs from the first four, higher plants, were extracted from plants collected in the late summer of 1983. It should be noted here that yields of OFAs from <u>E. microcarpa</u> were higher during the early Summer months when maximum growth was occurring.

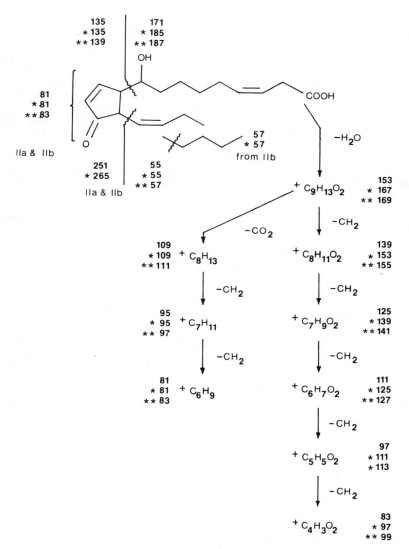

Figure 4. Low and high resolution mass spectral fragmentation patterns of Component II.

Table II. Carbon Thirteen NMR Correlation

IIa

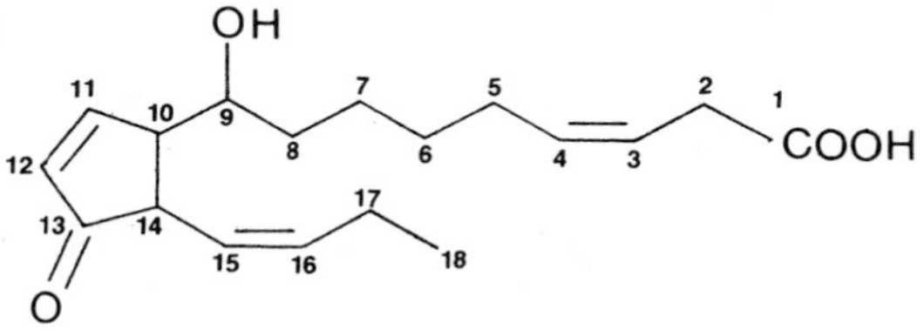

IIb

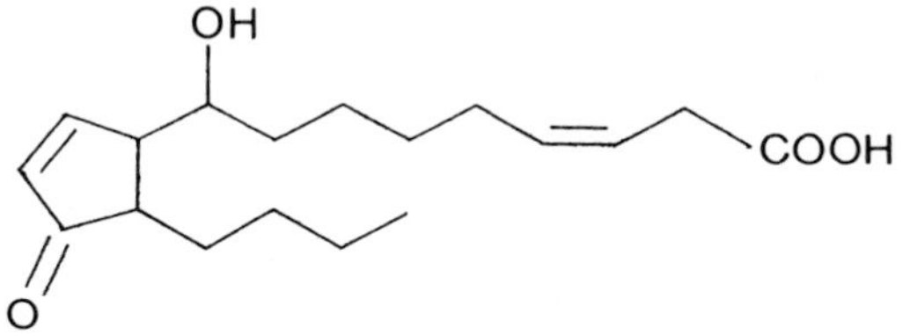

Carbon
No.   Chemical Shifts   Attached Proton Test

	IIa	IIb	IIa	IIb
1	176.5	176.5	UP	UP
2	36.9	36.9	UP	UP
3	132.0	131.8	DOWN	DOWN
4	138.2	138.1	DOWN	DOWN
5	33.8	34.0	UP	UP
6	29.4	29.2	UP	UP
7	28.8	28.7	UP	UP
8	40.8	40.4	UP	UP
9	73.1	71.7	DOWN	DOWN
10	44.2	44.2	DOWN	DOWN
11	129.1	128.9	DOWN	DOWN
12	142.5	142.4	DOWN	DOWN
13	208.2	207.9	UP	UP
15	141.0	29.3	DOWN	UP
16	130.0	29.7	DOWN	UP
17	24.7	24.0	UP	UP
18	13.7	13.9	DOWN	DOWN

Table III.   OFA Occurrance by HPLC

Source	I	II
Potamogeton sp. (Pond Weed)		++
Najas sp.	+	+
Thalassia sp. (Turtle Grass)	++	
Ruppia sp. (Widgeon Grass)	+	++
Chara sp.	+	
Oscillatoria sp.[1]	+	
Chlorella sp.[1]	+	
Pond A		++
Pond B	+	
Pond C	+	++
Pond D	++	++

Legend: Pond A — normal diversity of phytoplankton
          Pond B — Microcystis sp. bloom
          Pond C — very little phytoplankton
          Pond D — dense growth E. microcarpa
          1 — laboratory cultures
          (++) major peak, (+) minor peak

When Winter senescent plant material was harvested and   extracted,
yields of OFAs were greatly reduced. OFAs from the four ponds were
extracted easily from 20-50 liters of pond water by passing the water
through a tube filled with ODS silica. These ponds produced between
1 and 1.5 ppm of OFAs after elution with methyl formate. The origin
of I, IIa and IIb in these ponds is uncertain at this time.

Discussion

OFAs in E. microcarpa and other higher aquatic plants, as well as in
waters where these plants grow, appear to exert definite effects on
the phytoplankton community in eutropic systems by showing selective
inhibition against blue-green algae. The large number of these metabolites is surprising, but perhaps should be expected in view of the
variety of chain-lengths and unsaturated centers found in fatty acid
constituents of aquatic plants (19). At least one enzyme system from
a higher plant, flax, has the ability to convert C_{18}, C_{20}, and
C_{22} trienoic fatty acids to cyclopentenone fatty acids (17).
      Recent work in these laboratories indicates that small amounts
of the OFA mixture from E. microcarpa, less than 1 ppm, stimulate
many species that are inhibited at higher concentrations. Also, individual components of the mixture may inhibit, have no effect on, or
stimulate different algae.
      The presence of I and smaller amounts of other OFAs in algae
(Table III) is interesting in view of autoinhibition of algae cultures (7, 20, 21). Autoinhibition occurs as cultures begin to age at

high numbers of cells/liter, presumably because of cellular metabolites being released into the medium. Pratt showed that autoinhibitors from Chlorella vulgaris also inhibited C. pyrenoidosa and other algae (22). He also found, as we did with OFAs, that low concentrations of the autoinhibitors stimulated growth (23). Spoehr studied the antibiotic chlorellin produced by C. pyrenoidosa, the same substance studied by Pratt (24), and found evidence that the material was a complex mixture of oxidized unsaturated fatty acids (8). He concluded that the mixture resulted from photooxidation, since inhibition of Staphylococcus aureus was increased by exposing the inhibitors to air and sunlight. It seems possible, however, since OFAs were detected from Chlorella sp. in these laboratories, that his inhibitors were similar to ours when isolated but became more oxidized on treatment with air and sunlight. This possibility is being investigated.

As stated earlier, the effects of individual OFAs on the growth of algae are complex. Liquid chromatograms of plant extracts, including the three algae, while having the same general appearance, have components in differing amounts as well as components with different retention times. These differing patterns of constituents may relate to algal succession in eutropic systems. Boyd, in a study involving two membered cultures of blue-green and green algae, found that many green algae were inhibited by blue-green algae but that none of the blue-green algae were inhibited by green algae. He also found one green alga to be stimulated by a blue-green alga. He found inconsistent patterns, however, when filtrates of water from bloom ponds were used in growth media and attributed pattern differences to extracellular substances from genera other than the dominant bloom alga (25).

It remains to be demonstrated that OFAs are extracellular metabolites of algae and that they are prominant factors among the many determinants of succession and dominance patterns of phytoplankton. It appears that these compounds are worthy of further structural study, and that the effects on individual algae show promise of providing insight into the complex interactions in aquatic ecosystems.

Summary

A relatively large number of OFAs have been detected in several aquatic plants in small amounts. Bioassays of the OFA mixture against various algae showed them to have good specificity against blue-green algae. Three principal components of the group were identified from E. microcarpa and were also indicated in pond water inhabited by this plant. Bioassay of these components indicate a similar specificity against blue-green algae. Further work is necessary before the overall effects of individual OFAs on dominance and inhibition patterns can be determined with reasonable certainty.

Literature Cited

1. Hutchenson, G. E. Am. Nat. 1961, 95, 137-45.
2. Dugelate, R. C. Limnol. Oceanogr. 1967, 12, 685-95.
3. Tilman, D.; Kilham, S. S.; Kilham, P. Ann. Rev. Ecol. Systems 1982, 13, 349-72.

4. Whittaker, R. H.; Feeny, P. P. Science 1971, 171, 757-70.
5. Putnam, A. R. Chem. Eng. News 1983, 61(14), 34-35.
6. Keating, K. I. Science 1977, 196, 885-87.
7. Proctor, V. W. Limnol. Oceanog. 1957, 2, 125-29.
8. Spoehr, V. W. "Fatty Acid Antibacterials From Plants"; Carnegie
    Institution, Washington, D.C., Publication No. 586, 1949; p. 24.
9. Hasler, A. D.; Jones, E. Ecology 1949, 30, 359-64.
10. Fleming, A. N. M.S. Thesis, University of Southern Mississippi,
    Hattiesburg, Mississippi, 1973.
11. "Experimental Treatment of Catfish Ponds with Algal Inhibitors,"
    National Marine and Fisheries Service, 1983.
12. Tison, D. L.; Pope, D. H.; Cherry, W. B.; Fliermans, C. B.
    Journal of Applied Environmental Microbiology 1980, 39, 456-
    459.
13. Clark, L. R. Ph. D. Thesis, University of Southern Mississippi,
    Hattiesburg, Mississippi, 1980.
14. Rogers, V. A. M.S. Thesis, University of Southern Mississippi,
    Hattiesburg, Mississippi, 1983.
15. van Aller, R. T.; Clark, L. R.; Pessoney, G. F.; Rogers, V. A.
    Lipids 1983, 18, 617-22.
16. Rogers, V. A.; van Aller, R. T.; Pessoney, G. F.; Watkins,
    E. J.; Leggett, H. G. Lipids 1984, 19, 303-305.
17. Zimmerman, D. C.; Feng, P. Lipids 1978, 13, 313-16.
18. Vick, B. A.; Zimmerman, D. C. Lipids 1979, 14, 734-40.
19. Shorland, F. B. In "Chemical Plant Taxonomy"; Swain, T., Ed.;
    Academic Press: New York, 1963; pp. 255-57.
20. Pratt, R. Amer. J. Bot. 1940, 27, 52-56.
21. Pratt, R. Amer. J. Bot. 1940, 27, 431-36
22. Pratt, R. Amer. J. Bot. 1943, 30, 404-08.
23. Pratt, R. Amer. J. Bot. 1942, 29, 32-33.
24. Pratt, R.; Spoehr, H. A. Science 1944, 99, 351-2.
25. Boyd, C. E. Weed Science 1973, 21, 27-37.

RECEIVED August 14, 1984

Spikerush (*Eleocharis* spp.): A Source of Allelopathics for the Control of Undesirable Aquatic Plants

F. M. ASHTON[1], J. M. DI TOMASO[1], and L. W. J. ANDERSON[2]

[1] University of California, Department of Botany, Davis, CA 95616
[2] U.S. Department of Agriculture, Aquatic Weed Research Laboratory, Davis, CA 95616

Axenic cultures of dwarf spikerush (Eleocharis coloradoensis) were established in 4 L aspirator bottles containing quartz sand and a synthetic culture medium. These were periodically drained and the effluent subjected to fractionation and bioassays. This crude leachate was passed through a C_{18} cartridge to separate polar from nonpolar compounds. The nonpolar fraction was eluted from the cartridge with acetone and the solvent evaporated with N_2 gas. The polar fraction was lyophilized. Both fractions were subjected to exclusion chromatography using G-15 Sephadex for further fractionation. Four bioassays were developed to determine the biological activity of the several fractions. Two assays used explants, excised parts, of the target species hydrilla (Hydrilla verticillata) and sago pondweed (Potamogeton pectinatus) The other two assays used tomato cell cultures or lettuce seedling root growth. The most active allelopathic fraction was peak 1B of the polar portion. The molecular weight of the phytotoxic compound(s) appears to be between 600 and 1000. It will be further purified using TLC and characterized by GC-MS, NMR, and IR. A review of the literature on the allelopathic aspects of Eleocharis spp. is also presented.

Spikerush (Eleocharis spp.) is primarily aquatic in nature and includes a few species that are competitive with certain aquatic species. These competitive Eleocharis species are low growing and do not interfere with water management in reservoirs, irrigation canals, or drainage ditches. However, their competitive nature appears to reduce the growth of troublesome tall-growing aquatic weeds such as Potamogeton spp., Elodea spp., Hydrilla verticillata (L.f.) Royle, and certain algae. The competitive nature of spikerush on higher plants was reported by Oborn et al. (1) in 1954. In 1970, Yeo and Fisher (2) reported on their research with Eleocharis relative to its rate of migration, methods of establishing stands,

0097–6156/85/0268–0401$06.00/0

growth habits, and degree of controlling submersed aquatic weeds.
Subsequently, there has been additional research on the biology of
Eleocharis and its application as an aquatic weed control agent.
Since these aspects are not within the theme of this symposium, they
are not given further consideration.

Research on the mechanism of the competitive influence of
spikerush on other aquatic plants suggests a chemical phenomenon,
allelopathy. Frank and Dechoretz (3) found that the production of
new shoots of American pondweed (Potamogeton nodosus Poir.) and sago
pondweed (P. pectinatus L.) was reduced significantly when winterbuds and tubers of these species were planted in dwarf spikerush
[Eleocharis coloradoensis (Britt.) Gilly] sod. Although these experiments did not demonstrate allelopathy, they encouraged these
investigators (3) to conduct a definitive experiment. In this
experiment, dwarf spikerush was planted in one container and allowed
to become established for 3 months. At this time three winter-buds
of American pondweed and three tubers of sago pondweed were planted
in separate areas of a second container. These two containers were
connected by plastic tubing and an inserted needle valve controlled
the gravity flow of the leachate from the dwarf spikerush container
into the pondweed container at a rate of 0.5 L per day. After 16
weeks dwarf spikerush reduced the number of newly formed sago
pondweed plants by about three-fold (Figure 1). Similar results
were obtained with American pondweed, but the inhibition was only
about two-fold. Fresh weight of both species were reduced by dwarf
spikerush and both were found to be chlorotic. Additional studies
by these investigators (3) and others (4) have confirmed the
allelopathic effects of dwarf spikerush.

Stevens and Merrill (5) isolated, identified, and synthesized
the phytotoxic compound dihydroactinidiolide (DAD) from dwarf spikerush (Figure 2). Mowed vegetation was dried and threshed to remove
seeds. The plant material was steeped for 1 week at room temperature in 40% aqueous ethanol. The ethanol was removed in vacuo at
50 C and liquid-liquid extraction of the aqueous residue with ethyl
acetate yielded a dark-brown oil. The ethyl acetate extract was
chromatographed on silica gel and eluted with chloroform to give
material showing phytotoxic activity. This material was further
purified by fractional crystallization, preparative TLC, rechroma-
tography on TLC plates, crystallization, and GC. Mass spectral
analysis suggested that the phytotoxic compound was dihydroactinidiolide. This identification was confirmed by its NMR, IR, and UV
spectra. The compound was later synthesized. The active fractions
were identified during the purification by measuring root growth of
germinating watercress (Nasturtium officinale R. Br. in Ait.) seeds.
Pure dihydroactinidiolide reduced root length of watercress approximately 30% between 10 and 20 ppm. It also inhibited germination of
radish (Raphanus sativus L.) 50% at 5 ppm and no germination
occurred at >50 ppm.

Previous reports (6,7) suggested that Eleocharis microcarpa
Torr. may be a natural algal inhibitor. Keating (8) indicated that
certain fatty acids may have significance to algal succession. A

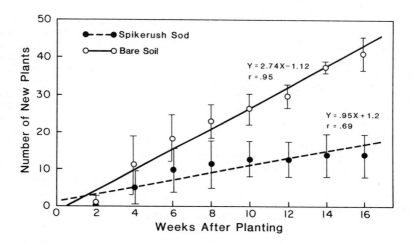

Figure 1. Production of sago pondweed in tubs supplied with water from tubs containing spikerush sod and bare soil. Reproduced with permission from Ref. 3. Copyright 1980, Weed Science Society of America.

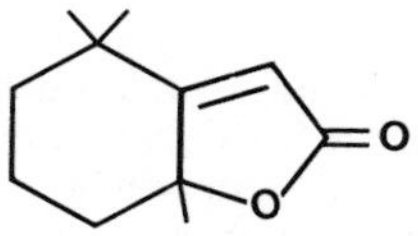

Figure 2. Structure of dihydroactinidiolide, an allelopathic compound isolated from <u>Eleocharis</u> <u>coloradoensis</u>.

C_{20} cyclic trihydroxy unsaturated fatty acid was isolated and characterized as a representative member of a group of oxygenated fatty acids from E. microcarpa (9). The compound was characterized as 11-hydroxy-14-(3,5-dihyroxy-2-methycyclopentyl)-tetradec-9-ene-12-yneoic acid (I). It inhibited the growth of the blue-green alga Anabaena flos-aquae (Lyngb.) DeBrebisson. This organism was used as a bioassay to follow the isolation and final purification. In addition to (I) above, 33 TLC bands were biologically active. In a very recent paper (10), these same investigators described a rapid method of isolating a relatively pure fraction of oxygenated fatty acids from plants and natural waters. However, in contrast to the low-growing species, dwarf spikerush and slender spikerush (E. acicularis), E. microcarpa is a tall-growing species and therefore would not be a suitable replacement for tall-growing aquatic weeds. Details of the research with E. microcarpa are presented by VanAllen in another paper of this symposium.

Initial investigations (3) with dwarf spikerush were carried out in open non-sterile containers. Though results suggested a phytotoxic response, it is difficult to determine whether this effect was due to associated microorganisms or spikerush leachate. To alleviate this problem, Wach (11) established axenic spikerush cultures in terraria containing an autoclaved soil mixture. The effluent from terraria with established spikerush cultures could be compared to sterile control terraria. Under these conditions, the occurrence of novel substances in spikerush leachate could be directly attributed to the presence of spikerush. However, the organic compounds obtained from the soil mixture caused difficulty in isolating specific compounds derived from spikerush. In the research reported here, we used quartz sand rather than the soil mixture as a substrate to eliminate extraneous organic compounds from the soil mixture.

Materials and Methods

Establishment of Terraria. Dwarf spikerush tubers were sterilized in 10% hydrogen peroxide for 20 min and transferred to petri dishes containing 3% sucrose, 4.3 g/L Murashige and Skoog salts (12) and 0.8% agar. Tubers were grown in petri dishes for three weeks under continuous fluorescent light at approximately 100 $Ei/m^2.s$.

Terraria consisted of 4 L Pyrex aspirator bottles fitted with a short piece of Tygon tubing over the glass nipple at the bottom. Each terrarium contained 650 g washed quartz sand and 500 ml deionized water with 3% sucrose and 2.2 g Murashige and Skoog salts. The bottles were closed at the top with a rubber stopper fitted with cotton plugged 'N' shaped glass tubes to allow for passage of gasses. Terraria were autoclaved at 5 lbs/in², 110 C for 1 h.

Three-week-old spikerush seedlings were asceptically introduced into cooled terraria and placed under continuous fluorescent light at 100 $Ei/m^2.s$. Temperatures ranged from 22 to 28 C. After two months of undisturbed growth, a thick sod of spikerush had developed from rhizomatous daughter plant production. Each terrarium was drained through the bottom outlet tube, the medium discarded and replaced with fresh sterile medium containing 500 ml deionized water and 2.2 g Murashige and Skoog salts. At two month intervals, for approximately eight months, terraria were drained and replaced with fresh Murashige and Skoog salt medium. Effluent was saved for further fractionation and bioassays.

Leachate Fractionation. Crude leachate was filtered through C_{18} cartridge (Sep-Pak) in order to separate polar from nonpolar compounds. Nonpolar fraction was eluted from C_{18} cartridge with acetone, evaporated under nitrogen gas and stored at -20 C. Polar leachate was lyophilized and stored at -20 C.

Both polar and nonpolar extracts were further fractionated by exclusion chromatography on a K50 Pharmacia column containing G-15 Sephadex gel. The effluent, eluded with degassed distilled water, passed through a Hitachi 034 UV-Vis Effluent Monitor. Samples absorbing at a wavelength of 220 nm were collected on a Gilson Micro-Fractionator at 1.5 min intervals. The volume of each sample was approximately 7 ml. Samples corresponding to each peak were combined, lyophilized, and stored at -20 C.

Bioassays. Four bioassays were developed to test the phytotoxic affect of spikerush leachate. Two assays involving target plants (hydrilla and sago pondweed) were tested at the explant level. Two additional assays were developed, one involving whole plants [lettuce (Lactuca sativa L.) seedlings], and the other cultured cells [tomato (Lycopersicon esculentum Mill.) cell suspension].

In the first bioassay, two-node sections of hydrilla stems were allowed to initiate axillary shoots in 1% Hoaglands solution (13). Ten shoot-forming explants were transferred to 125 ml erlenmeyer flasks containing 50 ml deionized water and 500 or 250 ppmw crude polar extract (5 reps/treatment). The flasks were placed under continuous light at 25 C for seven days. New shoot length, adventitious root number, and root length were recorded.

In a second assay, shoots attached to sago pondweed tubers were excised at the base of the tubers. Ten shoots were placed in each 125 ml erlenmeyer flask containing 50 ml deionized water and 200 or 400 ppmw of peak 4 (5 reps/treatment). The flasks were placed in a growth chamber on an orbital shaker at 100 rpm and subjected to alternating 12 h light and dark periods at 27 C. Following one week of treatment; shoot length, number of rooting shoots, number of roots per shoot, and length of roots were measured.

Tomato cell suspensions in modified Murashige and Skoog medium (12) were assayed for change in cell volume at varying concentrations of polar and nonpolar extracts. Twenty-five ml erlenmeyer flasks (5 reps/treatment) containing 10 ml of sterile medium, tomato cell suspension, and extract were allowed to settle for 20 min in sidearm test tube attachments. Relative height of cell volume fraction to medium was recorded each day for 11 days. Cell suspensions were incubated in the dark at 27 C on an orbital shaker at 125 rpm. The doubling time of control cell suspensions were approximately 2.25 days.

In a fourth experiment, lettuce seeds (var. 'greenhart') were assayed for root growth inhibition after 3 days of exposure to varying concentrations of polar and nonpolar exclusion chromatographic peaks. Twenty-five lettuce seeds were placed on Whatman #1 filter paper in 15 x 60 mm petri dishes. Two milliliters of deionized water containing nonpolar or polar extract were added to each dish (4 reps/treatment). Petri dishes were placed in a growth chamber at 25 C and alternating 12 h periods of light and dark. Root length, percent germination, and obvious symptoms were recorded.

Results and Discussion

Fractionations. Following lyophilization of the polar leachate fractions, separations by exclusion chromatography suggest three major molecular weight regions, labelled peak 1A, peak 1B and peak 2 (Figure 3). Retention times on a G-15 Sephadex gel indicate molecular weights of peaks 1A and 1B compounds to be between 600 and 1000.

Medium from chromatographed control terrarium suggests that peak 2 is represented by organic components of Murashige and Skoog salts. A subsequent run of aqueous Murashige and Skoog salts also demonstrated a similar curve. Nevertheless, peak 2 was collected, lyophilized, and tested in each bioassay in order to determine whether additional phytotoxic compounds were present.

The brownish, oily nonpolar fractions was chromatographed on the same column as was the polar leachate. One distinct peak, labelled peak A, was observed (Figure 4). An additional peak comprised of several smaller peaks was collected and labelled peak B. Estimates of molecular weight suggest that peak A and B are also between 600 and 1000.

Bioassays. Two-node hydrilla explants were treated with either 250 or 500 ppmw crude polar spikerush extract. One week after treatment, axillary shoot length at both 250 and 500 ppmw exhibited no significant difference from that of control (Table I). In each treatment, shoot lengths increased by approximately 500%. However, the mature stem sections in both treated samples appeared chlorotic and flaccid, often with leaves senescing. When developing axillary shoots were excised from two-node stem sections and subject to 0, 250, and 500 ppmw crude polar spikerush extract, virtually no growth was observed in either the treated or the control samples. Presumably the growth and elongation of the developing shoots depend on compounds produced by mature stem sections.

Though the number of shoots developing adventitious roots was not significantly different between treated samples and control, the average length of adventitious roots declined significantly at 500 ppmw. At this concentration, roots appeared somewhat brownish while the tips were slightly swollen.

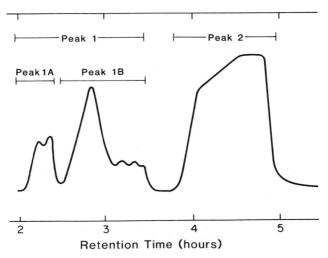

Figure 3. Exclusion chromatography of polar leachate through G-15 Sephadex gel. Retention time at flow rate of 4.8 ml/min in K50 Pharmacia column.

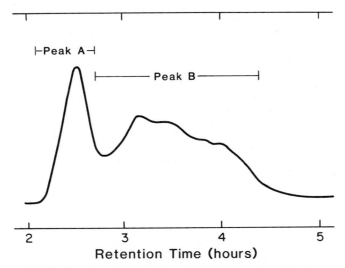

Figure 4. Exclusion chromatography of nonpolar leachate through G-15 Sephadex gel. Retention time at flow rate of 4.8 ml/min in K50 Pharmacia column.

Table I. Hydrilla Root Growth Assay Using Two-node Explants Treated
          with Crude Polar Spikerush Extract

	Treatment (N=40)		
	control (mm)	250 ppmw (mm)	500 ppmw (mm)
Initial axillary shoot	3.0	3.3	3.6
Axillary shoot, one week	14.9(100)[a]	18.7(125)	17.5(117)[b]
Adventitious roots	35.7(100)	32.7(92)	12.7(36)[b]
Shoots developing adventitious roots	58%(100)	59%(102)	55%(95)

[a]Percent of control in parenthesis.
[b]Only value significantly different from the control at the 5% level
according to Duncan's multiple range test.

     Excised sago pondweed shoots treated with 1000 ppmw crude polar
spikerush extract also displayed no significant reduction in length
from that of control samples with one week exposure.   However, root
development was considerably influenced by the treatment (Table II).
The number of shoots forming adventitious roots declined by 86%,
whereas the average number of adventitious roots per rooting shoot,
And the average length of adventitious roots were 33% and 40% of
control, repsectively.

     When peak 1 (combination of 1A and 1B) of the polar extract was
tested against sago pondweed shoots, the number of shoots producing
adventitious roots was reduced to 38% of control.   Neither average
root length, nor the number of adventitious roots per root-
producing-shoot were significantly reduced. It is possible that the
addition of a high concentration of non-sterile organic compounds
(1000 ppm crude polar extract) to the medium may affect root growth
by microbial contamination.

Table II. Sago Pondweed Excised Shoot Assay After One Week Exposure to Polar Spikerush Extracts

	1000 ppmw crude polar extract	peak 1 400 ppmw	peak 1 200 ppmw
	Treatment (N=40)		
	% control		
Shoot length	99	117	101
Shoots developing adventitious roots	14[a]	38[a]	38[a]
Number of adventitious roots per rooting shoot	33[a]	70	89
Root length	49[a]	124	103

[a]Value significantly different from control at the 5% level according to Duncan's multiple range test.

Nonpolar extracts were not tested in either the hydrilla explant or the sago shoot assay due to the limited supply of material.

Cultured cell assays are widely used in investigations designed to measure phytoxicity of herbicides (14,15). Its application has many advantages over the explant assays previously discussed (11). Experiments can be designed so that only milligram amounts of test material are required. In addition, cell populations are relatively homogeneous, translocation of test substance is not a problem since the cells are surrounded by the medium, and potential problems with microorganism interactions are eliminated under axenic conditions of cell suspensions. However, there are three disadvantages to this approach. First, all manipulations of cells and the addition of test substances must be aseptic. Second, non-differentiated cell suspensions may be insensitive to toxins which only manifest their actions on differentiated cell types or developmental processes. Thus, there exists the possibility that a population of cultured cells and the plant species from which they were cultured may not have the same physiological response to an applied toxin (16). Finally, the species of the tested cell suspension may not typify the degree of selectivity to a compound that is characteristic of the target species, e.g. aquatic weeds.

Axenic cultures of suspended tomato cells were treated with polar and nonpolar spikerush extracts. Results suggest that growth in cell volume after one week is severely reduced at both 500 and 1000 ppmw crude polar extract (Table III). The inhibition observed in the crude polar extract is apparently the result of the higher molecular weight compound(s) in peak 1. Though only slight inhibition was demonstrated at 250 ppmw, almost complete cessation of cell growth occurred at 500 ppmw (Table III, Figure 5). In addition to reduction in cell volume, cell suspensions in peak 1 treatments developed a brownish coloration within 24 h of exposure.

Table III. Tomato Cell Suspension Volume Assay at Various Fractions
and Concentrations of Spikerush Leachate

Treatment	ppmw	% control after 7 days
Crude polar extract	1000	11
	500	40
Peak 1 - polar	500	4
	250	73
Peak 2 - polar	1000	108
Murashige & Skoog salts	1000	96
Crude nonpolar extract	350	89

Seeds of lettuce and other species have frequently been used to
bioassay for the allelopathic activity of plant exudates (17, 18, 19).
As with the use of cell suspensions, there are certain advantages
and disadvantages to this methodology. The experimental simplicity,
small amounts of material required and short time frame are
certainly attractive qualities. However, species used in such
bioassays quite often do not represent the actual target species
under consideration. This is especially true when terrestrial crop
species are substituted for weeds of aquatic systems. Nevertheless,
information obtained from such experiments are often valuable when
used in conjunction with results of other assays.

The effects of 250 ppmw polar peaks 1A, 1B, and 2, nonpolar
peaks A and B, and Murashige and Skoog salts are represented in
Table IV. Though polar peak 2, nonpolar peak A, nonpolar peak B,
and Murashige and Skoog salts demonstrate a slight decrease in root
growth; polar peak 1B resulted in significantly greater reduction
than all other treatments. When polar peak 1B was applied to
lettuce seeds at 50, 100, 150, 200, 250, and 500 ppmw; the growth
curve obtained suggests that the concentration of peak 1B required
to inhibit root growth by 50% is approximately 100 ppmw (Figure 6).

In addition to reduction in root length, seedlings exposed to
all concentrations of polar peak 1B displayed both an increase in
root hairs and chlorosis at the tips of the cotyledons. The
severity of these symptoms increased with the dosage.

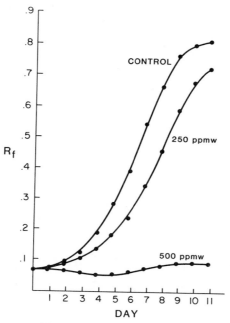

Figure 5. Tomato cell suspensions exposed to peak 1 polar extract of spikerush culture. Plotted as relative fraction of cell volume over volume of medium vs. time.

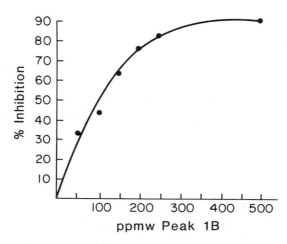

Figure 6. Affects of polar peak 1B on root growth of lettuce seedling following three day treatments.

Table IV. Effects of 250 ppmw Murashige and Skoog Medium, Polar and
Nonpolar Peaks from Axenic Cultures of Spikerush on Lettuce Seedling
Root Growth[1]

Treatment (250 ppmw)	mean (mm)	% inhibition
Polar peak 1A	31.1	3 [a]
Polar peak 1B	4.9	85 [c]
Polar peak 2	20.8	35 [b]
Murashige & Skoog salts	21.4	34 [b]
Nonpolar peak A	24.7	23 [b]
Nonpolar peak B	22.5	30 [b]
Control	32.2	0 [a]

[a]
Values followed by the same letter are not significanlty different
at the 5% level according to Duncan's multiple range test.

[1]
Assessment 3 days following treatment to dry lettuce seed.   Four
repetitions with 25 seeds per petri dish.   Growth chamber at 25 C,
12 h light and 12 h dark periods.

Conclusion

From the results acquired to date, the following conclusions and
postulations can be made:

a) Peak 1 or, in particular, peak 1B of the polar spikerush
leachate appears to contain the most active allelopathic activity
observed.

b) The molecular weight of the phytotoxic compound(s) of
interest is apparently between 600 and 1000.

c) Results from the hydrilla explant, sago pondweed shoot tip,
and lettuce seed assays suggest that the phytotoxin may act, in
part, as an inhibitor of root growth and development.

d) The phytotoxic compound(s) may translocate via the apoplast, as evidenced by the chlorosis in cotyledon tips of treated lettuce seed.

e) The allelopathic activity of the polar leachate of spikerush has been demonstrated on two associated aquatic weeds, hydrilla and sago pondweed. This is in agreement with field tests, which note a suppression of daughter plant production in both these target weeds.

Present fractionation techniques have been achieved by utilizing differences in molecular size, e.g. exclusion chromatography, and polarity, e.g. C_{18} cartridge. Further purification and identification will be accomplished on the basis of differential polarity through the use of thin layer chromatography (TLC), high performance liquid chromatography (HPLC), gas chromatography—mass spectroscopy (GC—MS), nuclear magnetic resonance (NMR), and infrared (IR) spectra techniques.

In cooperation with USDA—ARS, isolated compounds demonstrating phytotoxic activity will be synthesized and tested to determine lethal concentrations, mode of action, and mechanism of action.

The identification of an allelopathic compound(s) and its mode of action may eventually permit the economical use of spikerush as a biological control measure in aquatic systems too fragile for chemical control, e.g. recreational water, drinking water, and certain irrigation canals.

Acknowledgments

This research was supported by a cooperative agreement with the U.S.D.A., contract number 58-9AHZ-9-432.

Literature Cited

1. Oborn, E. T.; Moran, W. T.; Greene, K. T.; Bartley, T. R. Joint Laboratory Report SI-2, U.S.D.I.; Bureau of Reclamation, U.S.D.A., A.R.S.: Denver, Colorado, 1954; 84 pp.

2. Yeo, R. R.; Fisher, T. W. F.A.O. International Conference on Weed Control; Weed Sci. Soc. Amer., 1970; pp. 450-463.

3. Frank, P. A.; Dechoretz. N. Weed Sci. 1980, 28, 499-505./

4. Ashton, F. M., unpublished data.

5. Stevens, K. L.; Merrill, G. B. J. Agric. Food Chem. 1980, 28, 644-6.

6. Spoehr, H. A. 1949, Carnegie Institution, Washington; Publication No. 586.

7. Proctor, V. W. Limnol. Oceanog. 1957, 2, 125-9.

8. Keating, H. A. Science 1977, 196, 885-7.

9. VanAller, R. T.; Clark, L. R.; Pessoney, G. T.; Watkins, E. J.; Rogers, V. A. Lipids 1983, 18, 617-22.

10. Rogers, V. A.; VanAller, R. T.; Pessoney, G. T.; Watkins, E. J.; Leggett, H. G. Lipids (in press)

11. Wach, M. J. M.S. Thesis, University of California, Davis, CA. 1981.

12. Murashige, T.; Skoog, F. Physiol. Plant. 1962, 15, 473-97.

13. Hoagland, D. R.; Arnon, D. I. Calif. Exp. Stn. Circ. 347: Berkeley, 1950; 32 pp.

14. Ellis, B. E. Can. J. Plant Sci. 1978, 59, 775-8.

15. Zilha, S.; Gressel, J. Plant Cell Physiol. 1977, 18, 815-20.

16. Chaleff, R. S.; Parson, M. F. Genetics 1978, 89, 723-8.

17. Tang, C-S; Young, C-C Plant Physiol. 1982, 69, 155-60.

18. Lehle, F. R.; Putnam, A. R. Plant Physiol. 1982, 69, 1212-6.

19. Forney, D. R.; Foy, C. L. Weed Sci. Soc. Am. Abstr. 1984, 56, 147.

RECEIVED June 27, 1984

Improvements in the Synthesis of Strigol and Its Analogs

A. B. PEPPERMAN, JR., and E. J. BLANCHARD

Southern Regional Research Center, Agricultural Research Service, U.S. Department of Agriculture, New Orleans, LA 70179

Conditions have been found for improving the yield of products from some of the steps in the synthesis of strigol reported by Sih. For several of the other steps, such as closure of the B-ring and C-ring, efforts to improve the yield were unsuccessful and some consistently gave lower yields than reported. Several of the simplest analogs of strigol, alkoxy-butenolides containing only the D-ring, were prepared by etherification of a hydroxybutenolide with an alcohol. Conditions and limitations of the reaction were determined. Preparation of two-ring strigol analogs have been carried out by a modification of the literature procedure.

Witchweed [Striga asiatica (L.) Kuntze] is an economically important root parasite affecting many warm season grasses, including such important crop members of the Gramineae family as corn, grain sorghum, and sugar cane (1). The parasite has long been a problem in South Africa (2) and is now widespread throughout the Eastern Hemisphere. Since its discovery in the Western Hemisphere in North and South Carolina in the 1950s, it has been the object of quarantine, control, and eradication programs by the United States Department of Agriculture (3). Although the quarantine program has been effective, and the control program has permitted significant acreage to be removed from quarantine, the problem still exists and further research is needed.

The ability of viable witchweed seed to remain dormant in the soil for many years, only to germinate when favorable conditions prevail, makes eradication difficult (2). Usually, seed will not germinate unless pretreated in a warm, moist environment for several days before exposure to a chemical exuded from the roots of a host plant or some non-host plants (4). An active chemical in the root dates of cotton was isolated in 1966 (5) and identified in 1972 (6) by Cook and co-workers as compound I.

The compound (I) was given the trivial name strigol and it has been shown to be a very potent witchweed seed germination stimulant.

I

Total synthesis of (±)-strigol was reported in 1974 by Sih and co-workers (7) and the details of their scheme and resolution of (±)-strigol was reported in 1976 (8). Raphael and co-workers reported the synthesis of strigol by a different method also in 1974 (9) and 1976 (10).

Cook and co-workers suggested that strigol may be representative of a new class of plant hormones and that other biological effects should be examined (6). In 1976 Johnson and co-workers (11) reported the synthesis and testing of several strigol analogs, some of which were excellent seed germination stimulants for both Striga and Orobanche species. In 1982, Pepperman and co-workers tested a number of precursors and analogs of strigol and showed that several compounds which contained only part of the strigol molecule were active as witchweed seed germination stimulants (12).

A project has been undertaken at the Southern Regional Research Center which has as its goal the preparation of sufficient quantities of strigol and its analogs to permit the broad spectrum of tests necessary to understand the role of these compounds in the germination, growth, and reproduction of witchweed, and to determine their potential as control agents for witchweed when applied to infested fields. For a scale-up of the synthesis of strigol, the first approach was to modify one of the existing synthetic sequences to give overall better yields. Sih and co-workers (8) described two routes to strigol, one starting with α-cyclocitral (2,6,6-trimethyl-cyclohex-2-ene-1-carboxaldehyde) and the other with β-cyclocitral (2,6,6-trimethylcyclohex-1-ene-1-carboxaldehyde). The two routes converged after several steps and the remainder of the synthesis was identical. The α-cyclocitral scheme appeared to us to present fewer experimental problems than the β-cyclocitral scheme and was selected for further study.

Results and Discussion

There are 11 steps in the build-up of the A-B-C ring portion of strigol according to Sih's synthetic scheme. These are shown in Figure 1. By repeating these steps in our laboratory both on a small scale and on the same scale used by Sih, we found that we could not

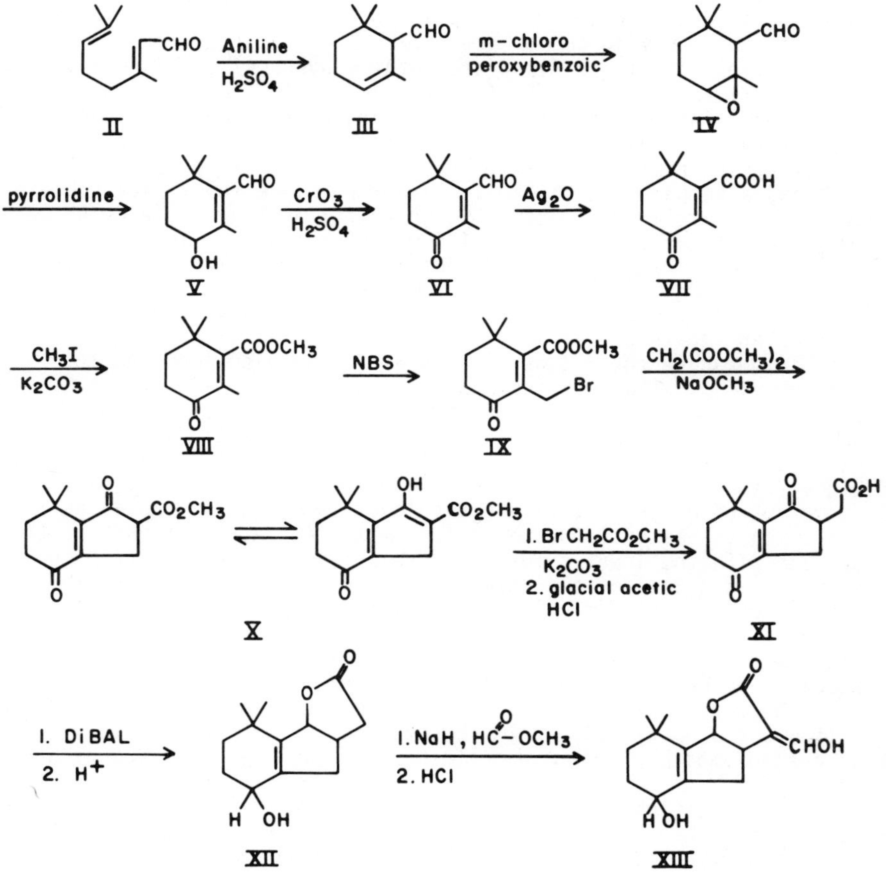

Figure 1. Construction of A–B–C rings in strigol.

duplicate some of the yields reported. A material balance for a
large scale synthesis of strigol using these steps and the yields
reported by Sih showed that starting with 8.3 kg of citral, which is
about the upper limit for our pilot plant equipment since such a
large excess of H_2SO_4 is required, one would obtain only 417 g of
XIII with the proper configuration to produce strigol. However, the
coupling step also produces 4'-epistrigol and the total yield of
strigol would only be 156 g.

Examination of the first step of the synthesis, ring closure of
citral (II) to a mixture of cyclocitrals, indicated that temperature
control and molar ratio of reactants were of critical importance to
obtain maximum yield. Gedye and co-workers (13) reported a yield of
57% of cyclocitrals using a 10 to 1 molar ratio of H_2SO_4 to citral
anil and a temperature of -20°. In our work, we found that the use
of a 10 to 1 molar ratio of H_2SO_4 to citral anil gave yields of only
about 34% of cyclocitrals. A 14.5 molar excess of acid upped the
yield to 54%. Maximum yield of 66% was obtained with an 18.7 molar
excess of H_2SO_4. This increase of 9% over the literature yield
gives us 16% more α-cyclocitral to go into the next step or 750 g at
pilot-plant scale.

Temperature control at -15° to -25°C was also required for
maximum yield. The best results were obtained by maintaining a
temperature of -20 to -25°C during the addition of citral anil to
the acid and at -15°C for the duration of the reaction. At this
temperature range, the formation of α-cyclocitral (III) is favored.
Higher temperatures caused excessive polymer formation and favored
formation of β-cyclocitral whereas lower temperatures caused a
reduction in the yield of the citral mixture. At least part of the
problem with the lower temperature reaction was the fact that the
sulfuric acid tended to freeze around the inside of the reaction
vessel causing the effective molar ratio of acid to anil to be
reduced. These lower temperature reaction mixtures were also
lighter in color which indicated less polymer formation but this was
accompanied by a lower yield of cyclocitrals.

The next step of the synthesis, the epoxidation of α-cyclocitral
(III) to 2,3-epoxy-2,6,6-trimethylcyclohexane-1-carbox aldehyde (IV)
proceeded smoothly and in excellent yield (96%) as reported by Sih
(8).

Isomerization of the epoxide (IV) with pyrrolidine was carried
out as described by Sih (8) and consistently gave yields of 35-60%
rather than the 73% reported. Changes in experimental conditions
including longer reaction times at lower temperatures, use of freshly
distilled pyrrolidine, use of NaOH dried pyrrolidine, and use of
distilled epoxide (IV) had little effect on the yield. The only
variation that improved the yield was to allow the reaction to
proceed at ambient temperature for a longer period of time than the
recommended 3 hours. Allowing the reaction to proceed for 40 hours
provided a maximum 67.5% yield. Other bases such as sodium
carbonate, triethylamine, diethylamine 1,5-diazabicyclo[4.3.0]non-
5-ene(DBN), and sodium methoxide all gave lower yields of distilled
product than pyrrolidine. It is important to use the hydroxyaldehyde
(V) as soon as possible since it is a very unstable material.

The next step in the Sih sequence involved oxidation of the
allylic alcohol and the α,β-unsaturated aldehyde to the keto acid
VII, which was accomplished by simultaneous oxidation with the Jones

reagent (14). While admitting that there were problems with this
reaction, Sih et al. reported yields of 45-55% of crude product.
Our results were not as good, with yields only in the range of
10-30%. The 30% yields were obtained only when a double (or triple)
distillation of crude V was carried out to give hydroxy aldehyde of
95% or greater purity for use in the oxidation step.

The acid (VII) formed in the Jones oxidation is readily
separated from the nonacidic organic material by extraction of the
ether layer with NaHCO$_3$ solution. The material not extracted by
NaHCO$_3$ (neutral organics) was shown by NMR spectroscopy to consist
of primarily the keto aldehyde VI.

In our experience, the sterically hindered aldehyde group was
largely untouched by the Jones reagent. The long reaction time (10
h) used by Sih et al. was required to cause oxidation of the
aldehyde group since oxidation of the alcohol proceeds readily and
rapidly. However, this long exposure to the Jones reagent caused
byproduct formation as was readily apparent from the NMR spectrum of
the crude acidic products, since several different methyl groups
were evident which were not attributable to VII. Based on these
results, a two-step procedure for the oxidation to VII, gave a
substantial improvement of yield. In work reported previously (15),
alkaline silver (I) oxide gave the crude acid in 72% yield. The
reaction was exothermic and rapid and the crude product had only a
small percentage of byproducts as shown by NMR. The facility of
oxidation of VI by Ag$_2$O was unexpected since earlier workers (16)
had obtained low yields (30-40%) in oxidations of α,β-unsaturated
aldehydes, and we had observed yields of about 20% in the room
temperature oxidations of β-cyclocitral and V with Ag$_2$O. The
extended conjugation provided by the keto group appears to be the
activating influence for the facile oxidation of VI by Ag$_2$O. The
relative purity of the product was anticipated since it is known
that Ag$_2$O does not attack carbon-carbon double bonds. Using the
two-step procedure and combining the acid produced in the Jones
oxidation step with that produced by Ag$_2$O oxidation, total yields of
70-85% of acid VII were obtained.

Esterification of the acid (VII) to produce VIII proceeded
smoothly as described by Sih (8) and no attempt was made to modify
this satisfactory procedure. Bromination of VIII to produce IX was
quantitative as described by Sih and no further modification was
necessary.

Closure of the B-ring in strigol involves the use of dimethyl
malonate to give the keto-enol mixture X which we obtained in only
30-55% yields rather than the 86% reported by Sih (8). Changes in
experimental conditions involving time and temperature did not
improve the yield past 55%. Preparation of XI typically produced
50-60% yields rather than the 72% reported by Sih (8). Closure of
the C-ring and concomitant reduction of the A-ring keto group is a
complicated reaction, giving a mixture which requires chromatography
to separate the isomers from the byproducts. For preparation of
XII, we have not yet realized the yields reported by Sih.

Preparation of XIII proceeded as described by Sih (8) and
yields of about 85% (reported 93%) were obtained. The coupling step
of XIII with the bromobutenolide (XIV) (see Figure 2) proceeded to
form a mixture of strigol and epistrigol in approximately the yields
reported by Sih (8). An attempt to use the sodium enolate with the

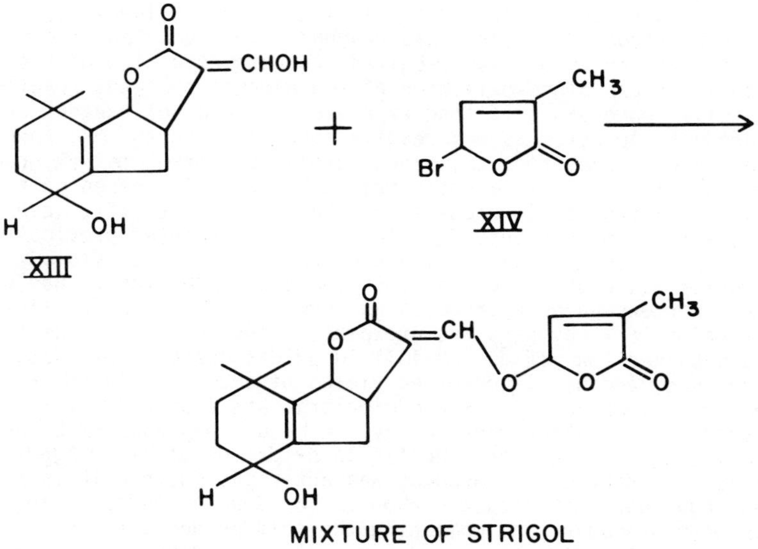

Figure 2. Last step in strigol synthesis.

bromobutenolide in the manner of Cassady and Howie (17), resulted in a very low yield of the strigol isomers, despite the fact that this method is more effective for formation of the analogs.

The D-ring of strigol is prepared in a six step sequence. The first three steps are Organic Synthesis preparations (18) and involve; (1) conversion of 4,4-dimethoxy-2-butanone to methyl-5,5-dimethoxy-3-methyl-2,3-epoxypentanoate by the action of methyl chloroacetate and sodium ethoxide; (2) conversion of the epoxide to methyl-3-methyl-2-furoate by prolonged heating at 160° with distillation of released methanol; and (3) hydrolysis of the ester to 3-methyl- 2-furoic acid (XV) by the action of sodium hydroxide followed by acidification. The next three steps involve; (4) photooxygenation (19) of XV in ethanol (or methanol) to give the alkoxybutenolide (XVI); (5) hydrolysis by concentrated HCl to the hydroxybutenolide (XVII); and (6) bromination of XVII by the use of carbon tetrabromide in the presence of triphenylphosphine (8) to form the bromobutenolide (XIV).

In our laboratory, we found that reaction of XV with other alcohols such as methanol, propanol, iso-propanol, and butanol gave yields of 60-80% of XVI which were comparable to ethanol and required about the same reaction time of 35-45 hours. For preparation of XVI, none of these alcohols offered any particular advantage over ethanol, but the hydrolysis (in dilute HCl) of the methoxybutenolide (XVIb) proceeded more rapidly (4-8 hrs.) than that of the ethoxybutenolide (XVIa) which can require up to 24 hrs for complete hydrolysis. Also in the hydrolysis of XVIb the product was cleaner thus giving a better yield of XVII. In earlier work (12), it was shown that activity as a witchweed seed germination stimulant occurred with just the D-ring of strigol, in particular for XVIa. It was of interest then to prepare several alkoxybutenolides (XVI) for testing. Some can be prepared by photo-oxygenation as stated

above. Some that could not, such as the benzyl, allyl, and lauryl
deratives, were prepared from the hydroxybutenolide (XVII) by
reaction with the appropriate alcohol in benzene in the presence of
catalytic amounts of p-toluenesulfonic acid. The etherification
reaction (or pseudoesterification reaction) proceeds with the
splitting out of water which azeotropes with the benzene and is

XVII   **XVI**

collected in a Dean-Stark trap. The following alcohols were successfully employed in this reaction: n-propyl, iso-propyl, cyclohexyl,
lauryl, allyl, and benzyl. Phenol, triphenyl carbinol, and t-butyl
alcohol did not give the desired alkoxybutenolides. Triphenyl
carbinol and t-butyl alcohol are evidently too bulky sterically and
will not form XVI. The t-butyl derivative can be formed by the
photooxygenation reaction. The use of trifluoroacetic anhydride as
a specific catalyst for esterification with phenols (20) was
attempted but was unsuccessful.

In some of the etherification reactions, the butenolide dimer
XVIII was observed as a byproduct in yields up to 40%. The dimer

XVIII

was also observed in the hydrolysis of XVIb when the volume of HCl
became too low and the mixture overheated. The dimer was first
observed in our efforts to mesylate XVII, but only one isomer was
isolated. In both the hydrolysis experiment and the etherification
reactions, a mixture of two isomers of XVIII resulted which were
separable by silica gel column chromatography. The lower melting
isomer (LMI) had a mp of 141-142° whereas the previously isolated
higher melting isomer (HMI) had a mp of 180-182°. Elemental
analyses were satisfactory for both isomers. When the NMR spectra
of the isomers were taken in $CDCl_3$, clear differences were
observable since the H^2 proton appeared at $\delta6.97$ for LMI and at
$\delta6.87$ for HMI while the H^1 protons differed by 0.18 ppm ($\delta6.02$ for
LMI vs. $\delta6.20$ for HMI). When the NMR spectra of the two isomers
were taken in d_6-acetone no differences in chemical shift of these
protons were observed.

In an effort to form the dimer as the major product and to
determine what effect the temperature of reaction had on the ratio
of HMI to LMI, the hydroxybutenolide XVII was heated to reflux in
benzene (bp = 80.1°C), toluene (bp = 110.6°C), and xylene (mixture
bp = 137-144°C) in the presence of catalytic p-toluenesulfonic acid

while monitoring water evolution with a Dean-Stark trap. The yield of dimer XVIII was 50% in benzene, 98% in toluene, and 99% in xylene. The percentages of the HMI and LMI in the reaction mixture were determined by NMR spectroscopy in CDCl$_3$ and found to be relatively constant at 55-60% HMI and 40-45% LMI for all three solvents. The higher temperatures did cause more complete reaction but had little effect on the ratio of dimers.

The literature preparation (11) of the two-ring analog of strigol (2-RAS) involved reacting the sodium enolate (XIX) with the mesylate (XX) to form 2-RAS (XXI). In our work we found it very difficult to prepare and purify the mesylate. Low yields of 2-RAS contaminated with XVIII resulted.

XIX   **XX**   **XXI**

In our experience, better results were obtained using the method of Cassady and Howie (17) than by the mesylate route. Thus the sodium enolate (XIX) is reacted with the bromobutenolide (XIV) in acetonitrile to form the 2-RAS (XXI) in 50-60% crude yields which after column chromatography provided 20-30% of XXI. Similar yields were obtained when tetrahydrofuran (THF) was used as the solvent,

XIX   **XIV**

but improved yields and cleaner products resulted when the reaction was conducted in hexamethyl phosphoric triamide (HPT) or other polar aprotic solvents such as dimethyl sulfoxide (DMSO) or dimethyl-formamide (DMF). Crude yields could not be determined since it is very difficult to remove the last traces of these high boiling solvents, but column chromatography of the oily residues provided 30-40% of relatively pure 2-RAS which was then amenable to recrystal-lization. The polar aprotic solvents offered the advantages of a cleaner product which required less chromatography and the resulting higher yields of 2-RAS.

Another two ring analog of strigol was prepared in a similar manner from the sodium enolate of α-valerolactone (XXII) and XIV. There is a methyl group at the 5-position of the ring which would correspond to the C-ring of strigol, so the shorthand for this compound is Me-2-RAS (XXIII). Johnson and co-workers (21) had

reported this compound as an oil. We were able to isolate it as a
solid with a wide mp range which gave a good elemental analysis,
which is indicative of a mixture of disasteomers. Efforts to reduce
the mp range have thus far been unsuccessful. Polar aprotic
solvents did not help in increasing yield or providing a cleaner
product. Attempts to prepare several other 2-RAS derivatives have
thus far been unsuccessful.

Summary

In preparation for scale-up of the strigol synthesis described by
Sih (8), efforts were made to improve the yield of some of the seven
steps involved in the scheme. Of these steps, nine are satisfactory
from the standpoint of yield and experimental conditions. For three
of the steps, we have improved the yield and/or experimental conditions such that the yield of (+)-strigol would be raised to 2.85%
overall from citral rather than 1.53% based on Sih's procedure and
reported yields. Improvements were developed preparation of
α-cyclocitral (III), the oxidation of the hydroxyaldehyde (V) to the
ketoacid (VII), and for the preparation of the hydroxybutenolide
(XVII). For the remaining five steps, our attempts to change experimental conditions have failed to improve, and in most cases to even
obtain, the yields reported in the literature (8). We have
considered the preparation of strigol analogs and determined the
conditions and limitations for the preparation of a series of
alkoxybutenolides (XVI) and a butenolide dimer (XVIII). Modification of the literature procedure (11) to eliminate the use of the
mesylate (XX) and the use of polar aprotic solvents gave better
yields of the 2-RAS (XXI).

Acknowledgments

The authors wish to acknowledge the able technical assistance of
Lynda H. Wartelle.

Literature Cited

1. Shaw, W. C.; Shepard, D. R.; Robinson, E. L. and Sand, P. F.
    Weeds, 1962, 10, 182.
2. Saunders, A. R. Union of South Africa Dep. Agric. Sci. Bull.
    No. 128, 56 pp.

3. Pavlista, A. D. Weeds Today, <u>1980</u>, **11(2)**, 19.
4. Brown, R. Encycl. Plant Physio., <u>1965</u>, **15**, 925.
5. Cook, C. E.; Whichard, L. P.; Turner, B.; Wall, M. E. and Egley, G. H. Science, <u>1966</u>, **154**, 1189.
6. Cook, C. E.; Whichard, L. P.; Wall, M. E.; Egley, G. H.; Coggan, P.; Luban, P. A. and McPhail, A. T. J. Am. Chem. Soc., <u>1972</u>, **94**, 6198.
7. Heather, J. B.; Mittal, R. S. D.; and Sih, C. J. J. Am. Chem. Soc., <u>1974</u>, **96**, 1976.
8. Heather, J. B.; Mittal, R. S. D.; and Sih, C. J. J. Am. Chem. Soc., <u>1976</u>, **98**, 3661.
9. MacAlpine, G. A.; Raphael, R. A.; Shaw, A.; Taylor, A. W. and Wild, H. J. J. Chem. Soc., Chem. Commun., <u>1974</u>, 1834.
10. MacAlpine, G. A.; Raphael, R. A.; Shaw, A.; Taylor, A. W. and Wild, H. J. J. Chem. Soc., Perkin T. I, <u>1976</u>, 410.
11. Johnson, A. W.; Rosebery, G. and Parker, C. Weed Res., <u>1976</u>, **16**, 223.
12. Pepperman, A. B; Connick, W. J. Jr.; Vail, S. L.; Worsham, A. D.; Pavlista, A. D., and Moreland, D. E. Weed Sci., <u>1982</u>, **30**, 561.
13. Gedye, R. N.; Arora, P. C. and Deck, K. Can. J. Chem., <u>1971</u>, **49**, 1764.
14. Bowden, K.; Heilbron, I. M.; Jones, E. R. H. and Weedon, B. C. L. J. Chem. Soc., <u>1946</u>, 39.
15. Pepperman, A. B. J. Org. Chem., <u>1981</u>, **46**, 5039.
16. Thomason, S. C. and Kubler, D. G. J. Chem. Educ., <u>1968</u>, **45**, 546.
17. Cassady, J. M. and Howie, G. A. J. Chem. Soc., Chem. Commun., <u>1974</u>, 512.
18. Burness, D. M. Org. Syn., Coll. Vol. IV, N. Rabjohn, Ed. John Wiley and Sons, Inc., New York, <u>1963</u>, 628, 649.
19. Farina, F. and Martin, M. V. An. Quim., <u>1971</u>, **67**, 315.
20. Bourne, E. J., Stacey, M., Tatlow, J. C. and Tedder, J. M. J. Chem. Soc., <u>1949</u>, 2976.
21. Johnson, A. W.; Gowda, G.; Hassanali, A.; Knox, J.; Monaco, S.; Razave, Z. and Rosebery, G. J. Chem. Soc., Perkin I, <u>1981</u>, 1734.

RECEIVED August 6, 1984

An Improved Partial Synthesis of (±)-Strigol

OLIVER D. DAILEY, JR., and SIDNEY L. VAIL

Southern Regional Research Center, Agricultural Research Service, U.S. Department of Agriculture, New Orleans, LA 70179

Ethyl 3-hydroxy-2,6,6-trimethylcyclohex-1-ene-1-carboxylate (Xa) has been prepared in four steps from ethyl 4-oxo-2,6,6-trimethylcyclohex-2-ene-1-carboxylate (III), readily obtained by the condensation of mesityl oxide and ethyl acetoacetate. Compound Xa can be incorporated in the Sih synthesis of (+)-strigol (J. Am. Chem. Soc., 1976, 98, 3661), and its preparation offers a significant improvement over existing methodology in terms of lower cost of starting materials and reagents and applicability to large scale synthesis. Several routes from III to Xa were investigated, and notable results obtained in these approaches are discussed. Finally, the conversion of Xa to the hydrindan XIII is described.

(+)-Strigol (I), isolated from cotton root exudates, is a very effective germination stimulant for witchweed [e.g. Striga asiatica (L.) Kuntze] seed (1). Structure I depicts the relative stereochemistry of natural strigol but the absolute configuration has not been determined. The parasitic weeds of __Striga__ species are thought to germinate primarily in response to a chemical signal from the host plant. Corn, rice, sugarcane, and sorghum are the major crop plants affected by __Striga__ (__2__). However, the structure of the germination stimulant exuded from their roots remains unknown.

Three total syntheses (__3,4__) of (__+__)-strigol have been reported in the literature. Of these the Sih approach (__3__), using citral as starting material appears to be the most applicable to large-scale preparation. Several partial syntheses of (__+__)-strigol have been reported, offering alternative approaches to intermediates in the Sih synthesis (__5-7__) or improvements in individual steps (__8,9__). Most recently, Brooks (__7__) has reported the synthesis of methyl 3-oxo-2,6,6-trimethylcyclohex-1-ene-1-carboxylate (__IIb__) in five steps from α-ionone in 48% overall yield, and other investigators have reported the one-step preparation of the corresponding acid __IIc__ from a mixture of α- and β-cyclocitrals (__9__).

CH₃ CH₃

> **CH₃ CH₃**
> CO₂R   __IIa__, R = CH₂CH₃
>   __b__, R = CH₃
> CH₃   __c__, R = H
> O

To date, probably less than five grams of (__+__)-strigol has been synthesized. For extensive field studies, a large quantity (in excess of 100 g) will be required. Although three total syntheses and several partial syntheses of strigol have been reported, the need for a synthesis adaptable to large scale preparation still exists. The goal of the present study is to achieve a preparation of strigol, suitable for large-scale production, based upon inexpensive starting materials and reagents and requiring a minimum of chromatographic purification.

Results and Discussion

__Preparation of the Starting Material III.__ The starting material utilized in these investigations directed toward an improved partial synthesis of (__+__)-strigol is ethyl 4-oxo-2,6,6-trimethyl-cyclohex-2-ene-1-carboxylate (__III__, Scheme I). This compound possesses the methyl groups and ester functionality in the appropriate positions for elaboration to strigol according to the Sih route starting with the key intermediate ethyl 3-hydroxy-2,6,6-trimethyl-cyclohex-1-ene-1-carboxylate (__Xa__). Compound III has been prepared by the zinc chloride or boron trifluoride catalyzed condensation of mesityl oxide with ethyl acetoacetate (__10,11__) or by the boron trifluoride catalyzed condensation of ethyl acetoacetate with acetone (__11,12__). These reactions give a product mixture of __III__ and __IV__ with __III__ as the major product. In addition, isophorone has been isolated as a product (__10__). Presumably, isophorone is formed by the preferential

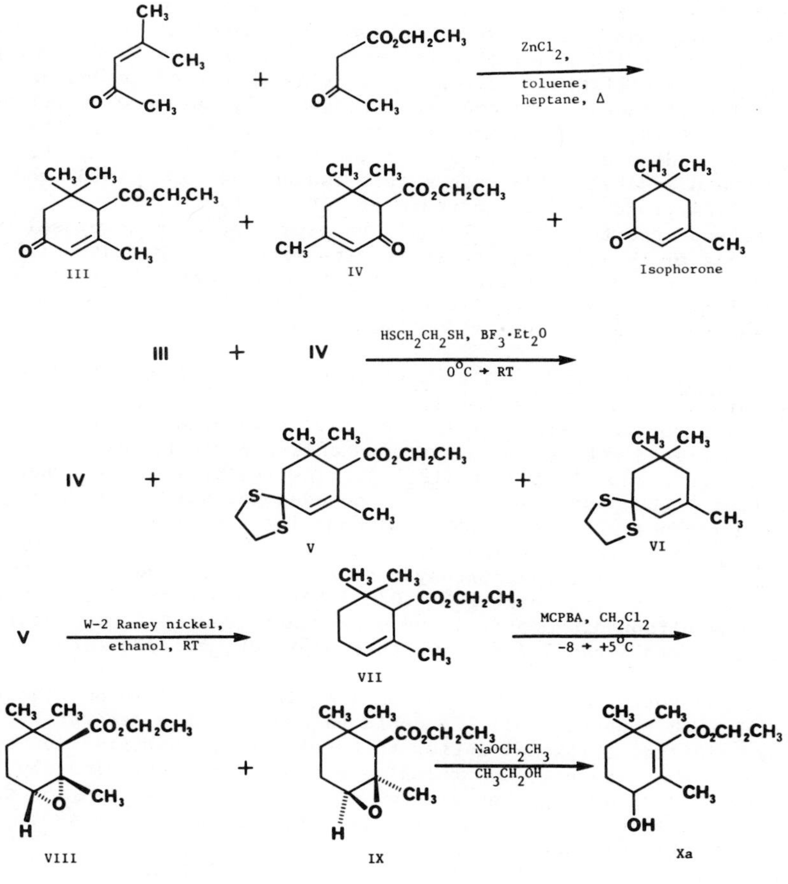
ZnCl2, toluene, heptane, Δ
III
IV
Isophorone
HSCH2CH2SH, BF3·Et2O
0°C → RT
V
VI
W-2 Raney nickel, ethanol, RT
VII
MCPBA, CH2Cl2
-8 → +5°C
VIII
IX
NaOCH2CH3
CH3CH2OH
Xa

Scheme 1

hydrolysis and decarboxylation of IV, thus accounting for the predominance of III in the product mixture.

A number of methods for the preparation of III were examined. Best results were obtained using a modification of the procedure of Surmatis, et al. (10): A mixture of 1.00 mole of mesityl oxide, 1.00 mole of ethyl acetoacetate, and 0.18 mole of zinc chloride in 200 ml of heptane and 100 ml of toluene was heated at reflux for 67 hours to give a 37% yield of a 7:1 mixture of III and IV following workup and distillation. Boron trifluoride catalyzed condensation of ethyl acetoacetate and acetone afforded a 7:3 mixture of III and IV in 29% yield, in excellent agreement with literature results (11,12). Sodium ethoxide catalyzed condensation of ethyl aceto-acetate and mesityl oxide gave mostly isophorone (50% yield) and only a 5.2% yield of a 6:1 mixture of III and IV.

It was reported that III could be separated from IV by selective hydrolysis of IV (10), but these results were later disputed (12). Indeed, in our hands purification of III by selective hydrolysis of IV to isophorone using the literature method (10) failed. There was no observed change in the product mixture.

Reduction of Enone III to Olefin VII. Compound III was obtained in pure form by column chromatography for use in investigating direct conversion to the olefin VII (Scheme I). A solution of pure III in ether saturated with gaseous hydrogen chloride was treated with activated zinc at 0°C for 2h (13,14). The isolated crude product consisted of two uncharacterized compounds, neither of which was the desired product VII. Additional methods for the direct conversion of III to VII were investigated: treatment with p-toluenesulfon-hydrazide and sodium cyanoborohydride in a 1:1 mixture of dimethyl-formamide and sulfolane at 100-110°C (15); conversion of III to its p-toluenesulfonhydrazone (16) and subsequent treatment with sodium borohydride in ethanol (17). Neither approach gave promising results.

A successful approach to alcohol Xa is outlined in Scheme I. Liu (12) had reported that III could be conveniently separated from IV by selective thioketalization to V followed by purification by column chromatography. Accordingly, a mixture of approximately 70% III and 30% IV (60.9 g) was treated with 1,2-ethanedithiol (50 ml) and boron trifluoride etherate (6 ml). The crude product was distilled under reduced pressure to yield 45.1 g (54.4%, 77.6% based upon III as starting material) of crude V, sufficiently pure for the next step. Column chromatography of lower boiling impure fractions furnished the dithioketal VI (10% yield), unreacted IV, and additional V.

Approximately 60 g of Raney nickel activated catalyst (supplied as a 50% slurry in water, pH 10 by Aldrich Chemical Company) was washed twice with ethanol (150 ml) and added to a solution of 15.9 g of V in 500 ml of ethanol. The mixture was stirred at room temperature for two hours. An additional 18 g of Raney nickel was added, and the mixture was stirred at room temperature for three hours. Following workup, 9.11 g (83.4%) of crude ethyl 2,6,6-trimethylcyclohex-2-ene-1-carboxylate (VII) sufficiently pure for the next step was obtained.

Raney nickel desulfurization is not generally considered practical for large scale synthesis in that large excesses (tenfold or more) are customarily used. Although only a fivefold excess (potentially as little as a two- or threefold excess) of Raney nickel was required in the conversion of V to VII, a further improvement could be achieved using conditions more suitable for large scale preparations. To this end, the treatment of V with metallic sodium in liquid ammonia was investigated (18). The reaction was run twice, and a single product was obtained in each case. The NMR spectrum of the crude product was consistent with the structure XI (yields of 56% and 65%, Equation 1). Although the undesired, but not unexpected, reduction of the ester functionality took place, the observed facile desulfurization lends feasibility to the investigation of alternate routes incorporating the desulfurization step later in the synthesis.

$$(1)$$

Synthesis of Allylic Alcohol Xa. A 3.84 g sample of olefin VII was treated with m-chloroperoxybenzoic acid (MCPBA) in dichloromethane for 1.5 hours at 0°C and 2.5 hours at 20°C. The NMR spectrum of the crude product indicated a mixture of approximately 75% epoxide VIII and 25% IX (structural assignments based upon assumed epoxidation preferentially from the less hindered side). Purification by column chromatography furnished 0.61 g of IX and 2.58 g of VIII. The separation was performed for characterization purposes; the crude epoxidation mixture was suitable for subsequent transformations.

Two reagents were investigated as less expensive replacements for MCPBA in the epoxidation of VII: (a) 30% hydrogen peroxide in acetic acid at 90–100°C; (b) 30% hydrogen peroxide and acetonitrile in ethanol in the presence of potassium bicarbonate (19,20). The first reaction gave a mixture of VIII and an unidentified product. In the second case no reaction occurred.

However, peracetic acid does appear to be a suitable inexpensive substitute for MCPBA. Treatment of 275 mg (1.40 mmol) of VII with 0.40 ml (2.1 mmol) of 35% peracetic acid (FMC Corporation) containing 32 mg of sodium acetate afforded 283 mg (95%) of product consisting of ca. 75% VIII and 25% IX.

Treatment of 1.99 g of VIII with sodium ethoxide in refluxing ethanol for 1.5 hours yielded 1.88 g (94.5% of crude allylic alcohol Xa. A small sample of IX was also converted to Xa under these reaction conditions.

Conversion of Xa to the Hydrindan XIII. The ultimate synthetic target of these investigations is a mixture of the tricyclic alcohols XIV (Scheme II). In the Sih synthesis (3) the methyl ester

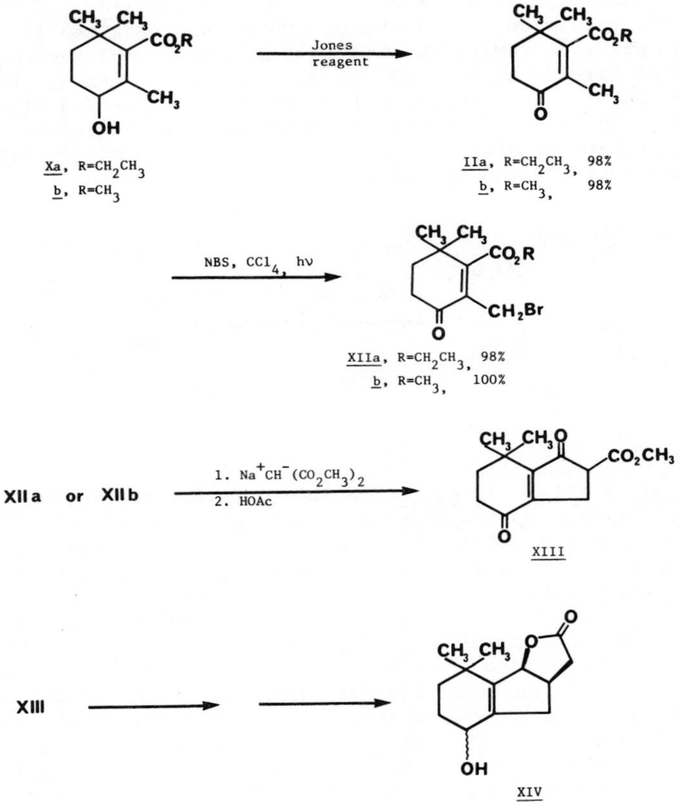
Jones reagent
Xa, R=CH2CH3
b, R=CH3
IIa, R=CH2CH3, 98%
b, R=CH3, 98%
NBS, CCl4, hν
XIIa, R=CH2CH3, 98%
b, R=CH3, 100%
XII a or XII b
1. Na+CH−(CO2CH3)2
2. HOAc
XIII
XIV

Scheme 2

Xb was utilized in the elaboration of the hydrindan XIII and subsequently compound XIV. We expected that Xa could be converted to XIII in the same manner as was in the Sih synthesis. This was confirmed in practice (Scheme II). Oxidation of 569 mg (2.68 mmol) of Xa with pyridinium chlorochromate in methylene chloride furnished 532 mg (94%) of enone IIa (7,21). Treatment of 130 mg of Xa with a slight excess of Jones reagent (3,22) afforded 126 mg (98%) of IIa. Allylic bromination of IIa with a 20% excess of N-bromosuccinimide (NBS) in refluxing carbon tetrachloride provided XIIa in 98% yield.

Sih reported the preparation of XIII by the reaction of the sodium salt of dimethyl malonate with the methyl ester XIIa at room temperature in methanol for 24 hours followed by heating at reflux under nitrogen for 6 hours. Recent investigations have revealed that the Sih procedure gives inconsistent results (yields ranging from 14 - 86%) and that best results are obtained with 1:1 methanol/tetrahydrofuran as solvent and by eliminating the reflux step (23). Accordingly, 238 mg (0.82 mmol) of XII was treated with a fourfold excess of the sodium salt of dimethyl malonate under argon at room temperature. The reaction was complete after only one hour, affording 107 mg (55%) of XIII following work-up and recrystallization from ethyl acetate. The proton NMR spectrum indicated that the product existed primarily in the keto form. The somewhat low yield can be attributed to the fact that the reaction was run on a small scale and residual dimethyl malonate interfered with the recrystallization.

Finally, the number of steps in the strigol synthesis could be reduced by the direct conversion of Xa to XIIa. To this end, Xa was treated with 1.2 equivalents of NBS in refluxing carbon tetrachloride under the irradiation of a 150-watt lamp. Purification of the crude product mixture by thin layer chromatography afforded two products: IIa (27%) and the unanticipated bromoketone XV (31%). The reaction

was repeated using 2.2 equivalents of NBS. The major products were IIb and XV with only a trace of XIIa being formed. There was no significant change in product composition upon using 3.3 equivalents of NBS under the same reaction conditions.

Since it has been established that treatment of IIa with NBS under conditions identical to those described above gives a near quantitative yield of XIIa, it is highly unlikely that bromoketone XV arises from IIa. In the reaction of Xa with NBS, it appears that a common intermediate is responsible for the formation of both IIa and XV. A possible intermediate is shown by structure XVI. It has been postulated that a similar compound is formed in the reaction of NBS with the pyranylate of an allylic alcohol akin to Xa (11).

$$CH_3 \quad CH_3$$
$$CO_2C_2H_5$$
$$CH_3$$
$$Br \quad OH$$

XVI

Conclusions

The feasibility of utilizing ethyl 4-oxo-2,6,6-trimethylcyclohex-2-
ene-1-carboxylate (III) as starting material in the elaboration of
(±)-strigol (I) has been established. The results reported in this
paper provide the basis of an attractive alternative for a large
scale preparation of I in terms of cost and practicality. On a
molar basis, the cost of the starting material for the original Sih
synthesis (3), citral, is 8.5 times greater than the combined cost of
mesityl oxide and ethyl acetoacetate, used in the preparation of III.
However, the overall yields of the analogous intermediates IIa and
IIb are comparable. Assuming an optimal 60% conversion of citral to
α-cyclocitral and utilizing the yields reported in the literature
(3), the overall yield for IIb in six steps from citral through the
intermediacy of α-cyclocitral is 20%, and in seven steps through the
intermediacy of β-cyclocitral the overall yield is 23%. The overall
yield for the six step synthesis of IIa reported in this paper is
20%.

In conclusion, a practical route form III to the hydrindan
intermediate XIII has been established, requiring no chromatographic
separations or purifications and generally utilizing inexpensive
reagents. However, several additional refinements in the general
route are being investigated, and results will be reported in a
comprehensive paper which will include a complete experimental
section describing all new compounds reported herein.

Mention of a trademark, proprietary product or vendor does not
constitute a guarantee or warranty of the product by the U. S.
Department of Agriculture and does not imply its approval to the
exclusion of other products or vendors that may also be suitable.

Literature Cited

1. Cook, C. E.; Whichard, L. P.; Wall, M. E.; Egley, G. H.;
    Coggon, P.; Luhan, P. A.; McPhail, A. T. J. Am. Chem. Soc.,
    1972, 94, 6198.
2. Riopel, J. L.; In "Vegetative Compatibility Responses in
    Plants"; Moore, R., Ed.; Academic Press: New York, 1983; pp
    13-34.
3. Heather, J. B.; Mittal, R. S. D.; Sih, C. J. J. Am. Chem.
    Soc., 1976, 98, 3661.
4. MacAlpine, G. A.; Raphael, R. A.; Shaw, A.; Taylor, A. W.;
    Wild, H. J. J.C.S. Perkin Trans. 1, 1976, 410.

5. Dolby, J.; Hanson, G. J. Org. Chem., 1976, 41, 563.
6. Cooper, G. K.; Dolby, L. J. J. Org. Chem., 1979, 44, 3414.
7. Brooks, D. W.; Kennedy, E. J. Org. Chem., 1983, 48, 277.
8. Pepperman, A. B. J. Org. Chem., 1981, 46, 5039.
9. Sierra, M. G.; Spanevello, R. A.; Ruveda, E. A. J. Org. Chem., 1983, 48, 5111.
10. Surmatis, J. D.; Walser, A.; Gibas, J.; Thommen, R. J. Org. Chem., 1970, 35, 1053.
11. Rubenstein, H. J. Org. Chem. 1962, 27, 3886.
12. Liu, H. J.; Hung, H. K.; Mhehe, G. L.; Weinberg, M. L. D. Can. J. Chem., 1978, 56, 1368.
13. Yamamura, S.; Hirata, Y. J. Chem. Soc. (C), 1968, 2887.
14. Toda, M.; Hirata, Y. J. Chem. Soc. Chem. Commun., 1969, 16, 919.
15. Hutchens, R. O.; Maryanoff, B. E.; Milewski; C. A. J. Am. Chem. Soc., 1971, 93, 1793.
16. Bamford, W. R.; Stevens, T. S. J. Chem. Soc., 1952, 4735.
17. Caglioti, L.; Graselli, P. Chem. Ind. (London), 1964, 153.
18. Ireland, R. E.; Wrigley, T. I.; Young, W. G. J. Am. Chem. Soc., 1958, 80, 4604.
19. Payne, G. B. Tetrahedron, 1962, 18, 763.
20. Payne, G. B.; Deming, P. H.; Williams, P. H. J. Org. Chem., 1961, 26, 659.
21. Corey, E. J.; Suggs, J. W. Tetrahedron Lett., 1975, 2647.
22. Bowden, K.; Heilbron, I. M.; Jones, E. R. H.; Weedon, B. C. L. J. Chem. Soc., 1946, 39.
23. D. W. Brooks, personal communication.

RECEIVED August 8, 1984

Strigol: Total Synthesis and Preparation of Analogs

DEE W. BROOKS, EILEEN KENNEDY, and H. S. BEVINAKATTI

Department of Chemistry, Purdue University, West Lafayette, IN 47907

An improved total synthesis of racemic strigol (1) and 4'-epistrigol (25) from alpha-ionone (2) is described which is applicable on a multigram scale. The preparation of several analogs representing A-D-ring (30 and 31), A-B-D-ring (34), and A-B-C-D-ring (35 and 36) partial structures of strigol (1) are outlined to demonstrate the flexibility and potential of utilizing both early and advanced synthetic intermediates prepared enroute to strigol for access to compounds which can be tested for biological activity and evaluation of key structure-activity relationships.

Strigol (1) is a potent seed germination stimulant for witchweed, a harmful parasitic plant acting upon important crops such as corn, sorghum, sugar cane, and rice.(1,2) The fact that strigol is not readily available from natural sources, motivated efforts to develop total synthetic schemes. Several partial (3-6) and two total syntheses (7,8) have been reported. The interest in field testing strigol as a control agent in witchweed infested areas provides further impetus to develop improved synthetic routes applicable on a multigram scale. Our efforts toward this objective will be described.

In addition, the elucidation of structure-activity relationships might provide a lead to the preparation of simpler analogs with biological activity. The use of synthetic intermediates encountered enroute to strigol can be used for the preparation of several analogs representing partial strigol structures. Our endeavors on this theme will be outlined.

Synthetic Studies of Strigol

The excellent total synthesis of strigol developed by Sih and co-workers (7) formed the basis for our efforts to devise an improved synthesis. The general synthetic plan involves consecutive A+B+C+D-ring formation as shown.

Commercially available alpha-ionone (2) appeared to be a reasonable starting point for a strigol synthesis as it contained the required carbon framework and workable functionality for an A-ring intermediate. A series of five reactions, involving no chromatographic purifications, provided an efficient preparation of 3-oxo-2,6,6-trimethylcyclohex-1-ene-1-carboxylate (7) (6), the A-ring intermediate used previously by Sih (7) for the synthesis of strigol. Regioselective epoxidation of 2 with m-chloroperoxy-benzoic acid gave a 6:1 mixture of isomeric epoxides 3a,b which were subsequently treated with sodium metaperiodate and a catalytic amount of potassium permanganate in aqueous 2-methyl-2-propanol, resulting in cleavage of the enone functionality to provide the isomeric mixture of epoxycarboxylic acids 4a,b. Esterification with iodomethane in acetone containing excess potassium carbonate followed by treatment with sodium methoxide in methanol resulted in epoxide cleavage providing the allylic alcohol 6, which was readily oxidized with pyridinium chlorochromate to the enone 7 (48% overall yield from alpha-ionone).

2   3a,b   4a,b R=H   6 X=OH,Y=H
                        5a,b R=CH$_3$ 7 X,Y=O

Further studies were conducted to improve this scheme to the A-ring unit, particularly, with respect to cost and scale-up practicality. Replacement of expensive meta-chloroperoxybenzoic acid by 30-40% peroxyacetic acid in acetic acid (four equivalents at 0°C) gave a 94% yield of epoxides 3a,b in a 9:1 ratio.

The oxidative cleavage by the procedure of Lemieux (9) was not convenient on a large scale due to the dilute reaction conditions, (7.5 L of 2:1 water:2-methyl-2-propanol) and the need for nine equivalents of relatively expensive sodium metaperiodate

(232g) for optimum reaction with the enone substrate **3a,b** (25g). Several alternatives were investigated. Ruthenium tetraoxide catalyzed oxidation with four equivalents of sodium metaperiodate in a carbon tetrachloride, water, acetonitrile solvent mixture (10) provided the isomeric epoxyaldehydes **8a,b** in 76% yield along with small amounts (10-15%) of the acids **3a,b**. Attempts to oxidatively cleave the enone functionality in **3a,b** with potassium permanganate under phase transfer conditions (11-13) resulted in complex product mixtures. A process involving ozonolysis (14) followed by various oxidative workup methods to access 4a,b was examined. Treatment of **3a,b** with excess ozone (ozonized oxygen was generated by a Wellsbach Model T-23 ozonizer operating at an oxygen pressure of 8 p.s.i. and a flow rate of 0.015 cu.ft./min.) at -78°C in methanol followed by treatment with sodium metaperiodate (15), with chromic acid (16), or with alkaline aqueous hydrogen peroxide (17), gave low yields (less than 20%) of the desired acids **4a,b**. Ozonolysis of **3a,b** in ethylacetate at 0°C followed by treatment with periodic acid or chromic acid gave mainly the tricarboxylic acid **9**.

Since various ozonolysis procedures with oxidative workups were uniformly unsuccessful, the desired target **4a,b** was changed to the corresponding aldehydes **8a,b**, which could be derived via reductive workup methods. Ozonolysis in methanol at various temperatures followed by treatment with dimethyl sulfide (18), gave a complicated mixture including **8a,b** in 28% yield. Ozonolysis of beta-ionone (10) in methanol at -30°C followed by treatment with zinc in acetic acid was reported to give a 90% yield of beta-cyclocitral (11). (19) Application of this procedure to **3a,b** gave a 50% yield of **8a,b**. This procedure was optimized simply by conducting the ozonolysis step at -78°C and the subsequent reductive workup at -10°C, provided **8a,b** in 86% yield (distilled product).

$$HO_2C(CH_2)_2C(CH_3)_2CH(CO_2H)_2$$
9

8a,b   **10**   **11**

Sih and coworkers (7) had studied the application of **8a,b** in their strigol synthesis. Their method of epoxide cleavage with pyrrolidine in ether was duplicated and proceeded to give the hydroxyaldehyde **12** in 90% yield. Oxidation of **12** to the ketoacid **14** with Jones' reagent (20) gave unsatisfactory yields. A two step procedure was developed by Pepperman (5) involving first Jones' oxidation of **12** to the ketoaldehyde **13** followed by treatment with alkaline silver(I) oxide to provide **14** in 70-85% yield. We found that the Jones' oxidation of **12** could be controlled to give the ketoaldehyde **13** in 92% yield, similar to Pepperman's results. In addition, we observed that a cold (0°C) carbon tetrachloride solution of **13** left exposed to air resulted in gradual oxidation over two weeks to provide ketoacid **14** which crystallized out of solution in 95% yield. Air oxidation of a neat film at room temperature gave lower yields and several side products. Thus, a practical synthesis of the A-ring unit represented by the ketoacid **14** in five steps from alpha-ionone in 63% overall yield has been achieved.

With the A-ring unit readily available, we directed our attention to the formation of the B-ring. At first, we duplicated the five step scheme reported in Sih's strigol synthesis involving; 1) esterification of the acid **14**, 2) allylic bromination with N-bromosuccinimide (NBS) to **15**, 3) condensation with the sodium salt of dimethyl malonate to **16**, 4) alkylation with methyl bromoacetate to **17**, and 5) acid catalyzed hydrolysis and decarboxylation to the acid **18**.

After numerous trials, we realized that the yield of **16** from the condensation reaction of the bromoester **15** with the sodium salt of dimethyl malonate was very dependant upon the reaction conditions and varied from 15-85%. Also, the use of excess dimethyl malonate complicated the isolation of pure product. Systematic optimization of the reaction conditions and amounts of reagents employed finally provided a reproducible one-pot procedure for the preparation of the diester **17** from **15**. The bromoester **15** (1 equiv) was added dropwise to a mixture of sodium hydride (2.5 equiv) and dimethyl malonate (1.2 equiv) in tetrahydrofuran at -10°C and the mixture was then stirred under nitrogen for 24h at 25°C. Aqueous acidic workup of an aliquot from the reaction verified the formation of the ketoester **16** which in the reaction mixture would exist as the corresponding sodium salt, therefore, methyl bromoacetate (1.5equiv) was added and the mixture was stirred 48h at 25°C. Aqueous acidic workup gave crude material which was purified by chromatography (silica gel, 10-30% ethyl acetate in hexane) to give the diester **17** in 82% yield. On large scale reactions the crude material was used without purification in the next step. Acid catalysed hydrolysis and decarboxylation of crude **17** gave the acid **18** in 64% yield after recrystallization from ether.

We were interested in obtaining the ester **19** directly from the crude diester **17** and found two suitable methods. The first method involved treatment of **17** with lithium iodide hydrate in dimethylformamide (<u>21</u>) at reflux for 1 hr., providing a 66% yield of **19** (purified by chromatography or distillation). The second method used was derived from the use of lithium chloride in hexamethylphosphorous triamide (suspected carcinogen) (<u>22</u>) but substituting less toxic N-methylpyrrolidone (NMP) as the aprotic solvent. Heating **17** at 100°C in NMP with lithium chloride (2.5 equiv.) for 6 h gave **19** in 68% yield.

A further improvement in the synthetic scheme was realized when the ketoaldehyde **13** was treated with 2.1 equiv. of NBS resulting in formation of the acid bromide **20** which was directly converted to the methyl ester **15** in 83% yield. The oxidation of aldehydes to the corresponding acid bromides with NBS has been reported.(<u>23</u>) This result provides a one-pot conversion of the ketoaldehyde **13** to the ester **15** in excellent yield, thus circumventing the oxidation step of aldehyde to acid and subsequent esterification.

20

The preparation of racemic strigol (**1**) and 4'-epistrigol (**25**) from keto-acid **18** was initially completed according to the procedures of the Sih synthesis (<u>7</u>), involving reduction of **18** to give **21a,b**, and hydroxymethylenation of **21a** to **22** by treatment with sodium hydride and methyl formate followed by O-alkylation with the bromobutenolide **23** using potassium carbonate in hexamethylphosphoric triamide. Reduction of **18** with diisobutylaluminum hydride provided a 1:1 mixture of isomeric hydroxylactones in 45% yield. Numerous attempts were tried to devise an improved method for reduction of either the acid **18** or the ester **19** to the desired lactone **21**. Reduction of the cyclohexyl carbonyl proceeded smoothly with $NaBH_4$, $LiBH_4$ and 9-BBN but attempts to further reduce the cyclopentanone either failed or resulted in complicated mixtures upon prolonged treatment. We were very pleased to finally obtain a simple solution to this problematic reduction involving the use of a reducing system of $NaBH_4$ and $CeCl_3$.(<u>24</u>) When the homogeneous aqueous solution of the sodium salt of the acid is mixed with an aqueous solution of 1 equivalent of $CeCl_3$ a yellow precipitate is formed. Subsequent addition of excess $NaBH_4$ (6 equiv) with stirring effects clean reduction within 2 h at room temperature to provide after acidification with 1 N HCl and extraction with dichloromethane a 75% yield of a 1:1 mixture of lactones **21a,b**. The action of the $CeCl_3$ in promoting this reduction is interesting. The free acid was not cleanly reduced by this procedure. The desired isomer **21a** can be selectively crystallized from an ether solution of the mixture.

Hydroxymethylene formation proceeded in the standard fashion (7) to give a 93% yield of **22**. Carcinogenic hexamethyl-phosphoric triamide was replaced by N-methylpyrrolidone as the solvent for the condensation reaction of **22** with the bromo-butenolide **23** in the presence of excess K_2CO_3 and gave a mixture of racemic strigol (**1**) (35%) and 4^4-epistrigol (**25**) (39%), which were readily separated by chromatography on silica gel.

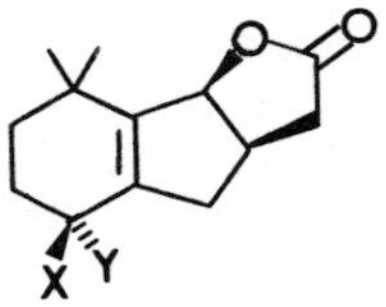

21a X=OH, Y=H
21b X=H, Y=OH

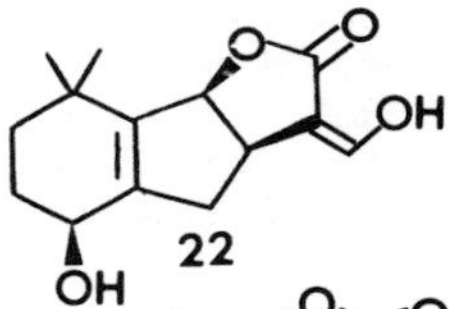

      22

23

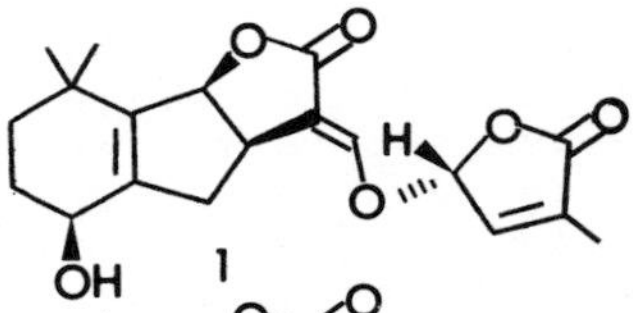

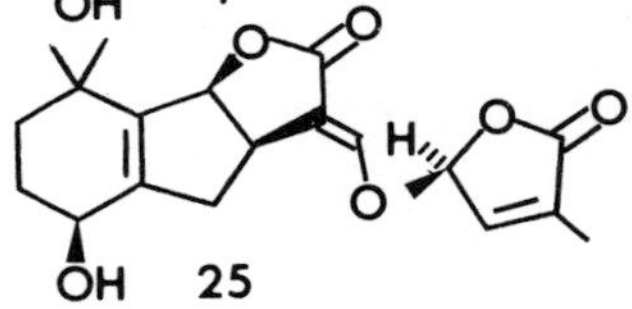

25

   1

In summary the synthesis of diketoacid **18** or diketoester **19**, representing a A-B-ring unit for strigol, has been accomplished in seven steps and 35-40% yield from alpha-ionone, using reagents and conditions appropriate for large scale synthesis and no required chromatographic purification. A simple efficient reduction of the diketoacid **18** to a mixture of hydroxylactones **21a,b** has been developed and the coupling reaction of the butenolide unit has been improved. From 100g of alpha-ionone, approximately 10g of **21a** and 10g of **21b** can be obtained, and the 10g of **21a** will provide about 5g of racemic strigol and 5g of 4'epi-strigol.

Preparation of Analogs of Strigol

Several studies directed toward the preparation of analogs of strigol have been reported. (7,25-30) Compounds **26** and **27** exhibit seed germination activity for both Striga and Orobanche species.(28) The racemic A-ring aromatic analog **28** showed 2% activity compared to natural strigol and the diastereomer **29** was 100 times less active than **28**.(27) These results are consistent with the observation by Sih (7) who reported that racemic strigol was 10,000 times more active than racemic 4'-epistrigol (**25**).

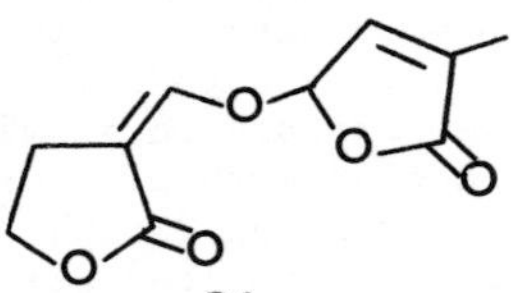

26

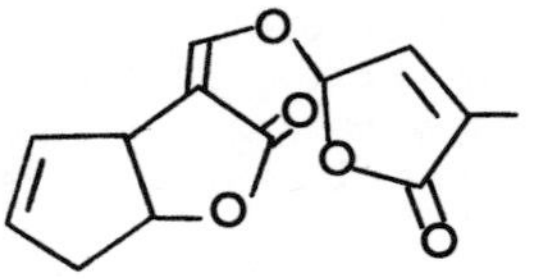

27

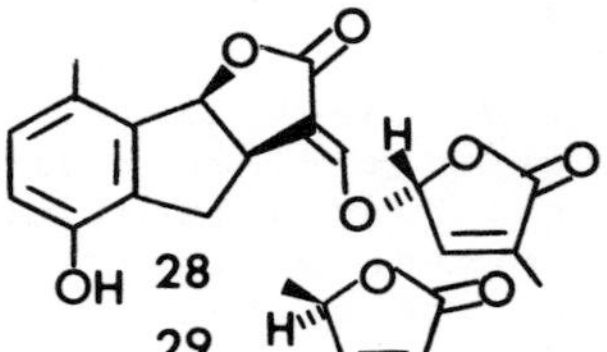

28
29 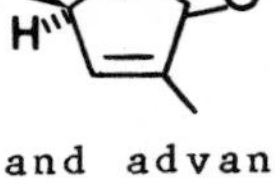

We are interested in applying both early and advanced synthetic intermediates developed enroute to strigol for analog preparations. Examples of the diverse analogs which are being prepared for structure-activity evaluation are described as follows.

The A-D-ring analog **30a,b** (mixture of epimers) has been prepared from the epoxide **3a,b** by base catalyzed epoxide cleavage, hydroxymethylenation, and O-alkylation of the butenolide unit using standard conditions. Hydroxymethylenation of keto-ester **7** followed by butenolide addition provided the A-D-ring analog **31**.

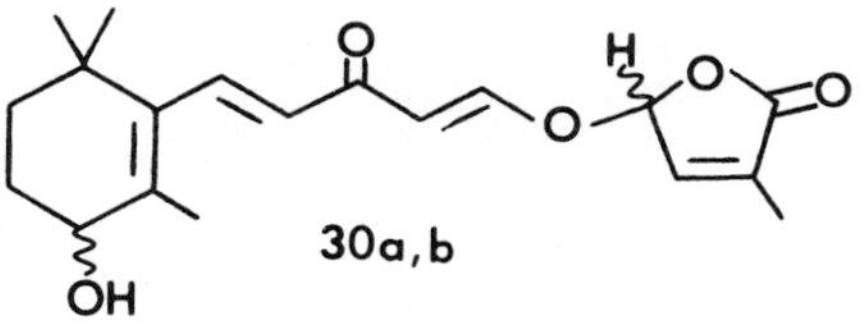

30a,b

Decarboxylation of **16** using the previously described NMP, lithium chloride method provided the dione **32**. Selective reduction of the least hindered carbonyl was readily effected using sodium borohydride providing **33**. Hydroxymethylenation followed by O-alkylation of the butenolide unit by standard procedures provided the A-B-D-ring analog **34a,b** (racemic mixture of epimers).

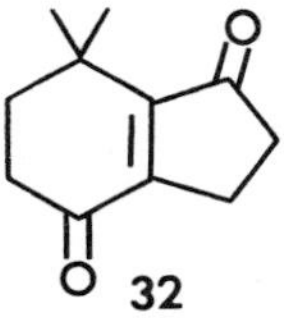

32 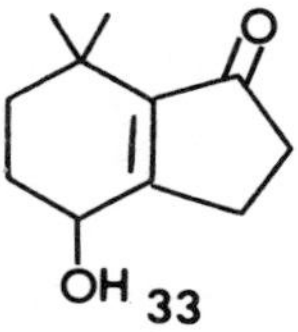   **33** 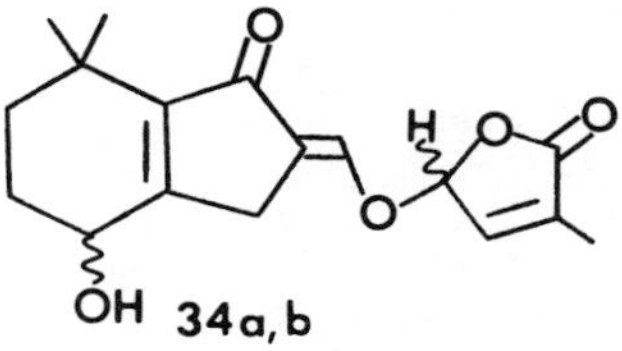   **34a,b**

As far as we know, a study of simple changes in the butenolide portion of strigol has not been reported. It would be interesting to identify whether the position of the methyl group influences the biological activity. This can be readily examined by preparing the strigol analogs **35** and **36** from the advanced A-B-C-ring intermediate **21a**.

35 $R_1=R_2=H$

36 $R_1=CH_3$ $R_2=H$

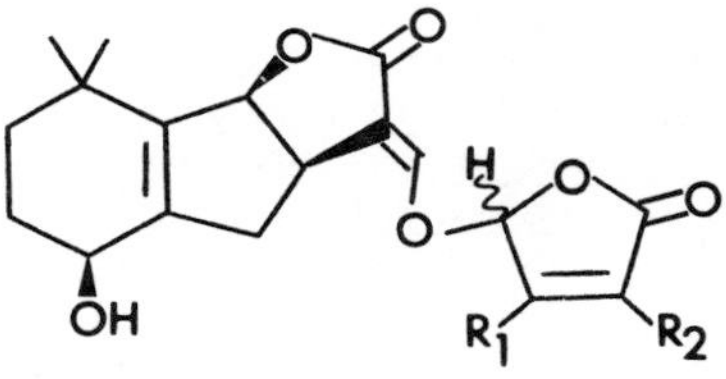

The synthetic studies which have been described have resulted in significant improvements in the preparation of racemic strigol and have also provided access to several analogs which will subsequently be tested for seed germination activity in order to elucidate key structure-activity relationships. These results and further investigations will hopefully lead to effective synthetic compounds for the control of witchweed and related parasitic plants.

Acknowledgments

We express our appreciation to the United States Department of Agriculture and Purdue University for financial support and Drs. A. Pepperman and S. Vail for helpful discussions during the course of this work.

Literature Cited

1. Cook, C.E.; Whichard, L.P.; Turner, B.; Wall. M.E.; Egley, G.E. Science (Washington, D.C.) 1966, 154, 1189.
2. Cook, C.E.; Whichard, L.P.; Wall, M.E.; Egley, G.E.; Croggon, P.; Luhan, P.A.; McPhail, A.T. J. Am. Chem. Soc. 1972, 94, 6198.
3. Dolby, L.J.; Hanson, G. J. Org. Chem.1976, 41, 563.
4. Cooper, G.K.; Dolby, L.J. J. Org. Chem. 1979, 44, 3414.
5. Pepperman, A.B. J. Org. Chem. 1981, 46, 5039.
6. Brooks, D.W.; Kennedy, E. J. Org. Chem. 1983, 48, 277.
7. Heather, J.B.; Mittal, R.S.D.; Sih, C.J. J. Am. Chem. Soc. 1976, 98, 3661.
8. MacAlpine, G.A.; Raphael, R.A.; Shaw, A.; Taylor, A.W.; Wild, H.J. J. Chem. Soc., Perkin Trans. 1976, 1, 410.
9. Lemieux, R.U.; von Rudloff, E. Can. J. Chem. 1955, 33, 1701.
10. Carlsen, P.H.J.; Katsuki, T.; Martin, U.S.; Sharpless, K.B. J. Org. Chem. 1981, 46, 3936.
11. Starks, C.M. J. Am. Chem. Soc. 1971, 93, 195.
12. Ogino, T.; Mochizuki, K. Chem. Letters 1979, 443.
13. Lee, D.G.; Chang, V.S. J. Org. Chem. 1978, 43, 1532.
14. Bailey, P.S. Chem. Rev. 1958, 58, 925.
15. Meyer, W.L.; Cameron, D.D.; Johnson, W.S. J. Org. Chem. 1962, 27, 1130.
16. Narula, A.S.; Dev, S. Tetrahedron Letters 1969, 1733.
17. Fremery, M.I.; Fields, E.K. J. Org. Chem. 1963, 28, 2537.
18. Heldeeveg, R.F.; Hogeveen, H.; Schudde, E.P. J. Org. Chem. 1978, 43, 1912.
19. Muller, N.; Hoffmann, W. Synthesis 1975, 781.
20. Bowden, K.; Heilbron, I.M.; Jones, E.R.H.; Weedon, B.C.L. J. Chem. Soc. 1946, 39.
21. Krapcho, A.P. Synthesis 1982, 805.
22. Muller, P.; Siegfried, B. Tetrahedron Letters 1973, 3565.
23. Cheung, Y.-F. Tetrahedron Letters 1979, 3809.
24. Luche, J.-F. J. Am. Chem. Soc. 1978, 100, 2226.
25. Cassady, J.M.; Howie, G.A. J.Chem. Soc. Chem. Comm. 1974, 512.
26. Johnson, A.W.; Roseberry, G.; Parker, C. Weed Research 1976, 16, 223.
27. Kendall, P.M.; Johnson, J.V.; Cook, C.E. J. Org. Chem. 1979, 44, 1421.
28. Johnson, A.W.; Gowda, G.; Hassanali, A.; Knox, J.; Monaco, S.; Razavi, Z.; Rosebery, G. J. Chem. Soc. Perkin I 1981, 1734.
29. Connick, W.J.; Pepperman, A.B. J. Agric. Food Chem. 1981, 29, 984.
30. Pepperman, A.B.; Connick, W.J.Jr.; Vail, S.L.; Worsham, A.D.; Pavlista, A.D., Moreland, D.E. Weed Science 1982, 30, 561.

RECEIVED July 11, 1984

Strigol Syntheses and Related Structure–Bioactivity Studies

SIDNEY L. VAIL, OLIVER D. DAILEY, JR., W. J. CONNICK, JR., and A. B. PEPPERMAN, JR.

Southern Regional Research Center, Agricultural Research Service, U.S. Department of Agriculture, New Orleans, LA 70179

The ability of some chemicals to stimulate seed germination is well known. The synthesis of strigol, its precursors, and analogs has attracted considerable attention because of the very high activity of strigol as a germination stimulant for seeds of some parasitic weeds. Strigol has been isolated only from the root exudate of a nonhost plant (cotton) and does not occur commonly in nature. Therefore, there is no known natural source at a useful level for isolation and purification of the compound. Considerable effort has been expended on alternate synthetic routes and improvements in the methods of preparation of amounts needed for field and laboratory research. Compounds other than strigol from host plants produce allelopathic effects that stimulate parasitic seed germination and ultimately result in extensive crop damage from the parasitic weeds. None of these other naturally occurring allelopathic compounds have been isolated and identified. However, some correlations between chemical structure and biological activity can be made on the basis of prior work. Strigol is a very useful compound for basic research on seed germination and plant growth, but there are many more promising options for solution of practical problems related to the induced germination of dormant seeds of parasitic weeds.

There are numerous examples of allelopathy involving seed of parasitic weeds that require a chemical signal for germination. Allelopathic chemicals are generally released by the host plant at times favorable for the germination, growth, and reproduction of the parasite. Since parasitic weeds can significantly reduce crop yields, a number of agricultural practices have been devised to reduce this loss. Some of the more effective methods to eradicate or reduce the effects of these parasites are to plant nonhost crops which stimulate suicidal germination or to apply synthetic germination stimulants to the field at an inappropriate time for the growth of the parasites (1-3).

Isolation, Identification, and Early Synthetic Schemes for Strigol

(+)-Strigol is a natural product with unusually strong properties as a germination stimulant for parasitic weeds of the genera Striga (witchweed) and Orobanche (broomrape).

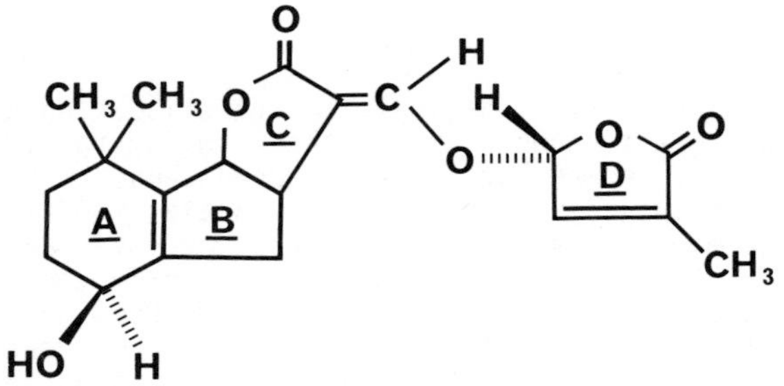

The structure of strigol is shown above with the rings identified as A, B, C, and D for further reference. Because the total synthesis of the compound for the production of even small amounts has proven to be exceedingly difficult, it is estimated that only about five grams of pure material had been isolated prior to 1984.

The syntheses of strigol, its precursors, and analogs have thus attracted considerable attention, and several papers presented in this symposium discuss various facets of the problem. The compound, originally isolated from the root exudate of cotton (a nonhost plant) (4), has not been found to occur commonly in nature. A total synthesis appears to be necessary to obtain useful amounts. The known (5,6) synthetic paths are not completely satisfactory for production of the required amounts for field testing as a germination stimulant or for other uses. Considerable effort has been expended on alternate synthetic routes, improvements in the original synthetic scheme, and in the synthesis of analogs. Many of these analogs, their precursors, as well as other organic compounds are also effective germination stimulants and have been studied in both the field and in the laboratory. Such studies with strigol and related compounds for allelopathic properties have been limited primarily to their effects upon selected parasitic weed seeds (7-10). The comments in this paper concerning bioactivity of the compounds are similarly limited, but assays of their effects upon other crop and weed seeds with varying germination requirements are in progress.

Effects of Various Chemicals on Germination of Selected Seed

Striga asiatica, one of many species of Striga found commonly in the Eastern Hemisphere, was also discovered in North and South Carolina in 1956. The seed may remain dormant in the ground for many years until stimulated by a chemical or chemicals released by certain plants. The germinated seed rapidly develops a radicle, receives a second chemical signal from the host, and in a relatively short period of time parasitizes the host plant. The host, e.g. corn, sorghum, sugarcane, rice or numerous forage crops and weeds belonging to the grass family, generally appears drought-stricken and often dies if the parasitic plant is not removed. Crop losses often have approached 100% in heavily infested fields (1-3).

The germination stimulant or stimulants from host plants have not yet been identified, but research on isolation and identification of these allelopathic compounds continues. Other nonhost plants, such as cotton, also release chemicals which stimulate the germination of witchweed seed and these crops can replace the cereal crops in witchweed-infected fields. If no acceptable host is present, the witchweed plant is unable to mature and produce seed. The importance of cereal crops as a staple food in underdeveloped countries makes growth of nonhost crops only partially acceptable, and there are numerous wild hosts that allow the witchweed to germinate, mature, and produce more seed (several thousand seeds can be produced by a single plant). Nevertheless, application of either natural or synthetic stimulants in the absence of a host plant is an effective way of reducing and eventually eliminating the witchweed problem.

In a review, Moreland, et al. (11) reported that thiourea, allylthiourea, D-xylulose, sodium diethyldithiocarbamate, L-methionine, n-propyl-di-n-propylthiolcarbamate, twelve 6-(substituted) purines, and two coumarin derivatives stimulated Striga seed germination. None of these compounds has been reported in natural stimulant preparations. Similarly, Riopel (2) in a more recent review noted that many compounds promote or inhibit seed germination. Ethylene is an effective germinator, but its use in under developed countries is minimal.

Many authors note that test conditions have a strong influence on results, and activity varies widely with a variation in concentration of the germination stimulant. A relatively high concentration of a stimulant often causes inhibition of germination. Preconditioning the seed in the presence of a germination stimulant can also cause inhibition of germination.

The apparent necessity of a second signal for haustorial initiation provides another possibility for chemical control. Riopel and coworkers (2,12-14) observed a few years ago that gum tragacanth, a foliar extract, was a very potent haustorial initiator. Two active compounds were isolated (Xenognosin A shown below, and a flavonoid with similar structure). The structural features required for haustorial formation include a meta relationship of hydroxyl and methoxyl groups and an alkyl branching ortho to the methoxy substituent.

A number of analogs were prepared to arrive at the above relationship. However, the identity of the chemicals needed for haustorial initiation and produced by any cereal crop remains unknown. The possibility of using a chemical to suppress or confuse haustorial formation has been recognized, but only a little work has been done on this aspect of control. The germinated seed will die in a short time if haustorial initiation and attachment are not realized.

Studies to correlate molecular structure of the compound with bioactivity in seed germination in the soil are difficult. It is uncertain whether the activity is elicited from compounds in an exudate of the plant or metabolites of compounds in the exudate. In general, it is assumed that the unaltered exudate contains the active compound. The situation is further complicated by the presence of numerous compounds in the exudate that may play only a secondary role (but a necessary one), yet possess no activity when separated from the exudate. Studies have confirmed that some compounds active in the laboratory do not possess the indicated activity in the field. Results from field studies are difficult to assess because the state of the seed in the soil varies and germination is less likely to occur if seed are not properly conditioned. Germination is inhibited by materials or conditions that are difficult or impossible to control in field studies and, as noted previously, high concentrations of germination stimulants often result in inhibition of germination. Further, seed counts in soil are very difficult, especially because of the small size of Striga seed. Thus, some theoretical questions involving germination in the soil may never be answered satisfactorily even though solutions to the practical problems of control have or will be achieved.

Resistant Varieties of Cereal Crops. Several varieties of cereal crops possess resistance to some species of witchweed which is attributed to the absence of an effective chemical stimulant in the root exudate of the crop plant. Williams (15) compared the effects of root exudate from a resistant and a susceptible variety of sorghum (grown in sand) on germination of Striga asiatica seed (designated Striga lutea by Williams). The aqueous eluents were compared at full strength and at diluted strengths. The full strength eluent from the resistant variety produced only a low level of germination and dilution reduced germination levels almost to zero. Similar studies with eluent from the susceptible variety produced a high level of germination at full strength and after dilution.

These studies have been extended by Parker, et al. (16), and root exudates of 24 sorghum varieties were screened for activity in germination of S. asiatica seed. In addition, root exudates of 15 sorghum varieties were tested with S. hermonthica, S. asiatica, and S. densiflora seed. There appeared to be a good correlation between the germination responses of S. hermonthica and S. asiatica, but germination responses of S. densiflora exhibited little or no correlation with the other species.

Some successes have been achieved in breeding cereal crops that possess resistance to witchweed and are suitable for commercial use (1); however, induced germination of witchweed seed using synthetic chemicals for control or eradication continues as the more promising approach to the ultimate solution of the problem. However, as noted in the above discussions, the chemical structures of the natural products from host plants responsible for germination in nature remain unknown. The root exudates of host plants contain many compounds that may act singly or in combination with other materials present to produce germination. The problem is further complicated by the fact that the actual stimulant may be the product of a chemical or biological modification of a component or components of the root exudate in the soil.

Improvements in the Synthesis of Strigol and Its Analogs

In the total synthesis of strigol, the first series of steps involving the A-ring are of primary importance in the preparation of large quantities of the compound. Clearly, large quantities of these starting materials are needed to compensate for the losses in yields in the later steps. Several separate investigations (17-21), not including those in this symposium, have been concerned primarily with improvements in this phase of the synthesis.

The procedures described in this symposium by Brooks and by Dailey represent new synthetic paths with several advantages for scaled-up synthesis. The procedures described by Pepperman outline improvements or modifications in a path suggested by Sih (5). Strigol synthesis has been completed by both Brooks and Pepperman using the new or improved procedures.

Review of Coupling Reactions for C- and D-Rings. The coupling of the two lactone rings is a similar step in the synthesis of strigol and strigol analogs. Yields and isomer distribution are the primary considerations. The configuration about the double bond connecting the C- and D-rings of naturally-occurring strigol is E (4); therefore, it is essential that reaction schemes chosen to form the enol ether bond in synthetically-prepared strigol, its isomers, and its analogs preserve the E stereochemistry. The configuration can be easily determined by ^{1}H NMR where the alkoxymethylene vinyl proton of the E isomer of many of these compounds absorbs at δ7.3-7.6, whereas the more highly shielded Z isomer proton absorbs at δ6.6-6.9 (5,6,22,23).

Researchers in this field have usually begun ring coupling with the preparation of the sodium salt of the hydroxymethylene (sodium enolate) derivative of the lactone corresponding or analogous to the fused ring portion of the strigol molecule. This derivative has been prepared using sodium hydride and methyl formate in ether at room temperature under anhydrous conditions. An alternative method involves the reaction in ether of metallic sodium, methanol, ethyl formate, and the lactone (8,9). In many cases, the crude sodium enolate can be used in further reactions without isolation, but acidification often gives the hydroxymethylene lactone that can be purified by recrystallization.

The final ring coupling reaction is usually an O-alkylation of the sodium enolate with a methylsulfonate-, bromo-, or chloro-butenolide in acetonitrile or an ether solvent (8,22-24). Use of the methylsulfonate derivative is least preferred because of its poor stability (9,24). The isolated hydroxymethylene lactone can be allowed to react with the bromobutenolide using potassium carbonate in hexamethylphosphoric triamide (caution: a potential carcinogen).

These reactions are highly selective for the desired E isomer of strigol, its isomers, and analogs. It has been postulated (6) that the E isomer of the hydroxymethylene lactone is thermodynamically favored because it has the maximum separation of the negative oxygen centers.

Correlations of Molecular Structure with Bioactivity in Seed Germination

Earlier in this paper studies were reported that indicated correlation of the molecular structure of the compound with bioactivity in seed germination in laboratory tests, as compared to tests performed in the field, offer distinct advantages. Most of what we know on this subject was obtained from laboratory test procedures. Results from field tests are also dependent upon the stability of the compound and physical factors such as solubility and adsorption in the soil.

Most of the work on correlation of molecular structure with bioactivity in witchweed seed germination has been produced by two groups (7-10). Johnson, et al. (7-9) prepared and evaluated a large number of strigol analogs and many approached the activity of strigol. In many studies by others, the results of the bioassays are presented, but the compounds from Johnson are described only by GR-number. GR-7 and -24, probably the more promising of these compounds, have been used in extensive field studies, and their structures are known.

The work of Pepperman and coworkers (10) is taken as the basis for this structure-activity discussion because about thirty compounds were tested for their activity in a bioassay that involved only one type of seed - S. asiatica. These compounds have been divided into groups according to low, moderate, and high activity in the bioassay (Table I). Considerable activity is noted in the two- and three-ring analogs of strigol and in the precursors that are similar to the A- and D-rings in the strigol. 3-Hydroxy-2,6,6-trimethylcyclohex-1-ene-1-carboxaldehyde (AB-4) has an activity essentially equal to that of strigol. Unfortunately, this hydroxy-aldehyde is unstable and is readily oxidized to the less active acid (10).

GR-24 has the highest activity of all of Johnson's compounds (9). It is essentially structurally identical to strigol except that the A-ring has been replaced with a phenyl in the same position. The isomer of GR-24 with a phenyl ring attached to the other nonbridgehead atoms of the B-ring was found to be less active. Cook et al. (22) prepared an analog of GR-24 wherein one hydroxyl and one methyl group were on the aromatic ring and in positions identical to

Table I
Bioactivity of Strigol, Strigol Analogs[a], their Precursors, and
Related Compounds[a]

Number or Designation[b]	Structure	Activity[c]
Low Activity:		
AB-5		55% at 10^{-4}
SA-4		48% at 10^{-5}
SA-3		55% at 10^{-5}
D-7		60% at 10^{-5}
Moderate Activity:		
Epi-strigol		83% at 10^{-6}
AB-1		55% at 10^{-7}

Table I (continued)

2-RAS   61% at 10^{-7}

SA-2   62% at 10^{-7}

3-RAS (High Melting)   66% at 10^{-7}

High Activity:

  3-RAS (Low melting)   60% at 10^{-8}

  3-RAS (Isomer Mixture)   68% at 10^{-8}

  D-1   40% at 10^{-9}

  AB-3   60% at 10^{-9}

  D-2   72% at 10^{-10}

Table I (continued)

Strigol   50% at 10^{-11}

AB-4   60% at 10^{-11}

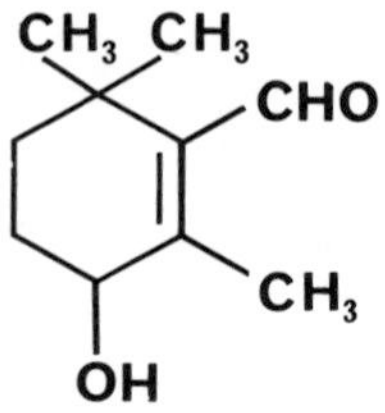

---

[a] *S. asiatica* seed were conditioned prior to the bioassay. Details on the bioassay and results of additional bioassays on these and other compounds were published (10).

[b] See Ref. 10. AB refers to precursors of the A-B rings of strigol. SA refers to precursors (or similar compounds) of strigol analogs. D refers to precursors of the D-ring of strigol. 2-RAS is a two ring analog of strigol and 3-RAS is a three ring analog. A mixture of isomers of 3-RAS is also known as GR-7.

[c] Numbers reported are a percent germination at the shown concentration (1 x 10M to the exponent shown). Many other values are shown in the referenced article (10), but the values recorded in this paper represent the highest activity at the lowest concentration. There is a great deal of variability in the results of germination tests as noted by many references in this text. See also ref. 25 and 26.

GR 24

those in the A-ring of strigol. The E-isomer was said to have only
2% of the biological activity of natural strigol.

Johnson and coworkers also prepared several two-ring analogs
and more three-ring analogs, but these compounds, except for GR-7
(3-RAS), have received little or no attention. Many of the three-
ring analogs had activities similar to GR-7 but reduced stability
and more difficult syntheses of the compounds appear to be the major
reasons for their limited use. For example, the isomer of GR-7,
wherein the double bond is moved to the other nonbridgehead atoms of
the B-ring, was found to be susceptible (as is GR-7) to degradation
by light and alkaline conditions.

There is no reported separation by Johnson and coworkers of geometric or optical isomers from their reaction mixtures. Connick,
Pepperman, and coworkers (10,24) and Cook et al. (22) separated
geometric isomers into their two diastereomeric racemates and noted
differences in activity for the geometric isomers. Many of the
reported activities of strigol-related compounds have been obtained
with mixtures of isomers.

Both applied and basic aspects of the use of chemicals to stimulate germination of seed continue. Strigol is primarily of basic
interest because it is so active in the germination of some parasitic
plant seeds. Because of quarantines on the seed and restriction of
much of the research to affected areas, progress in basic studies
has been limited. Efforts are underway to find bioassays other than
simple witchweed germination (27) so that the complete biological
and chemical significance of strigol and its related compounds can
be evaluated. Control of witchweed and other parasitic plants will
likely involve an integrated program of breeding resistant host
crops, selected agricultural practices, and effective use of
chemicals.

Much remains to be accomplished in the separation, isolation,
and identification of both naturally occurring and synthetic bioactive materials effective in the germination of parasitic weed seeds.
Structure-activity studies suffer from the lack of separation of
isomers in most synthetic samples. Strigol is an important tool in
basic studies on the effect of chemicals on seed germination, but it
is highly unlikely that the compound will meet practical field

requirements even if the A-ring synthesis options, discussed above, are optimized and totally suitable for large scale synthesis. More likely, smaller, active molecules and/or strigol analogs will be used in field procedures for eradication of plant parasites.

Literature Cited

1. Ramaiah, K. W.; Parker, C.; Vasudera Rao, M. J.; Musselman, L. J. 1983. "Striga Identification and Control Handbook", Information Bulletin No. 15, Patancheru, A. P., India: International Crops Research Institute for the Semi-Arid Tropics.
2. Riopel, J. L. In "Vegetative Compatibility Responses in Plants"; Moore, R. Ed.; Academic Press, New York, 1983, p. 13.
3. Shaw, W. C.; Shepard, D. R.; Robinson, E. L.; Sand, P. F. Weeds, 1962, 10, 182.
4. Cook, C. E.; Whichard, L. P.; Wall, M. E.; Egley, G. H.; Coggon, P.; Luhan, P. A.; McPhail, A. T. J. Am. Chem. Soc. 1972, 94, 6198.
5. Heather, J. B.; Mittal, R. S. D.; Sih, C. J. J. Am. Chem. Soc. 1976, 98, 3661.
6. MacAlpine, G. A.; Raphael, R. A.; Shaw, A.; Taylor, A. W.; Wild, H. J. J. Chem. Soc., Perkin Trans. 1, 1976, 410.
7. Johnson, A. W.; Rosebery, G.; Parker, C. Weed Res., 1976, 16, 223.
8. Johnson, A. W.; Rosebery, G. U. S. Patent 4 002 457, 1977.
9. Johnson, A. W.; Gowda, G.; Hassanali, A.; Knox, J.; Monaco, S.; Razavi, Z.; Rosebery, G. J. Chem. Soc., Perkin Trans. 1, 1981, 1734.
10. Pepperman, A. B.; Connick, W. J.; Vail, S. L.; Worsham, A. D.; Pavlista, A. D.; Moreland, D. E. Weed Sci., 1982, 30, 561.
11. Moreland, D. E.; Egley, G. H.; Worsham, A. D.; Monaco, T. J. In "Natural Pest Control Agents"; ADVANCES IN CHEMISTRY SERIES No. 53, American Chemical Society: Washington, D.C., 1966, p. 112.
12. Lynn, D. G.; Steffens, J. C.; Kamat, V. S.; Graden, D. W.; Shabanowitz, J.; Riopel, J. L. J. Am. Chem. Soc. 1981, 103, 1868.
13. Kamat, V. S.; Graden, D. W.; Lynn, D. G.; Steffens; J. C.; Riopel, J. L. Tetrahedron Lett., 1982, 23, 1541.
14. Steffens, J. C.; Lynn, D. G.; Kamat, V. S.; Riopel, J. L. Ann. Bot. (London), 1982, 50, 1.
15. Williams, C. N. Nature, 1959, 184, 1511.
16. Parker, C.; Hitchcock, A. M.; Ramaiah, K. V. Proceedings of the Sixth Asian-Pacific Weed Society Conference, Jakarta, Indonesia, 1977, p. 57.
17. Dolby, L. J.; Hanson, G. J. Org. Chem., 1976, 41, 563.
18. Cooper, G. K.; Dolby, L. J. J. Org. Chem., 1979, 44, 3414.
19. Pepperman, A. B. J. Org. Chem., 1981, 46, 5039.
20. Brooks, D. W.; Kennedy, E. J. Org. Chem., 1983, 48, 277.
21. Sierra, M. G.; Spanevello, R. A.; Ruveda, E. A. J. Org. Chem., 1983, 48, 5111.
22. Kendall, P. M.; Johnson, J. V.; Cook, C. E. J. Org. Chem., 1979, 44, 1421.

23. Cassady, J. M.; Howie, G. A. J. Chem. Soc., Chem. Commun.,
    1974, 512.
24. Connick, W. J.; Pepperman, A. B. J. Food Agric. Chem., 1981,
    29, 984.
25. Hsiao, A. I.; Worsham, A. D.; Moreland, D. E. Weed Sci., 1981,
    29, 98.
26. Musselman, L. J.; Worsham, A. D.; Eplee, R. E., Eds.
    Proceedings of the Second Symposium on Parasitic Weeds. North
    Carolina State University, Raleigh, N. C., 1979.
27. Bradow, J. M.; Fites, R. C.; Menetrez, M., Personal
    Communication.

RECEIVED July 24, 1984

INDEXES

Author Index

Subject Index

459

Production by Anne Riesberg
Indexing by Janet S. Dodd
Jacket design by Pamela Lewis

Elements typeset by Hot Type Ltd., Washington, D.C.
Printed and bound by Maple Press Co., York, Pa.